Biogenic Silica
Fundamentals and Applications

Biogenic Silica
Fundamentals and Applications

Edited by

Ashok Kumar Nadda
Jaypee University of Information Technology, India
Email: ashok.nadda09@gmail.com

and

Vijai Kumar Gupta
SRUC, University of Edinburgh, UK
Email: vijaikumar.gupta@dcu.ie

ROYAL SOCIETY
OF **CHEMISTRY**

Print ISBN: 978-1-83916-583-2
PDF ISBN: 978-1-83916-971-7
EPUB ISBN: 978-1-83916-972-4

A catalogue record for this book is available from the British Library

The Royal Society of Chemistry is a charity, registered in England and Wales, Number 207890, and a company incorporated in England by Royal Charter (Registered No. RC000524), registered office: Burlington House, Piccadilly, London W1J 0BA, UK, Telephone: +44 (0)20 7437 8656.

Visit our website at www.rsc.org

Printed in the United Kingdom by CPI Group (UK) Ltd, Croydon, CR0 4YY, UK

Preface

Biogenic silica can be known as silica or silicon dioxide, and is produced by living organisms. Understanding the importance of biogenic silica in diverse fields is demanding in terms of science. Its remarkable inherent qualities could be used to advance existing technologies. This book comprises nine important chapters that focus on biogenic silica, its characteristics, the advanced strategies for sustainably synthesizing biogenic silica, applications, future perspectives and challenges. In this book, Chapter 1 describes biogenic silica's sources, structure and properties. The second chapter explains the significance of recent biotechnological methods for biosynthesis of silica, its ecological importance, and possible applications. Chapters 3 and 4 explain the production of silica from agricultural waste and the characterization of biogenic silica by applying sustainable advanced technologies for the safety of environmental pollution and economic growth. Chapter 5 elaborates on biogenic silica nanoparticles and composites for supercapacitors. This chapter overviews the latest progress in the synthesis, fabrication and application of BSNPs and their composites for supercapacitor electrodes. Chapter 6 emphasizes the importance and potential of biogenic silica for biosensing and bioimaging applications. Chapter 7 highlights the molecular mechanisms and describes the application of biogenic silica composites in biomedical fields. Chapter 8 covers current research on the biological uses of biogenic silica, such as its usage in photonic applications, gene therapy, drug delivery systems and the detection or treatment of numerous diseased conditions. This book chapter also discusses fundamental advancements and exploratory development of biogenic silica in research for future biomedical applications. The last chapter discusses about biogenic silica as filler material for elastomers and possible uses in important applications such as biomedicine,

Biogenic Silica: Fundamentals and Applications
Edited by Ashok Kumar Nadda and Vijai Kumar Gupta

Published by the Royal Society of Chemistry, www.rsc.org

photonics, energy storage, separation technology, sensing, and energy conversion. In summary, the chapters mentioned above collectively offer a thorough presentation of the current knowledge and breakthroughs in the application of biogenic silica.

Ashok Kumar Nadda and Vijai Kumar Gupta

Editor Biographies

Dr Vijai Kumar Gupta is an Associate Professor at the School of Biotechnology. Vijai has almost 15 years of experience in teaching and research. His research interests include biomass valorisation, microbial engineering biotechnologies, functional microbiomes, advanced bioprocessing and fermentation technologies and bioactive natural products. Vijai is the Fellow of Linnaean Society, UK; Fellow of Mycological Society of India; Fellow of Indian Mycological Association; Fellow of the Academy of Microbiological Sciences, India; Pravasi Fellow of the National Academy of Agricultural Sciences, India; and Associate Fellow of the National Academy of Biological Sciences India. He has mentored/supervised many undergraduates, master's students, PhD researchers, and postdoctoral researchers. He has also edited a few books in his area of expertise for Elsevier, Wiley-Blackwell, Taylor and Francis, Springer Nature, and CABI, UK. His multidisciplinary research has yielded several high-quality scientific articles, and his work has been widely cited in the field.

Dr Ashok Kumar Nadda is working as an assistant professor in the Department of Biotechnology and Bioinformatics, Jaypee University of Information Technology, Waknaghat, Solan, Himachal Pradesh, India. He has completed his PhD in Biotechnology from the Department of Biotechnology, Himachal Pradesh University, Summer Hill, Shimla, India in 2013. He worked as a postdoctoral fellow in State Key laboratory of Agricultural Microbiology, Huazhong Agricultural University, Wuhan, China for two years, 2014–2016. He also worked as Brain Pool researcher/Assistant professor at Konkuk University Seoul South Korea for 2016–2017. Dr Ashok has a keen interest in enzyme immobilization, biomaterials, biopolymers, nanobiotechnology, biocatalysis, waste management, biomass degradation, biofuel synthesis, and biotransformation. Dr Ashok has published more than 150 research

Biogenic Silica: Fundamentals and Applications
Edited by Ashok Kumar Nadda and Vijai Kumar Gupta

Published by the Royal Society of Chemistry, www.rsc.org

papers in various internationally reputed journals 30 books, and 57 book chapters. He also contributes as editor and reviewer in various international journals. He is teaching Industrial enzyme technology, Metabolic Engineering, and Environmental Biotechnology to degree students. He is also supervising the doctorate, master's and bachelor's degree students in his laboratory.

Contents

Biogenic Silica: Fundamentals and Applications
Edited by Ashok Kumar Nadda and Vijai Kumar Gupta

Published by the Royal Society of Chemistry, www.rsc.org

CHAPTER 1

Biogenic Silica: Sources, Structure and Properties

BANDANA THAKUR,[a] RAHUL SHRIVASTAVA,*[a] VIJAI KUMAR GUPTA[b,c] AND ASHOK KUMAR NADDA*[a]

[a] Department of Biotechnology and Bioinformatics, Jaypee University of Information Technology, Waknaghat, Solan, Himachal Pradesh, 173 234, India; [b] Biodesign Europe, Dublin City University, Glasnevin, Dublin D09 K20V, Ireland; [c] School of Biotechnology, Dublin City University, Glasnevin, Dublin D09 K20V, Ireland
*Emails: juit.rahul@gmail.com; ashok.nadda09@gmail.com

1.1 Introduction

Biogenic silica (bSi) can simply be demarcated as silica or silicon dioxide that is produced by living organisms. The other terms that are frequently used instead of bSi are opal, biogenic opal, and amorphous opaline silica.[1,2] Chemically bSi is hydrated silica ($SiO_2 \cdot nH_2O$) and it is noteworthy to mention that many plants and animals depend on bSi for their survival. For instance, grasses and other plants contain minute Si content known as phytoliths. Also, silicifiers play a very important role in silica deposition. These organisms can absorb dissolved silicon from their surroundings and convert it into glassy skeletons.[3] Silicifiers, which include macro-organisms like siliceous sponges and other micro-organisms (silicoflagellates, diatoms, rhizarians) and several types of choanoflagellates, are among the most significant aquatic organisms.[4] On a global scale, silicifying organisms such as diatoms that depend on photosynthesis effectively use large amounts of silicon in addition to nitrogen, phosphorus, and inorganic carbon. This interconnection influences the biogeochemistry of these elements and plays

Biogenic Silica: Fundamentals and Applications
Edited by Ashok Kumar Nadda and Vijai Kumar Gupta

Published by the Royal Society of Chemistry, www.rsc.org

a crucial role in capturing atmospheric CO_2 within the ocean. The dissolved silicate, which is needed by the organisms to synthesize their amorphous silica frustules, strongly influences diatom productivity.[5–7] Rhizarians, choanoflagellates, and sponges that are heterotrophic organisms generate bSi without relying on the photoautotrophic processing of carbon (C) and nitrogen (N).

In aquatic ecosystems, the production of silicic acid arises from the weathering of the Earth's crust by rainwater enriched with CO_2. This process plays a pivotal role in regulating atmospheric CO_2 levels.[6,8] The primary sources of bSi include diatoms, sponges, and grasses/plants. Diatoms, are an important group of algae, and they are also known for their intricate Si cell walls. These microscopic algae extract Si from their environment and use it to construct protective outer shells or frustules.[9] In the case of sponges, certain types of sponges incorporate Si into their skeletons. Siliceous sponges are an example, and they often have intricate spicules made of Si.[10] Some plants, particularly grasses, can accumulate Si in their tissues. This is often seen in grasses growing in environments with high Si concentrations in the soil.[11] In 1989, Wilding *et al.* estimated that bSi content ranges between 3 to 20% in most soils. These numbers indicate that opal can frequently surpass the concentrations of potassium, calcium, magnesium, sodium, and phosphate and make up a sizeable portion of the Si component of various types of soils. Currently, sodium silicate (Na_2SiO_3), a silica precursor, is produced by smelting sodium carbonate (Na_2CO_3) with quartz or sand to extract Si. Although Si has been successful in applications such as electrical components, however this approach is less practical because large amounts of energy is needed.[12–14] Hence, to avoid the limitations associated with the conventional techniques of Si extraction, innovative green techniques that are not only environmentally friendly but also cost-effective need to be explored.

Moreover, bSi is an adsorbent to eliminate heavy metal ions (such as lead, mercury, zinc, and nickel) from wastewater streams. Additionally, bSi is employed as a catalyst in the synthesis of zeolites and mesoporous silicon, as well as in drug delivery systems.[15,16] Moreover, Si particles are needed for various industrial applications, which require high purity and controllable textural properties. Different degrees of purities of bSi are needed, depending on the intended application. Generally, it is sufficient to have Si purity of at least 97% db (dry basis) for use as pozzolan in concrete. On the other hand, high-tech applications like solar applications and electronics demand a purity of Si almost equal to 99.9 wt% db.[17,18] For example, achieving synthesis of advanced materials requires low carbon content and high silicon purity.[19] Similarly, for filler applications, desirable attributes include whiteness and suitable particle size,[20] while an amorphous structure and an optimized pore system (resulting in high specific surface area) are also important considerations.[19,21]

It is commonly known that Si serves as a precursor for a variety of applications, including chromatography, electronic coating, ceramics, concrete, and catalysts, optical materials as well as anticorrosion agents.[22–27] Also, there are many industrial uses for Si, such as anti-caking food additives, viscosity control, beverage clarification, anti-foaming agents, dough modifiers, and excipients in vitamins and medications.[28] Hence, there are several applications related to bSi but due to the high processing temperature, using high-purity Si in industrial applications is expensive. Therefore, using alternatives such as crop residues for Si synthesis has been identified as a good way to combat sustainability issues and lower the processing costs.[27] The subsequent sections address the possible sources, extraction methods and applications of bSi in detail, along with a brief discussion of the structure and general properties of bSi.

1.2 Sources of bSi

Numerous organisms, especially in the ocean, produce bSi, which is essential for marine ecosystems. Studying nutrient dynamics, carbon cycling, and ecosystem structure in both aquatic and terrestrial environments requires an understanding of the sources and cycling of bSi.[29,30] Furthermore, the accumulation of bSi in sediments adds to the geological record and can be examined to learn more about the environmental conditions that existed in ancient times. Both terrestrial and aquatic ecosystems are the sources of bSi. Diatoms are principally responsible for the production of bSi in aquatic ecosystems. Diatoms are a special kind of algae that have a Si cell wall known as a frustule.[31] These fragments can gather in sediments to form deposits called diatomaceous earth. Plants, especially some kinds of grasses, are important for the accumulation and deposition of bSi in terrestrial ecosystems. The procedure entails the roots of plants absorbing dissolved silica (DSi) from the soil, moving it throughout the plant, and then depositing it in plant tissues. Plants require Si to grow, and different crops store differing amounts of Si in their tissues. Phytoliths are the most prevalent form of bSi in plants. Plant cells give rise to rigid, microscopic structures called phytoliths made of Si. They give structural support to the plant and aid in defending it against diseases and herbivores.[32–34] Genetic variables, environmental factors, and soil silicon availability all affect the buildup of Si in plant tissues. Therefore, it is possible to categorize the sources of bSi as terrestrial or aquatic as depicted in Figure 1.1.

1.2.1 Biogenic Silica from Terrestrial Sources

Traditionally, it has been believed that soil/geological processes fuelled by Si weathering are primarily responsible for the transfer of Si from the land and also for the cycling of Si on land. Thus, the final export of DSi crucial for eutrophication and carbon sequestration in oceans and coastal zones is regulated by the comparatively gradual weathering processes involving both primary and secondary silicate minerals.[29,35] However, a growing body of research now indicates how crucial biological activities are for both global Si

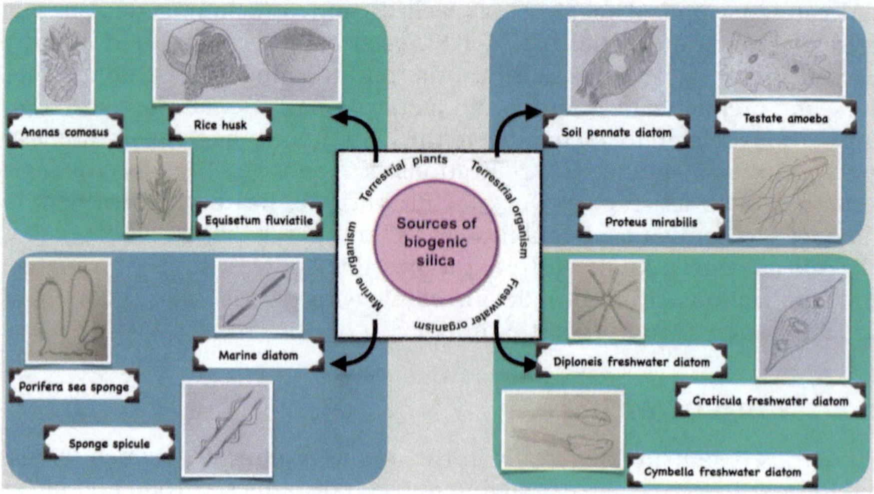

Figure 1.1 A diagrammatic illustration of different sources of bSi.

cycle and terrestrial Si mobilisation. Soil microbiota and land plants promote the weathering of silicate minerals. Consequently, vegetation takes up DSi as orthosilicic acid (H_4SiO_4), released through weathering. This leads to significant accumulations of silicon stored as amorphous bSi within ecosystem biomass and soils, predominantly in the structure of plant siliceous bodies known as phytoliths as shown in Table 1.1.[36,37] After plant decomposition and litterfall, the highly soluble bSi in biomass returns to the topsoil. Significant amounts of bSi have been discovered in wetlands, woods, and grassland soils.[38]

1.2.1.1 *Terrestrial Plants*

The realization that silicon is biologically necessary for plants is one of the foremost results of the data accumulated through numerous studies conducted on plants. Also, plants lacking in Si are frequently structurally weaker than those with sufficient Si, exhibit unusual growth, development, viability, reproduction, and are more vulnerable to abiotic and biotic stresses. Plants readily assimilate silicon in the dissolved form (*i.e.*, DSi), precipitating as solid amorphous silicon. Silicon is present in diverse tissues, including the epidermis of leaves, stems, roots, and reproductive organs of plants. Additionally, specialized silicon cells are found in grasses, along with hairs or trichomes.[36,39] These accumulations come in various forms and embellishments, some of which are unique in terms of taxonomy. These silicon structures, varying in size from a few to several tens of micrometers, are commonly known as opal phytoliths, plant phytoliths, opaline Si, or bSi. Silicon's function is similar to lignin's in plants as it is a structural element of the cell wall that provides resistance toward compression. It can be

Table 1.1 Sources of biogenic silica.[a]

Source name	Characteristic features	Si content	Ref.
Terrestrial plants			
Cynodondactylon	Leaf blades have a higher silica concentration than leaf sheaths and stems, according to a detailed examination of the LIBS spectra of the various plant parts.		42
RH	SiO_2 isolation was achieved through the use of fuoric technology. The specific surface area of the obtained samples was found to be 107 $m^2 g^{-1}$.	95–98%	43
RH and WH	Spherical nanocarriers made of rice husk exhibited a more crystalline structure with slit-shaped pores, whereas those made of wheat husk exhibited spherical nanoparticles with narrow cylindrical pores.	90–99.9% (RH) 65–75% (WH)	44
RHA	Chemical fuel pre-treatment is essential to reduce the risk of ash melting and enable the high-purity, sustainable production of bSi as a green chemical product.	88.6%	46
Rice hull ash	Depending on the production conditions, bSi powders derived from RHA could be inexpensive substitutes for the amorphous Si fillers used today.	93.40%	45
Equisetum fluviatile	The plant stems were used to extract the biogenic silica fibers, which had a diameter of approximately 120–150 μm and a length of several centimeters.		51
Ananas comosus	Rosette-like Si-based microparticles, ranging in size from 5 to 10 μm, which are formed by even smaller micro and nanoparticles that make up the pineapple's shell and bracts.		52
Terrestrial organisms			
Testate amoebae	The bSi that was preserved by living testate amoebae was only 0.45–1.57 kg SiO_2 ha^{-1}, the amount of biosilicification that occurred annually in different forest soils was estimated to be between 10 and 227 kg SiO_2 ha^{-1} $year^{-1}$.		57

Table 1.1 (*Continued*)

Source name	Characteristic features	Si content	Ref.
Testate amoebae	Euglyphids are preyed upon by a group of arcellinids known as hyalosphenids. Also, silica scales mineralized by euglyphids can be used by hyalosphenids to build shells.		60
Freshwater organisms			
Phytolith, diatom and sponge spicules	Main source of bSi in the rivers of Peninsular Malaysia appears to be phytoliths from terrestrial plants, which accounts for a sizeable amount of the budget for coastal silica.	92.8–98.3% (phytolith in Pahang River) 68.8% (diatom in Pontanian River) 79.3% (diatom in Endau River)	67
	Phytolith and diatom	The concentration of river bSi mainly from the Changjiang (Yangtze) and Huanghe (Yellow) rivers, change in response to in-stream primary production and terrestrial inputs.	68
Phragmites australis, *Impatiens glandulifera*, *Urtica dioica*, *Epilobium hirsutum* and *Salix* sp.	*P. australis* wetlands may play a crucial, albeit underappreciated, role as a sink for bSi in the biogeochemical cycling of Si.		70
Marine organisms			
Suberites *domuncula*	Four silicateins form a platform with serine clusters directed toward the center since silicatein-α is 4 times more abundant in the axial filament of *S. domuncula* than silicatein-β. The conaxially arranged silicateins' serines have an interaction with silicatein-β.		71

Deep sea sponges	Bio-polyP and bSi, produced by deep-sea sponges, have also been found in recent years to promote morphogenetics on both osteoblasts and osteoclasts.	72
Euplectella aspergillum	Deep-sea sponge-based bSi is less stiff than technical glass, but its architecture offers a significant amount of toughening due to both nanoscale and microscale structuring.	73
Diatom valves, sponge spicules	During the period when the Larsen ice shelf covered the sea surface, sponge spicules rather than diatoms were the primary source of bSi in the Eastern Antarctic Peninsula's sedimentary record.	75
Diatom	In the context of multifunctional supports for cell growth, using diatoms to produce bSi was suggested, which is a substitute source of mesoporous materials.	77
Navicula sp. and Rice straw hydrolysate (RSH)	It provides a mild condition, chemical-free method for recovering resources from RS in the form of nanostructured diatom bSi, an environmentally friendly step toward efficient management of rice straw waste.	79
Ostreococcus tauri and *Micromonas commoda*	This work demonstrates the widespread occurrence of biosilicification in both novel brackish picoeukaryote strains and common, well-studied marine species.	80
Radiolarians	The low-latitude South China Sea's bSi fluxes are primarily caused by radiolarians, as evidenced by the relatively strong bSi correlation for radiolarian abundance and weak for diatom abundance.	81

[a] Rice husk – RH; wheat husk – WH; rice husk ash – RHA.

incorporated with low energy expenditure and is essential for plant growth and energy-related processes. The enduring presence of fossil phytoliths in both terrestrial and aquatic paleoenvironments has been utilized to reconstruct historical ecosystems and human activities. Si is present in significant amounts in terrestrial plants, varying from less than 1 to 5% of the dry matter, and in some cases, up to 10% or even more.[36,40,41]

For example, Chauhan *et al.* analyzed the Laser-Induced Breakdown Spectroscopy (LIBS) spectra for various components (leaf blade, leaf sheath, and stem) of freshly harvested *C. dactylon* plants. The aim was to investigate the distribution pattern of silicon deposition across different plant segments. Leaf blades have a higher Si concentration than leaf sheaths and stems, according to a detailed examination of the LIBS spectra of the various plant parts. The findings from the LIBS analysis are compared with the density of phytoliths deposited in different locations of *C. dactylon*. In LIBS analysis, it is observed that leaf blades have the highest frequency of silicified cells, followed by leaf sheaths and stems.[42] Aside from calcium, aluminium, magnesium, sodium, and potassium oxides, rice residues such as straw and husk are recognized for containing SiO_2 in an amorphous state, with SiO_2 content being notably higher compared to other crops. Approximately 200 kg of rice husk (RH), yielding nearly 40 kg of ash with an 85–98% silicon content, is generated from one tonne of rice.[43] Using biomass as a renewable energy source for the production of biogenic materials has recently gained significant attention. Combining these techniques enables the production of high-quality bSi with purities exceeding 98 wt% using combustion residues, particularly from RH and rice straw (RS) characterized by elevated silicon contents in the fuel ash. In terms of global warming and CO_2 emissions, the entire process can be regarded as nearly neutral, but it can also solve the problems associated with disposing of RH and RS. Numerous cutting-edge uses for the resulting bSi are possible, including as drug delivery systems, catalysts, adsorbents, and more.[15]

In another study conducted by Porrang and others, bSi was extracted from cereals husk using acid leaching and it was further converted to sodium silicate as a Si precursor. For drug delivery, mesoporous Si nanocarriers were created in this study using natural materials like rice and wheat husk.[44] Similarly, in a different study bSi was obtained from rice hull ash (RHA). This study also suggests that RH, one of the many sources of bSi, is considered a valuable agricultural biomass material and an affordable source that can provide bSi for biomedical applications.[45] RH ash can yield extremely pure bSi if ash melting and agglomeration during combustion are prevented.[46] A different investigation supports the idea that RH is a valuable reservoir of biogenic silicon nanoparticles (Si NPs) characterized by remarkable physicochemical attributes and a wide range of potential applications. Several factors may affect the yield of extracted biogenic Si NPs. This study delves into the morphology and examines how the choice of solvent (sodium hydroxide, potassium hydroxide) and aging time (0, 6, 12, 18, 24 hours) influence the extraction yield and properties of specific biogenic Si NPs derived from Indonesian RH.[47] Using RHA as a source of bSi, the production of polycrystalline silicon was achieved, exhibiting impurity levels lower than those stipulated by the SEMI III standard for solar grade silicon feedstock, which is approximately 99.99% pure. To refine the RHA, cost-effective and energy-efficient techniques such as acid milling, boiling water wash purification, pelletization, and carbothermal reduction were employed. These processes were conducted using an experimental 50 kW electric arc furnace in batch mode, maintaining temperatures between 1700–2100 °C.[48] Setiawan and coworkers also demonstrated that RH is a good source of bSi.[49] In this work, three distinct molecular structure amine groups were used to functionalize bSi, which was extracted from RH. Grimm *et al.* endeavoured to propose an energy-optimized method for synthesizing metal oxide-supported bSi. Their proposed synthesis approach involves the integration of metal salts into the lignocellulosic matrix of biomass, aiming to reduce the pyrolysis steps. By calcining the modified RH or RS, the targeted metal oxide supported on bSi can be obtained directly.[50] Recently, a novel form of mesoporous, highly hierarchical fibrillar bSi was derived from the perennial plant *Equisetum fluviatile*. The inherent highly ordered structure of the plant stems can be largely preserved in the bSi skeleton by removing organic compounds through a combination of chemical and heat treatment.[51] It has also been proposed that Si, a physical barrier and component for improving mechanical properties, exists in plant tissues. Pineapples present in their shells, along with bracts that resemble rosette-shaped microparticles may be related to bSi. In a previous work, it is demonstrated that Si-based microparticles are co-purified while pineapple (*Ananas comosus*) nanocellulose is being extracted.[52] A different study focuses on the sophisticated textural characterization of bSi derived from various biomasses that accumulate Si, as well as agricultural waste materials like horsetail, rice, oat, and spelt husk. Thus, it is clear that plant-based biomass is a good source for bSi. Furthermore, the findings suggested that various plant species and origins would result in different textural characteristics and interconnections between the pores in bSi.[53] The fibrous material known as sugarcane bagasse is a great source of bSi. Additionally, sugarcane bagasse is an environmentally friendly precursor for synthesizing Si NPs with cost-effective production.[54] Moreover, it was found that the extracted bSi constituted between 0.01 and 5.9% of the dry biomass of leaves and wood in samples collected from 28 alpine plant species in the Swiss Alps (Valais). This observation extends to arborescent dicotyledons from 43 diverse families in temperate regions, showcasing a broad range of opal silicon content. Also, compared to woodlands, grassland communities produce significantly more opal Si, which increases the amount of bSi added to the soil.[55]

1.2.1.2 Terrestrial Organisms

Plants are not the only creatures that can biomineralize, that is, absorb minerals into their tissues. Certain fungi and insects that live on land are also capable of producing bSi. For instance, some ant species add Si particles to their exoskeletons, which increases the resilience and hardness of their outermost layer of defense. When discussing terrestrial organisms, testate amoeba, along with a few bacteria and fungi, are one of the main sources of bSi. Testate amoebae, a polyphyletic group of unicellular eukaryotes (protists) characterized by a test (shell) ranging in size from approximately 5 to 300 μm, constitute the primary origin of protozoic Si in soils.[56,57] Two supergroups can be used to classify testate amoebae: (1) Amorphea, which includes the order Arcellinida, and (2) TSAR, which includes the order Euglyphida. Arcellinida, an order encompassing testate amoebae with lobose pseudopodia and shells, exhibits shells formed through either the agglutination of foreign materials gathered from the environment (xenogenous shells) or a combination of agglutination and secretion (autogenous shells). Testate amoebae possessing filose pseudopodia are classified within the order Euglyphida, and most existing species in this order are characterized by siliceous shells comprised of self-synthesized Si platelets, also known as idiosomes.[56] Sommer *et al.* looked specifically at bSi, which is produced by testate amoebae and trees, while studying Si cycling in a beech forest in northeastern Germany. They observed that phytogenic Si pools amounted to 660 kg Si per hectare, with an average Si uptake by plants of 35 kg Si per hectare per year. Protozoic Si pools were significantly smaller, weighing 1.9 kg Si per hectare, but the Si uptake by trees was similar to that of the silicification process in disomic testate amoebae, reaching 17 kg Si per hectare per year.[58] It is known that at least certain fungi and bacteria, like *Proteus mirabilis*, can accumulate Si inside their cells. Additionally, by releasing acidic metabolites, these organisms can accelerate the dissolution of both crystalline and amorphous Si (bioweathering). To comprehensively grasp the importance of biota in Si cycling within agricultural plant–soil systems, there is a pressing need for additional research on bSi produced by sponges, fungi, and bacteria.[59,60]

1.2.2 Biogenic Silica from Aquatic Sources

Ocean surface planktonic organisms are the main producers of bSi. Within the top 100 meters of the water column, a sizable portion (60 per cent on average) of the bSi production is recycled through dissolution.[5] Also, one essential element of the global oceanic Si cycle is silicifying organisms. The silicic acid found in seawater can be used by diatoms, silicoflagellates, sponges, and siliceous rhizarians to construct complex skeletons that are thought to enhance vital functions like mechanical protection for the cell, armour against predators, an efficient pH buffer, or an enhancement for the absorption or accumulation of bio essential elements.[61,62] Previous research has indicated that the unique optical characteristics of the frustule may provide diatoms with an advantage.[63] Diatoms are thought to be the primary global contributors to the Si cycle, controlling the rate at which bSi is produced as well as its standing stock in the water column.[64] Hence, it is noteworthy to mention that over 20 billion tons of bSi produced worldwide each year are essential to the carbon cycle and have the power to influence atmospheric CO_2 concentrations. A quarter of this amount is produced in seawater by tiny photosynthetic algae called diatoms and, to a much lesser extent, by sponges and radiolarians of the zooplankton,[65] which is quite evident by looking at Table 1.1. Considering that diatoms make up between 35 and 75% of marine net primary production and roughly 20% of Earth's photosynthetic activity, it is clear that bSi is important to the planet's ecosystem. Various research initiatives, including the use of sediment traps and environmental molecular surveys, have further emphasized the importance of rhizarians in biogeochemical cycles and the transport of carbon and bSi to the deep continental ocean.[66]

1.2.2.1 *Biogenic Silica from Freshwater Organisms*

Rivers receive a significant amount of phytolith bSi from vegetation and soils. Thus, alterations in land use may impact the composition and concentrations of bSi in neighboring rivers. An investigation explored into the origin and characteristics of bSi present in rivers of the Asian tropical region, along with the phenomenon of reverse weathering observed in these fluvial systems. The study utilized samples collected from rivers in Peninsular Malaysia.[41,67] The findings revealed that the bSi samples obtained from sediments included phytoliths, diatoms, and sponge spicules. Elongate-form phytoliths constituted 92.8–98.3% of the total bSi in the Pahang River. Despite representing a minor fraction of the overall bSi volume, pennate diatoms dominated the diatom-derived bSi in this river. In contrast to the Pahang River, the Pontian and Endau rivers exhibited elevated concentrations of diatom-derived bSi, primarily of the Centricae form, making up 68.8% and 79.3% of the total bSi volumes, respectively. The bSi particulates had an average carbon content of 4.79%, ranging from 1.85% to 10.8%.[67]

According to the findings of Ran *et al.* for the rivers Changjiang (Yangtze) and Huanghe (Yellow), phytoliths and diatoms make up most of the riverine bSi. River bSi concentrations are dependent on in-stream primary production and terrestrial inputs.[68] If there was a bSi tracer, one could monitor the differences in bSi derived from freshwater and marine environments, establishing a connection between terrestrial and aquatic ecosystems. Tracking the origin of bSi in aquatic environments has traditionally involved identifying particles once bSi has been separated from other particles.[68,69] The vegetation and sediment in tidal marshes have the capacity to accumulate substantial quantities of bSi. Throughout the entire year, an investigation was conducted to assess the bSi content in prevalent freshwater marsh plants, including *Salix* sp., *Impatiens glandulifera*, *Urtica dioica*, *Phragmites australis*, and *Epilobium hirsutum*. It was seen that *P. australis* had the highest concentration of bSi (accumulating from 6 to 55 $mg\,g^{-1}$) among the living plants. The dead stems of *P. australis* exhibited the highest bSi content at 72.2 $mg\,g^{-1}$. In contrast, considerably lower levels of bSi were observed in *U. dioica* (<11.1 mg g^{-1}), *I. glandulifera* (<1.1 mg g^{-1}), *E. hirsutum* (<1.2 mg g^{-1}), and *Salix* sp. (<1.9 mg g^{-1}). Low concentrations of bSi (<6 mg g^{-1}) were found in underground biomass, with the exception of *P. australis* rhizomes (<15 mg g^{-1}). From the surface (9 to 10 mg g^{-1}) to deeper layers (5 to 7 mg g^{-1}), the sediment bSi content decreased. It was observed that in the biogeochemical cycling of Si, *P. australis* wetlands may be a crucial sink for bSi.[70]

1.2.2.2 Biogenic Silica from Marine Organisms

Among Si-forming organisms, siliceous sponges are unique in that they can enzymatically synthesize bSi, which has growing applications in nanobiotechnology, biomedicine, and biooptics.[71] The identification of an enzymatically active protein within the axial filaments of spicules (structural components of sponges such as demosponges and hexactinellids), responsible for synthesizing polymeric silicate known as bSi, marked a noteworthy progression in comprehending the formation of siliceous spicules. The enzyme, referred to as silicatein, has been observed to catalyze and polycondense bSi during both the axial and radial expansion phases of the spicules. In contrast to the deposition of bSi onto organic templates from a super-saturated solution, as observed in diatom frustules and plant phytoliths, the siliceous spicules of sponges undergo formation within an intra organism environment characterized by hypo-saturation. This occurs through an enzymatic mechanism that reduces the activation energy of the polycondensation reaction. Orthosilicic acid, also recognized as TEOS (tetraethyl orthosilicate), has been utilized as a substrate or precursor for bSi in enzymatic reactions within *in vitro* systems.[72] A different study suggests that the amorphous hydrated Si spicules that make up the skeleton of *Euplectella aspergillum* are held together by additional layers of Si.[73]

The bSi that reaches the seafloor primarily comprises siliceous hard parts from siliceous sponges, silicoflagellates, radiolaria, and diatoms. The main sources of bSi in marine sediments are typically diatom valves, radiolaria, and sponge spicules.[74] About half of the bSi deposited in the marine environment comes from sediments found in the Southern Ocean, and diatoms are mostly responsible for this siliceous material.[34] However, the findings of a different study suggest that, during the period when the Larsen ice shelf covered the sea surface, sponge spicules rather than diatoms were the primary source of bSi in this region's sedimentary record.[75] It has been documented that diatom-derived bSi possesses a unique chemical structure featuring exposed free hydroxyl groups that can be altered to attach chemical or biological entities.[76] A group of researchers recently revealed exciting potential applications of an alternate Si porous material derived from diatoms as a readily available, naturally occurring, multifunctional material for regenerative medicine.[77] Rembauville *et al.* demonstrated a robust correlation between wintertime bSi and particulate organic carbon export, primarily linked to *Fragilariopsis kerguelensis* in the Indian sector, employing sediment traps.[78] Another investigation proposes that valuable three-dimensional, nanoporous, hollow-structured bSi can be obtained alongside organic components such as lipids, proteins, and bioactives with minimal chemical input by utilizing diatoms as a tool.[79]

Despite not having silicon-dependent cellular structures, recent research has shown that some smaller marine organisms (such as the picocyanobacterium *Synechococcus*) can also absorb silicic acid, or DSi, and accumulate Si. That said, it was demonstrated that the five strains of picoeukaryotes grown in cultures supplemented with added dSi accumulate bSi. These strains include two marine species (*Ostreococcus tauri* and *Micromonas commoda*) and three novel isolates from the Baltic Sea.[80] A study proposed that the low-latitude South China Sea's bSi fluxes are primarily caused by radiolarians, as evidenced by the relatively strong bSi correlation for radiolarian abundance and weak for diatom abundance.[81] Open-ocean planktonic protists, known as siliceous polycystines and phaeodarians, are distributed all over the water column. These protists are characterized by intricate siliceous skeletons primarily developed through the absorption of silicic acid. While their role in the Si biogeochemical cycle remains somewhat obscure, these protists do play a part in the marine pools of organic carbon and bSi.[65]

1.2.3 Miscellaneous Sources

Other than the above-mentioned sources of bSi, it is perceivable that soil and vegetation are good sources of bSi. In certain regions of Brazil, lake bed sediments contain high concentrations of short fibers of bSi along with organic matter, sand, and clay.[82] On a global scale, there is a wealth of biomass residues readily accessible from agricultural food production. These biomass residues have important uses as feedstock for bioenergy and as starting points for producing materials with added value. The assessment involved the production of bSi from diverse biomass residues in Africa, including cornhusk, corncob, yam peelings, cassava peelings, and coconut husks.[83] The weathering of continental silicates is the primary source of DSi in rivers. Based on volume-weighted averaging, 16% of the suspended Si load from rivers is delivered to the ocean as bSi. In most cases, hot alkaline solutions are used to measure bSi in sediments.[41]

1.3 Structure and Properties of Biogenic Silica

It is notable to mention that bSi is characterized by various structural shapes, including sheet-like, globular, fibrillar, helical, tubular, and folded sheets. It also displays variations in density, content, and structure. There are many degrees of biosilica structural organization found in the natural world.[84]

1.3.1 Structure of Biogenic Silica

Due to its inherent amorphous nature, bSi is a biomineral with infinitely adaptable morphology. At locations of mineralization, the mechanisms governing structural organization and the resultant morphological forms involve spatial constraints, ordered particle aggregation, and chemical regulation.[85] The fundamental unit of most silicon-containing minerals is the SiO_4 tetrahedron as shown in Figure 1.2, where silicon is situated at its center, and oxygen atoms are in contact with each other. Additionally, it is known that minerals like stishovite and coesite, incorporating SiO_6 octahedral units, display sixfold coordination for silicon.[85,86] The Si–O bond measures approximately 0.162 nm in all these structures, significantly smaller than the sum of the covalent radii of silicon and oxygen atoms (0.191 nm). This size difference contributes to the remarkable stability of the siloxane bond. Central to the chemistry of silicon crystals is the arrangement of SiO_4 and SiO_6 units, along with their inclination to create three-dimensional framework structures.[87]

Nature exhibits various levels of hierarchically structured three-dimensional organization, ranging from the siliceous skeletons of marine glass sponges that can reach up to 2 m in height and 70 cm in width, to bSi nanogranules found in bacteria and mitochondria.[84] Also, most of the time, the nanostructure of bSi is described as nanospheres made of colloidal SiO_2 with diameters varying from 10 to 100 nm. The highly-branched polysilicic acid (general formula $[SiO_x(OH)_{4-2x}]_n$) that grows into these Si nanospheres is absorbed by the cells (microorganisms such as diatoms, radiolarians, silicoflagellates, and siliceous sponges) through active transport and functions as a Si nucleus.[87,88] The Si nanospheres frequently establish a three-dimensional network, giving rise to branched particle chains interconnected by siloxane linkages (Si–O–Si).[87,88] As is commonly observed in the diatom cell wall, flocculation (a process that almost closes and unites Si spheres and provides the surface a smooth appearance), fills the spaces between the nanospheres with Si in many situations. The Si deposited between the nanospheres usually dissolves first when the bSi dissolves, producing a porous structure. Both modern and ancient bSi have pore spaces with a mean diameter of 5–10 nm and an overall porosity range from 35–50%.[89]

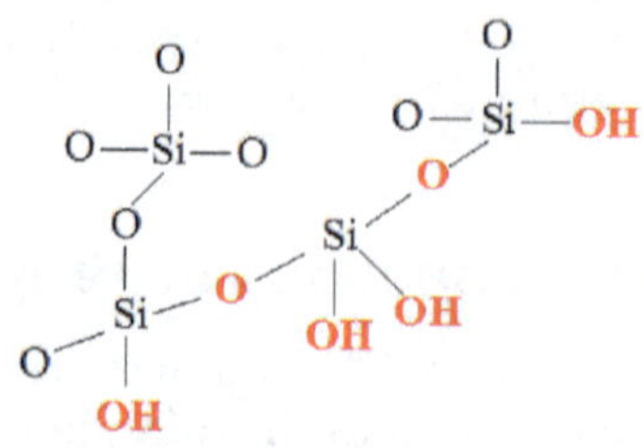

Figure 1.2 Structure of bSi.[86] Reproduced from ref. 86, https://doi.org/10.20517/2573-0002.2016.12, under the terms of the CC BY 4.0 license, https://creativecommons.org/licenses/by/4.0/.

1.3.2 General Properties of Biogenic Silica

The robust stability of Si–O–Si units significantly limits the reactivity of silicon in water at neutral pH, leading to the formation of condensed phases with potential crystalline or amorphous structures. Low temperature and pressure are preferred for amorphous phases where the Si–O–Si bond angle can vary greatly because of the large kinetic barrier to crystallization. Thus, one would anticipate that any bSi generated from scratch would be non-crystalline and subject to kinetic control regarding energetics.[84,85]

When discussing the marine environment, the mechanical strength of Si frustules of diatoms helps them fend off grazers.[90] Due to its enhanced solubility, bSi assumes a crucial role as a readily available source of Si (H_4SiO_4) for plants in soils. This aspect holds particular significance in agricultural biogeosystems.[91] In addition, bSi has exceptional absorption capacity, chemical inertness, thermostability, and a reasonably low cost.[92] Also, in comparison with synthetic Si, bSi is a much more biocompatible and nontoxic material which seems to be some of the most dependable properties of bSi.[93]

1.4 Extraction Methods for Biogenic Silica

Porous Si is typically created by precipitating alkaline silicates on an industrial scale. Alkaline silicates, or water glass, are generally produced by the very energy-intensive and high temperature (approximately 1400 °C) reaction of carbonate powders with Si sand.[15] Industrial processes include (i) hydrolyzing mineral precursors like TEOS, (ii) acidifying sodium silicate solutions, and (iii) smelting quartz sand with sodium carbonate at high temperatures (>1200 °C) to produce Si products like precipitated Si, fumed Si, arc Si, Si xerogels, and xerogels.[93] These synthesis routes all share energy-intensive characteristics and the emission of CO_2 along with the release of hazardous byproducts. To be precise the conventional method is not only costly but also environmentally dangerous since it produces 20 tons of wastewater, 0.23 tons of CO_2, and 0.74 tons of sodium sulfate for every ton of Si produced. This goes against the idea of sustainable development.[15,94] Because of these drawbacks, industrialized Si products are now less desirable in terms of cost and environmental impact, which restricts their widespread use in various industrial sectors. Thus, an alternative approach that is both environmentally friendly and commercially viable is needed to produce Si.[15,95] For this purpose, several different approaches such as biological, chemical and thermal methods are included.

1.4.1 Chemical Method

Chemicals including acids, bases, and other agents are employed in this process to raise the quality of the Si by eliminating impurities as shown in Figure 1.3(a) and (b).[96] In addition, contaminants can be changed into ions

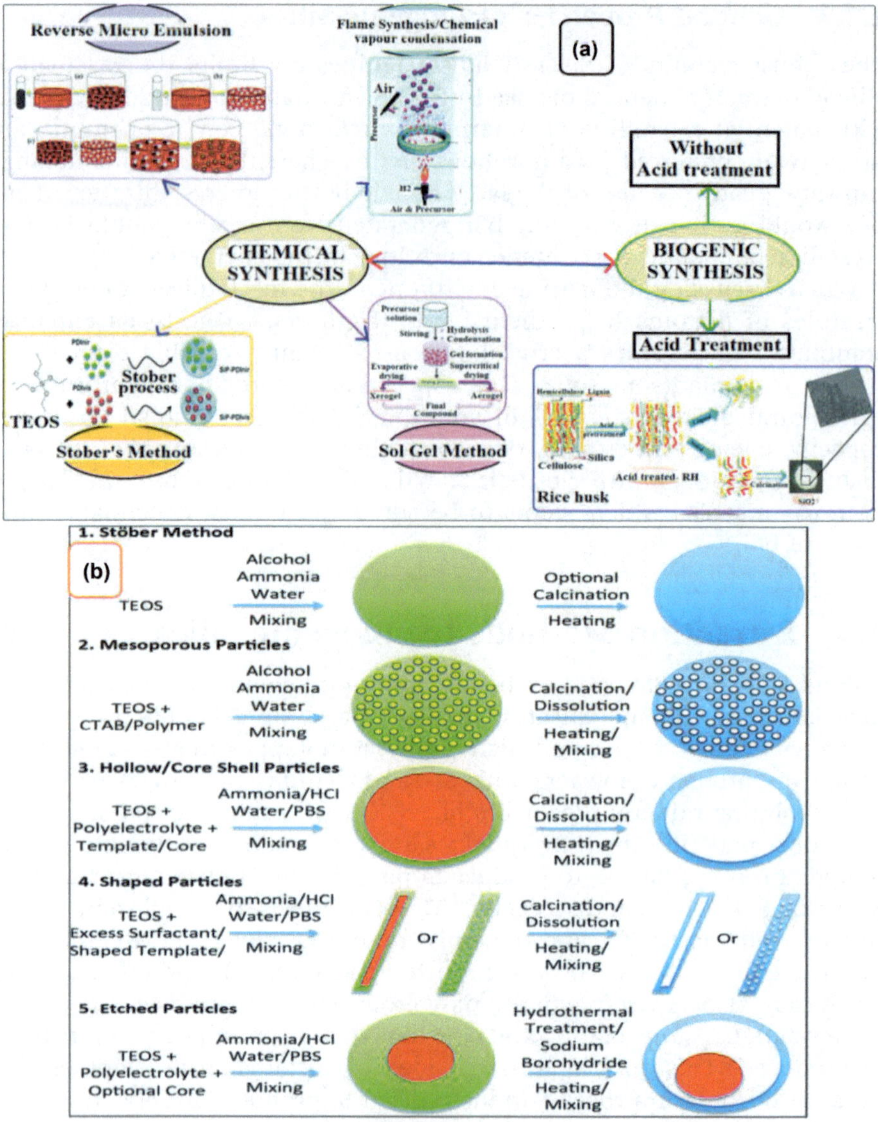

Figure 1.3 Depiction of (a) the synthesis of Si NPs through chemical and biogenic processes (b) typical techniques employed in the chemical synthesis of Si NPs.[96] Reproduced from ref. 96, https://doi.org/10.1039/D2RA01587K, under the terms of the CC BY 3.0 license, https://creativecommons.org/licenses/by/3.0/.

that can dissolve by acid leaching.[97] Given that RH is abundant in plants used for grain cleaning and can be utilized to produce various organic compounds like furfural, xylose, and polysaccharides, its potential as a renewable raw material for generating high-quality amorphous SiO_2 is highly promising.[98] One such study reported that chemical procedures were used to extract Si from RH. HFA was dissolved in 15% (1 : 4.5) NaOH solution by weight, then the mixture was kept at 100 °C for two hours before being filtered. Additionally, NaOH was added to separate the Si from the impurities and create a sodium silicate solution. This solution was then converted to hydrosol Si by using an ultrasonication-assisted method to get the pH of the solution down to 6 or 7. Consequently, the impact of pH on the purity of Si was generated, and upon dehydrogenation, gelation, and drying, powdered SiO_2 was acquired.[99]

Alyosef *et al.* exhibited two methods for producing amorphous Si from Egyptian RHs that are low-cost and least detrimental to the environment. The first method entails leaching with citric acid (CA), dry milling, and thorough washing of the husk sample with water. Two stages of the leaching procedure were used: first, for 180 minutes at 323 K and then for 60 minutes at 353 K. The leached sample underwent four successive stages of heat treatment in a muffle furnace following washing and drying.[100] All of the previously listed steps are included in the second route, with the exception of the CA leaching. The findings demonstrate that there are numerous benefits to CA leaching in terms of Si yield and porosity as well.[100] Utilising a combustion process in conjunction with pre-treatment techniques can yield superior bSi from RH. A study demonstrates the comparison of applicability of CA, and gluconic acid (GA) for the preparation of bSi from RH. The results showed that, with a small amount of Si loss, both GA and CA could remove the majority of the metal alkali impurities in RH. The results also indicate that GA is a great new environmentally friendly carboxylic acid that can be used to produce bSi from RH.[101]

1.4.2 Thermal Method

The thermal extraction process involves the valorization of agricultural biomass into ash using various methods such as a muffle furnace, fixed bed furnace, inclined step-grate furnace, cyclonic furnace, fluidized bed reactor, rotary kiln, or tubular reactor, with a specified time period typically ranging from 2 to 8 hours.[102] The thermal approach employs either pyrolysis or calcination as a form of heating. With the increasing interest in environmentally friendly methods for Si production, there is a demand for novel and dependable extraction techniques. One such technique was used by Park *et al.*, in which a pilot-scale one-pot alkali hydrothermal and ball-milling method was established for the continuous Si extraction.[103] In a different study, a thermal approach was used to extract Si from RHs, which effectively eliminated the majority of organic contaminants. The heating temperature was changed in this investigation to observe its impact on the percentage of Si that was achieved. Furthermore, the acid-treated RHs were utilized for the direct extraction of Si following dehydration with a NaOH solution. Then, Si is extracted from the sodium silicate that is left behind by adding the right quantity of mineral acid and heating it for 5–6 hours. This process produces

RHA, which is then pyrolyzed in a muffle furnace at temperatures ranging from 500 to 850 °C.[104]

Also, using residues from South African sugarcane and maize, biogenic amorphous Si was created through a thermo-chemical technique. To produce Si, various sugarcane fractions (such as the leaves, pith, and fiber) from South Africa were processed. The biomass samples were leached for two hours at 353 K with either 7 wt% H_2SO_4 or 7 wt% CA. Subsequently, the samples were subjected to a four-step program in a furnace, involving drying, rinsing, and burning, with temperatures ranging from room temperature to 873 K. Sugarcane leaves were used to produce high-quality amorphous bSi, with up to 99.3 wt% Si content and 0.1 wt% residual carbon. In addition to its outstanding textural characteristics, the produced bSi exhibited a surface area of up to 323 $m^2 g^{-1}$, a pore volume of 0.41 $cm^3 g^{-1}$, and an average pore diameter of 5.0 nm.[105]

1.4.3 Biological Method

Biological or microbiological approaches employ microorganisms like bacteria, fungi, cyanobacteria, chrysophytes like diatoms, and Porifera (also known as Californian red worms) to produce bSi in a more environmentally friendly and cleaner manner.[106] These organisms may utilize materials containing Si as a source of food, which causes a process called bio-silicification (the buildup of Si inside the cells). To obtain above 90% purity of bSi, the resultant product can be removed by disrupting the cells, then precipitating, purifying, and drying.[105,106] A study has reported employing enzymatic fluids comprising lipase, amylase, trypsinogen, and other digestive enzymes to extract Si from RHs that have been wet and digested in the mouths of California worms. Si NPs were produced with sizes ranging from 55 to 250 nm, and a Si recovery rate of 88% was achieved through the bio-digestion process of RH by worms after a 5-month processing period.[107] In a distinct study, Rohatgi *et al.* explored the potential use of white-rot fungus species, specifically *Cyathus*, to extract high-grade Si from RHs. The microbial fermentation process occurred over a span of 60 days, following which the material underwent drying at 80 °C, sieving to a specific size, and burning at 450 °C. According to the findings, this method resulted in diminished organic matter and a heightened concentration of inorganic material, specifically SiO_2.[108]

1.5 Applications

Scientists have been fascinated by bSi's remarkable structures and properties for a decade now, and they have been using it to create new devices and technologies. In addition to its chemical inertness, which makes bSi extremely demanding in terms of science, it also demonstrated some of its remarkable inherent qualities that could be used to advance technology.[76] Regarding applications in nanotechnology, bSi is superior to synthetic

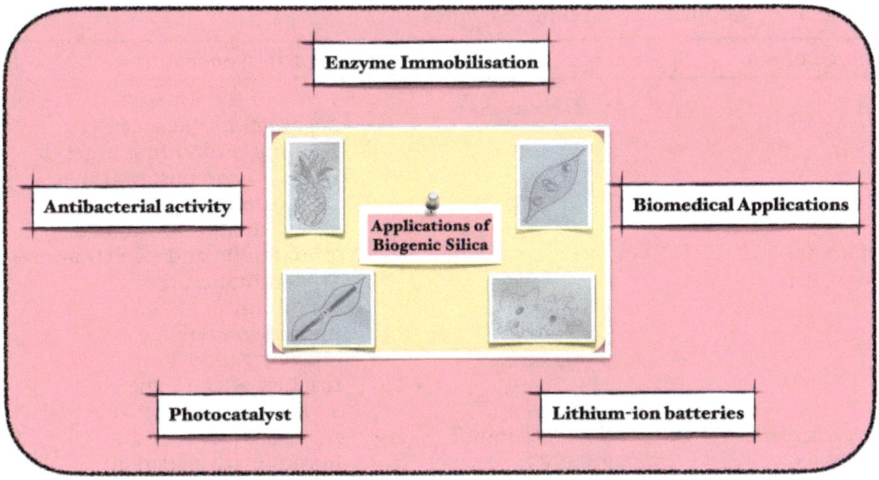

Figure 1.4 A schematic illustration of different applications of bSi.

nanoporous Si materials because it can be produced more affordably through *in vivo* biomineralization, which avoids the need for hazardous chemicals.[76,109] The following section highlights the various fields where bSi finds its application (Figure 1.4) and also gives a brief overview of the research work being conducted where bSi is the focal point as listed out in Table 1.2.

1.5.1 Photocatalysis

The unique properties of bSi enhance its applicability in catalysis and photocatalysis as shown in Figure 1.5 for various environmental applications, including the removal of organophosphate pesticides, extraction of heavy metals from wastewater, treatment of textile effluents, and decolorization of dyes.[109] Photocatalytic degradation of methylene blue (MB) using Si NPs derived from RS was explored in a particular research work conducted by Singh *et al.*[109] The potential and effectiveness of surface-modified polyacrylonitrile nanofibers/bSi composites for the photocatalytic degradation of malachite green dye were investigated.[110] Some studies have reported that forming nanostructures when Fe_3O_4 is dissolved into Si results in increased stability and photocatalytic activity enhancement.[111] On this basis Purwiandono *et al.* aimed to assess the photocatalytic potential of Fe_3O_4 dispersed in bSi (Fe_3O_4@SiO_2) derived from snake fruit (*Salaccazalacca*) leaf ash (SLA) from the standpoint of developing magnetite nanostructures and using agricultural waste. The study aimed to investigate the physicochemical characterization of Fe_3O_4@SiO_2 synthesized by SLA. Additionally, the research sought to provide mechanistic insights into the photocatalytic activity of Fe_3O_4@SiO_2 in the photocatalytic wet peroxidation and adsorption processes related to pollutants and Rhodamine B in batik wastewater.[112]

Table 1.2 Applications of biogenic silica.[a]

Source of bSi	Specific application	Characteristic feature	Ref.
RS	PCD of MB	66.26% of the dye was degraded in the first 10 minutes and first-order kinetics were observed in the degradation study, with a half-life of 6.79 minutes.	109
Diatomite and RH	PCD of MG	For diatomite and rice straw Si membranes, the maximum photodegradation efficiency of MG was reached after 15 and 25 minutes, respectively.	110
Salaccazalacca leaf ash	PC wet peroxidation of rhodamine B	Fe_3O_4@SiO_2 showed the maximum degradation efficiency (DE) of 99.9% after 60 min, and the reusability feature was expressed, as there was an insignificant change in the DE over the fifth cycle of use.	112
Diatom	PCD of methyl orange	The photocatalyst (bSi doped with doped with palladium(II) chloride) showed quick degradation (within 1 min), with an efficiency of 85% reaching over 98% after 75 minutes.	113
Diatom (*Pinnularia* sp.)	PC activity for acetaldehyde	Naturally occurring bSi–titania materials exhibit dependable immobilization of TiO_2 in the ordered structure of the diatom frustule and have high surface area (108 $m^2 g^{-1}$) and good porosity.	114
Bamboo leaves	PCD of MB	While other photocatalytic degradation using TiO_2/SiO_2 samples follows pseudo-first-order kinetics, the kinetics of photocatalytic degradation over 40 TiO_2/SiO_2 and 60 TiO_2/SiO_2 with the addition of H_2O_2 obey the modified Freundlich kinetic model.	115
RH	PCD of MB	In addition to BH-TiO_2/SiO_2, this technique can be used to create other element-self-doped TiO_2 in just one step, such as phosphorus- and iodine-containing loofah and kelp.	116

Table 1.2 (*Continued*)

Source of bSi	Specific application	Characteristic feature	Ref.
Sugarcane bagasse	BM application (biocompatibility with hLFCs)	bSi NPs' benefits may allow for a range of real-world uses, including bone tissue engineering, scaffolds for *in vitro* lung cell culture, precursors for nanocomposites.	119
RH	BM application (biocompatibility with hLFCs)	The bSi NPs may be safely used for bone tissue engineering in BM applications as well as an anti-caking agent in the food industry, according to preliminary research.	120
RH	Bone tissue engineering (biocompatibility with hMSc)	Amorphous Si is a suitable source of bSi NPs for solar cell applications, according to photoluminescence studies. Furthermore, there were no notable alterations in the nuclear morphology of hMSc caused by bSi NPs.	121
Sorghum bicolor seed head	Bone tissue engineering (biocompatibility with hMSc)	The bSiNs cause hMScs to undergo osteogenic differentiation in a time- and dose-dependent manner. Osteoblast formation may be stimulated by bSiNs, and scaffolds based on bSiNs may be employed in bone tissue engineering.	122
RH	Bone tissue engineering (biocompatibility with hMSc)	The materials are nontoxic and biocompatible, according to both the MTT assay results and the microscopy observations.	123
Allium cepa bulb	Antioxidant activity	The bSi NPs exhibited good thermal stability and were spherical, well-dispersed, and amorphous, with an average size of 18 ± 2 nm. The results of the DPPH assay demonstrate the radical scavenging ability of the bSi NPs, indicating their inherent biocompatibility.	124
RHA	Anti-bacterial activity	With the smallest particle size, maximum surface area, and best bacterial compatibility, the brown RHA-derived nanopowder is the most suitable material.	126

Table 1.2 (*Continued*)

Source of bSi	Specific application	Characteristic feature	Ref.
Pineapple peel	Bacterial anti-adhesive films	The high-density coverage coating created by BSR reduced the adhesion of *E. coli* while increasing the PDMS roughness and surface's water repellency. Because of its high porosity, this coating may also be used as a platform for encapsulating active compounds.	127
Punica granatum leaf extract	Anti-bacterial activity	The bSi NPs mediated by *P. granatum* leaf demonstrated strong antibacterial activity against the chosen pathogenic bacteria (*Salmonella* and *E. coli*).	128
Lantana camara	Anti-bacterial activity	When compared to TG hydrogels, the cytocompatible CTAB@bSiNP-TG hydrogels showed an approximately 10- to 12-fold increase in anti-bacterial activity against *Pseudomonas aeruginosa* and *Staphylococcus aureus*.	129
Equisetum myriochaetum	Enzyme immobilization	Horseradish peroxidase (HRP) and *Coprinus cinereus* peroxidase (CiP), two commercial peroxidases, were immobilized on the biogenic materials.	130
RH	Enzyme immobilization	When compared to the free enzyme, PersiXyn2@bis-AE@RKIT-6 improved the breakdown of lignocellulosic agricultural waste (RS) and reduced sugar production by up to 35%.	131
Oil palm leaves	Enzyme immobilization	Results point to the potential application of NC-SiO_2-PES/CRL to catalyze a quick and efficient production of PeVa, as well as the suitability of NC-SiO_2-PES for other enzyme activation.	132
Oil palm leaves	Enzyme immobilization	The immobilization parameters yielded maximum protein loading and specific activity of	133

Table 1.2 (*Continued*)

Source of bSi	Specific application	Characteristic feature	Ref.
		67.5 ± 0.72 mg g^{-1} and 320.8 ± 0.42 U g^{-1} of support, respectively, with 80% of the CRL (84.5 mg) initially offered immobilized onto the surface.	
Oil palm leaves	Enzyme immobilization	In comparison to free CRL (48.75 ± 0.70%), the CRL/SiO_2/Fe_3O_4/GO yielded 77.43 ± 1.04% of EV, demonstrating its suitability to hyperactivate and stabilize CRL for adequate EV production.	134
RH	Enzyme immobilization	Compared to octyl-SiO_2, the ester productivity of immobilized TLL on amino-SiO_2 was 1.3–3.1 times higher.	135
RH	LiBs	At a current density of 0.5 A g^{-1}, these SNCC spheres could provide a high reversible specific capacity of 1380 mA h g^{-1} while maintaining 1031 mA h g^{-1} after 100 cycles.	136
RH	LiBs	At a high applied current density of 1000 mA g^{-1}, the composite system made of bSi NPs produced an initial capacity of 1000 mA h g^{-1}.	137
RH	LiBs	Even at a rate of 16 A g^{-1}, a remarkably high delithiation capacity of 907 mA h g^{-1} can be maintained. After 200 cycles, a discharge capacity of 830 mA h g^{-1} at a current density of 1 A g^{-1} was achieved.	138
RH	LiBs	Porous C/SiO_2 composite shows good rate-capability, good cycling stability, and a high discharge specific capacity (approximately 1105 mA h g^{-1} at 0.1 A g^{-1}) when tested as anode material.	139
RH	LiBs	After 100 cycles, roughly 600 mA h g^{-1} steady state reversible capacity at 100 mA g^{-1} current density was achieved.	140

Table 1.2 (*Continued*)

Source of bSi	Specific application	Characteristic feature	Ref.
Diatom	Paleoproductivity and evolution of the environment	When utilizing bSi as a stand-in for diatom primary production, one should take the bulk sediment's grain size distribution into account.	144
Diatom	Waterproofing surface treatment for recycled concrete	Diatom biodeposition works well as a protective surface treatment for building materials made of cement.	146
RH	Gravity chromatography	The high surface area (196.2 $m^2 g^{-1}$) of highly pure mesoporous bSi NPs demonstrates superior performance for gravity chromatography.	147
RH	Silica-reinforced cellulose-based composite bioplastic films	After 21 days, the biodegradability of bioplastic films in the soil ranged from 25.5% to 55.7%, which was considerably higher than that of conventional plastics.	148

[a] Photocatalytic degradation – PCD; biomedical – BM; reactive oxygen species – ROS; biogenic silica nanostructures – bSiNs; biogenic silica rosettes – BSR; polyethersulfone – PES; *Candida rugosa* lipase – CRL; pentyl valerate – PeVa; ethyl valerate – EV; *Thermomyces lanuginosus* lipase – TLL; silicon/nitrogen-doped carbon/carbon nanotube – SNCC.

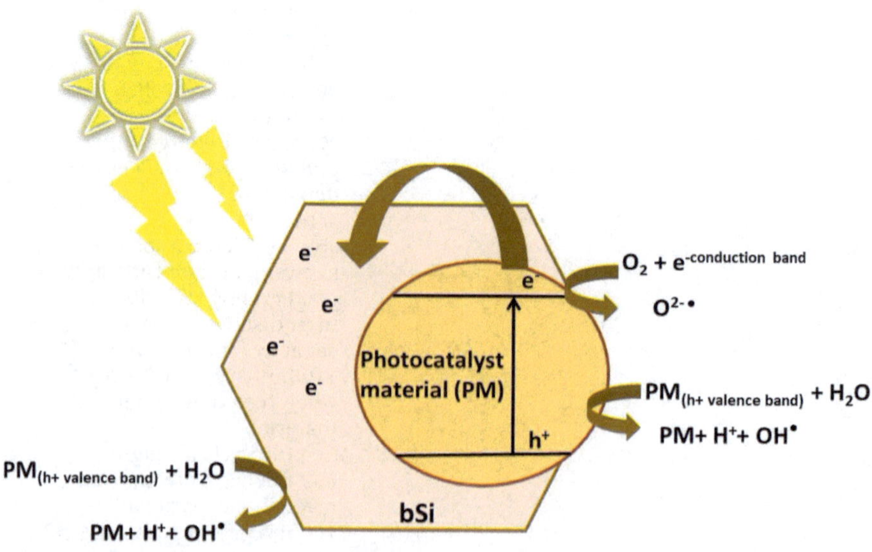

Figure 1.5 Role of bSi in photocatalysis.

A novel photocatalyst based on diatom bSi doped with palladium(II) chloride nanoparticles was produced and tested for effective methyl orange degradation in water solution.[113] The biological immobilization of titanium dioxide (TiO_2) in the porous 3D architecture of the frustules was achieved through a controlled culture of the diatom *Pinnularia* sp. on soluble titanium in a batch procedure. The extracted *Pinnularia* cells were treated with either high-temperature treatment or nitric acid (65%) to isolate the silica–titania frustules. The silica–titania material that is formed is assessed for its potential to reduce acetaldehyde (C_2H_4O) through photocatalysis.[114] In a different study set, bSi from bamboo leaves was extracted, and titanium tetraisopropoxide was used as a titania precursor using a sol–gel mechanism to create TiO_2/SiO_2 composite photocatalysts. The MB photocatalytic degradation process was used to assess the relationship between physicochemical parameters and photocatalytic performance under UV irradiation, both with and without adding H_2O_2 as an oxidant.[115] Similarly, RH was used in another study to synthesize biogenic hierarchical TiO_2/SiO_2 for the degradation of MB.[116]

1.5.2 Biomedical Applications

Because of their biological characteristics, Si NPs have drawn a lot of attention recently for a variety of biomedical applications, such as protein adsorption and separation, nucleic acid detection, molecular imaging, gene therapy, drug delivery, and scaffolds.[117] Particularly, tissue engineering and regenerative medicine applications use nanomaterials based on Si. Dietary intake was suggested by Jugdaohsingh and coworkers to significantly increase bone mineral density in men and premenopausal women.[118] In a different study, the authors used sugarcane bagasse as a bio-precursor to create Si NPs. The evaluation of Si NPs biocompatibility in human lung fibroblast cell (hLFCs) revealed that there are no appreciable cytotoxic effects of Si NPs. Additionally, Si NPs to some extent affect gene expression and the course of the cell cycle. According to this research, Si NPs and hLFC are biocompatible and can be used for bone tissue engineering, scaffolds for *in vitro* lung cell culture, along with other applications.[119] On a similar basis another study presents a novel and straightforward method to synthesize high-purity Si NPs from RHs and assess their biocompatible characteristics using *in vitro* cell-based techniques. Using a viability assay and evaluations of cellular morphological alterations, intracellular ROS production, mitochondrial transmembrane potential, and oxidative stress-related gene expression, the biocompatibility of the Si NPs with hLFCs was explored. According to the findings, there was no discernible incompatibility between the Si NPs in these *in vitro* cell-based methods. According to these initial results, Si NPs are biocompatible, may be the greatest substitute for synthetic Si, and can be used in biomedical applications.[120]

RH was used as a precursor to create Si NPs and using human mesenchymal stem cells (hMSc) and the MTT assay, the biological characteristics of bSi NPs were investigated. The results of the cell viability studies demonstrated the high level of biocompatibility between Si NPs and hMSc. These Si NPs can be used in bone tissue engineering, according to the findings of this study.[121] In another study, the authors valorized *Sorghum bicolor* seed head with the production of a biocompatible bSi nanostructure (BSN). Furthermore, because bone marrow hMSc can differentiate into a variety of lineages, the authors have evaluated the role of BSN on osteogenic differentiation using hMSc in an *in vitro* model.[122] In another study, RH was used as a Si source to create bSi–metal phosphate nanocomposites. A cell viability test and the examination of microscopy images were used to investigate the *in vitro* biological characteristics of the nanocomposites. The material was shown to be nontoxic and to have excellent biocompatibility with hMSc through cytocompatibility studies. These nanocomposites can be employed in bone tissue engineering, as per these findings.[123]

The biogenic polyphosphate (bio-polyP) and bSi, two naturally occurring polymers produced by deep-sea sponges, have also been found in recent years to promote morphogenetic on both osteoblasts and osteoclasts. The discoveries of this research have led to the development of novel bioinspired methods using 3D printing technology to create bone biomimetic templates.[72] In a different study, a green chemistry method was used to create the silica dioxide nanoparticles using the *Allium cepa* bulb aqueous extract. This work also indicates that the synthesis of Si NPs mediated by *A. cepa* may find use in biomedical applications.[124]

1.5.3 Antibacterial Activity

Tailored biogenic Si NPs have been shown by Capeletti *et al.* to have antibacterial qualities and less cellular toxicity when applied to cell lines.[125] Furthermore, human cell lines tend to exhibit biocompatibility with biogenic Si NPs.[121] The study conducted by Sharma *et al.* created bSi nanopowder from RHA using a chemical process and assessed their toxicity and compatibility with bacteria. Microorganisms classified as Gram-positive (*Escherichia coli*) and Gram-negative (*Staphylococcus aureus*) were used to test the bSi nanopowders' antibacterial activity. This work highlights the significance and utility of RHA-mediated synthesized bSi in the fields of medicine, biomedicine, clinical, and biology.[126] In a different study, the antibacterial activity of bSi rosettes derived from leftover pineapple peels against *Escherichia coli* as well as the bSi rosette films' anti-adhesive qualities on polydimethylsiloxane (PDMS) were assessed. The study also suggests that because of its high porosity, this coating may be used as a platform for encapsulating active compounds. Therefore, to enhance its antifouling properties, various bSi rosette arrangements could be investigated.[127]

Punica granatum leaf extract was used as a capping and stabilizing agent for the synthesis of Si NPs. The antibacterial activity of *P. granatum* assisted Si NPs was evaluated against two Gram-negative bacterial pathogens, namely *Salmonella* sp. and *E. coli*. The antibacterial experiments demonstrate the good antibacterial activities of Si NPs aided by *P. granatum*.[128] The invasive species *Lantana camara* was used to produce Si NPs which were then chemically altered utilizing nitrogen-containing moieties, such as CTAB and APTES. The modified Si NPs were employed as electrostatic cross-linking agents to create hydrogels made of tragacanth gum (TG). When compared to TG hydrogels, the cytocompatible CTAB@SiNP-TG hydrogels showed an approximately 10- to 12-fold increase in antibacterial activity against *Pseudomonas aeruginosa* and *Staphylococcus aureus*.[129]

1.5.4 Enzyme Immobilization

Enzymes are naturally occurring catalysts with various intriguing qualities, including high activity, specificity, stereo selectivity, *etc.* However, before they can be used on an industrial scale, some properties need to be modified.[76] Enzyme application in an industrial process is determined by its capacity to endure hazardous industrial circumstances and its ability to recover efficiently from catalytic reactions at industrial reactors. Soluble enzymes can be made more reusable, have improved thermal and storage durability, and have increased catalytic efficacy through immobilization.[76] The use of porous Si materials for enzyme immobilization has advanced significantly, but depending on the biomolecule, the Si source, and the intended application, a proper immobilization strategy is needed. In a recent study, the potential of bSi extracted from a rarely studied plant (*E. Myriochaetum*), its conversion to silicon and possible applications of both biogenic materials as supports for enzyme immobilization were investigated.[130] Also, using RH as a renewable source for synthesizing Si NPs, PersiXyn2 (recombinant xylanase enzyme) was immobilized to improve its stability and recyclability.[131]

In a different study, raw oil palm leaves (OPL) were used to extract nanocrystalline cellulose (NC) and SiO_2, which were then used as nano-fillers in polyethersulfone (PES) to create NC-SiO_2-PES, a platform for immobilizing *Candida rugosa* lipase (CRL).[132] Onoja *et al.* described the synthesis of butyl butyrate using a composite made of magnetite (Fe_3O_4) coated in nanosilica that was extracted from the ash of OPL as nanosupports for immobilizing CRL.[133] On a similar basis bSi was obtained from OPL and was used as an enzyme support. Its qualities could be enhanced by adding graphene oxide (GO) and Fe_3O_4 to produce a better ternary support for immobilizing enzymes, such as CRL which was further used to produce ethyl valerate.[134] Also, using a straightforward hydrothermal method, Si was extracted from RHs and functionalized with triethoxy(octyl)silane-OCTES (octyl-SiO_2) and (3-aminopropyl)-triethoxysilane-3-APTES (amino-SiO_2) in order to use it as a support for the adsorption-based immobilization of lipase from *Thermomyces lanuginosus*.[135]

1.5.5 Lithium-ion Batteries

Even though the as-synthesized silicon material's electrochemical performances still require development, producing silicon anode materials from biomass resources permits the battery industry to utilize subsidiary agricultural products effectively.[136] A group of authors report for the first time the fabrication of silicon/nitrogen-doped carbon/carbon nanotube (SNCC) nano/micro-hierarchical structured spheres using a simple electrospray method, using RH as the Si source. The inexpensive silicon composites derived from RH that were created using an easy-to-scale synthetic technique show promise for the next generation of rechargeable lithium-ion batteries (LiBs).[136] In a different work set, Si from RH was utilized to create Si NPs using a magnesiothermic reduction process. Moreover, the Si NPs were utilized to create a binder-free composite system with graphene and Si NPs for use as the anode material in LiBs.[137] Similarly, using a magnesiothermic reduction process, RHs were utilized as a source of raw materials to create porous Si. Also, a new technique was presented for creating a porous Si composite with reduced GO for LiBs anodes.[138]

$ZnCl_2$ was used as an activating agent during a simple carbonization process to create RHs into micro-sized porous C/SiO_2 composites in an Ar atmosphere. This porous C/SiO_2 composite shows good rate-capability, good cycling stability, and a high discharge specific capacity when tested as anode material for LiBs. This C/SiO_2 composite anode material's ease of use and inexpensive cost make it a viable substitute for the conventional graphite anode.[139] In a similar fashion a different study has investigated the use of SiO_x/C composite material as an anode for rechargeable lithium batteries, which is directly produced from agricultural RH byproducts using a financially feasible and ecologically friendly method.[140]

1.5.6 Other Applications of bSi

Because bSi is porous, it can be used in various ways. For example, it can be used as an affordable, non-traditional adsorbent and a good substitute for more expensive activated carbon in treating effluents to get rid of harmful metal ions or dyes.[141] In reference to this a study was conducted to develop a time and cost-effective method to increase the specific surface area and mesopore volume of bSi by partial pseudomorphic transformation in a microwave reactor using a low-cost surfactant.[141] Because of its high specific surface area and porous nature, bSi may find application in low-cost biocide delivery systems that provide controlled, sustained biocide release to protect wood and crops.[142] In addition, bSi can be used to make water glass at much lower temperatures (353–373 K) and under much milder conditions than in the traditional method of furnace roasting with sand and sodium carbonate (~1670 K) and additional hydrothermal treatment. Without a carbothermal reduction at 2173 K, high-purity Si can also be utilized to synthesis silicon tetrachloride ($SiCl_4$), silicon carbide (SiC), and silicon nitride (Si_3N_4).[142,143]

The sedimentary record of bSi and its generally good preservation efficiency all suggest that bSi is a potentially significant paleoproductivity proxy.[144] Consequently, variations in the abundance of bSi in sedimentary records are frequently very valuable for reconstructing paleoenvironments.

With significant worldwide ramifications, bSi records have been utilized over time to reconstruct paleoproductivity and paleoclimatic changes based on variations in their regional distribution.[81] And, since bSi is a major regulator of Si fluxes from terrestrial to aquatic ecosystems, it has been discovered to be a basic component of the global Si and carbon cycles. Since diatom reproduction is highly dependent on Si bioavailability (for the synthesis of diatom frustules), these Si fluxes regulate marine diatom production globally. Because marine diatoms make up to 54% of the biomass in the oceans, they can fix large amounts of CO_2 through photosynthesis.[64] The application of silicaceous biodeposition by diatoms as a waterproofing surface treatment for recycled concrete is the research subject for Maldonado and coworkers.[145]

One of the main biogenic elements of oceanic sediments bSi, can also be used to reconstruct historical fluctuations in the effectiveness of the biological pump of CO_2.[146] According to the research conducted by Schneider and others, bSi can be used to synthesize MCM-41/-48 and MFI-type zeolites, offering a sustainable method of producing high-performance materials for a variety of uses, including adsorption and catalysis.[147] Thus, either by its biological function or as a material, bSi offers a dual interest. A study demonstrated that bSi could be used to fabricate bioplastic films and ultimately in packaging. The goal of the study was to prepare Si-reinforced cellulose-based composite bioplastic films *via* a casting method, which could be evaluated in packaging applications, and to simultaneously extract cellulose fiber and crystal bSi from the same RHs.[148]

1.6 Conclusion

Living things have been chosen by nature to have the best potential to produce certain functional components that are essential to their survival, development, and procreation. These functional elements with highly replicable properties, have attracted a lot of interest from researchers as a very potential alternative to synthetic materials, particularly bSi. In particular, the unique properties of bSi, in terms of composition and structure have been used in a variety of applications.[76] Also, the wide diversity in the sources of bSi, be it terrestrial or aquatic has increased the bioavailability and utilisation of bSi in recent times. The use of agricultural biomass as a renewable source for the production of bSi has been highlighted in contemporary research which surely helps in resolving emerging issues of sustainability and waste management. In addition to this, the emerging extraction techniques of bSi are designed so that it is environmentally friendly and commercially viable.

Applications for bSi are abound, as they possess an unmatched diversity in form and structure. The primary benefit of bSi over other synthetic mesoporous materials is its ability to produce Si structures *in vivo* without the need for intricate chemical processes, making it an incredibly affordable material for analytical sciences. Scientists have been fascinated by bSi's remarkable structures and properties for a decade now, and they have been using it to create new devices and technologies. In addition to its chemical inertness, which makes bSi extremely demanding in terms of science, it also demonstrated some of its remarkable inherent qualities that could be used to advance technology.[76,149]

The number of research work carried out in recent times calls attention to the fact that bSi is fascinating in terms of its formation, structure, sources and application. Despite the fact that bSi has been investigated in a variety of fields, including the creation of novel materials and the development of new devices, it would not be surprising if their many characteristics were designed to demonstrate their utility in previously unexplored fields. An examination of the growing number of bSi applications may reveal new characteristics in them, leading to the development of fresh approaches to use them in unexplored fields.

References

1. D. J. Conley and C. L. Schelske, Biogenic silica, in *Tracking Environmental Change Using Lake Sediments: Terrestrial, Algal, and Siliceous Indicators*, 2001, pp. 281–293.
2. G. Bavestrello, R. Cattaneo-Vietti, C. Cerrano, S. Cerutti and M. Sará, Contribution of sponge spicules to the composition of biogenic silica in the Ligurian Sea, *Mar. Ecol.*, 1996, **17**(1–3), 41–50.
3. A. Petrucciani, N. Llopis Monferrer and M. López-Acosta, Silicifiers: The Glassy Creatures of the Ocean, *Front. Young Minds*, 2023, **11**, 1172756.
4. P. J. Tréguer, J. N. Sutton, M. Brzezinski, M. A. Charette, T. Devries, S. Dutkiewicz, C. Ehlert, J. Hawkings, A. Leynaert, S. M. Liu and N. Llopis Monferrer, Reviews and syntheses: The biogeochemical cycle of silicon in the modern ocean, *Biogeosciences*, 2021, **18**(4), 1269–1289.
5. S. Dixit, P. Van Cappellen and A. J. van Bennekom, Processes controlling solubility of biogenic silica and pore water build-up of silicic acid in marine sediments, *Mar. Chem.*, 2001, **73**(3–4), 333–352.
6. S. Loucaides, T. Behrends and P. Van Cappellen, Reactivity of biogenic silica: Surface *versus* bulk charge density, *Geochim. Cosmochim. Acta*, 2010, **74**(2), 517–530.
7. F. Fraysse, O. S. Pokrovsky, J. Schott and J. D. Meunier, Surface properties, solubility and dissolution kinetics of bamboo phytoliths, *Geochim. Cosmochim. Acta*, 2006, **70**(8), 1939–1951.
8. D. Zhu, J. N. Sutton, A. Leynaert, P. J. Tréguer, J. Schoelynck, M. Gallinari, Y. Ma and S. M. Liu, Revisiting the biogenic silica burial flux determinations: A case study for the East China seas, *Front. Mar. Sci.*, 2023, **9**, 1058730.
9. D. Liu, P. Yuan, Q. Tian, H. Liu, L. Deng, Y. Song, J. Zhou, D. Losic, J. Zhou, H. Song and H. Guo, Lake sedimentary biogenic silica from diatoms constitutes a significant global sink for aluminium, *Nat. Commun.*, 2019, **10**(1), 4829.

10. M. Maldonado, M. López-Acosta, C. Sitjà, M. García-Puig, C. Galobart, G. Ercilla and A. Leynaert, Sponge skeletons as an important sink of silicon in the global oceans, *Nat. Geosci.*, 2019, **12**(10), 815–822.
11. J. Clarke, The occurrence and significance of biogenic opal in the regolith, *Earth-Sci. Rev.*, 2003, **60**(3–4), 175–194.
12. L. A. September, N. Kheswa, N. S. Seroka and L. Khotseng, Green synthesis of silica and silicon from agricultural residue sugarcane bagasse ash–a mini review, *RSC Adv.*, 2023, **13**(2), 1370–1380.
13. H. Ahmad-Alyosef, H. Uhlig, T. Münster, G. Kloess, W. D. Einicke, R. Gläser and D. Enke, Biogenic silica from rice husk ash–Sustainable sources for the synthesis of value added silica, *Chem. Eng.*, 2014, **37**, 667–672.
14. F. C. Pa, A. Chik and M. F. Bari, Palm ash as an alternative source for silica production, *MATEC Web Conf.*, 2016, **78**, 01062.
15. H. Beidaghy Dizaji, T. Zeng, I. Hartmann, D. Enke, T. Schliermann, V. Lenz and M. Bidabadi, Generation of high quality biogenic silica by combustion of rice husk and rice straw combined with pre-and post-treatment strategies—A review, *Appl. Sci.*, 2019, **9**(6), 1083.
16. S. Kwon, R. K. Singh, R. A. Perez, E. A. Abou Neel, H. W. Kim and W. Chrzanowski, Silica-based mesoporous nanoparticles for controlled drug delivery, *J. Tissue Eng.*, 2013, **4**, 2041731413503357.
17. Y. Shen, Rice husk silica derived nanomaterials for sustainable applications, *Renewable Sustainable Energy Rev.*, 2017, **80**, 453–466.
18. N. Soltani, A. Bahrami, M. I. Pech-Canul and L. A. González, Review on the physicochemical treatments of rice husk for production of advanced materials, *Chem. Eng. J.*, 2015, **264**, 899–935.
19. S. Chandrasekhar, P. N. Pramada and L. Praveen, Effect of organic acid treatment on the properties of rice husk silica, *J. Mater. Sci.*, 2005, **40**, 6535–6544.
20. C. W. Foo, J. Huang and D. L. Kaplan, Lessons from seashells: silica mineralization *via* protein templating, *Trends Biotechnol.*, 2004, **22**(11), 577–585.
21. Y. Shen, P. Zhao and Q. Shao, Porous silica and carbon derived materials from rice husk pyrolysis char, *Microporous Mesoporous Mater.*, 2014, **188**, 46–76.
22. Y. Hu, T. Cai, H. Zhang, J. Chen, Z. Li, L. Zhao, Z. Li and H. Qiu, Two copolymer-grafted silica stationary phases prepared by surface thiol-ene click reaction in deep eutectic solvents for hydrophilic interaction chromatography, *J. Chromatogr. A*, 2020, **1609**, 460446.
23. Z. Xu, J. Zhong, X. Su, Q. Xu and B. Liu, Experimental study on mechanical properties of silica-based ceramic core for directional solidification of single crystal superalloy, *Ceram. Int.*, 2018, **44**(1), 394–401.
24. E. Mora, G. González, P. Romero and E. Castellón, Control of water absorption in concrete materials by modification with hybrid hydrophobic silica particles, *Constr. Build. Mater.*, 2019, **221**, 210–218.

25. J. M. Falcón, L. M. Otubo and I. V. Aoki, Highly ordered mesoporous silica loaded with dodecylamine for smart anticorrosion coatings, *Surf. Coat. Technol.*, 2016, **303**, 319–329.
26. S. Islam, H. Bakhtiar, M. S. Abd Aziz, M. B. Duralim, S. Riaz, S. Naseem, M. Bin Abdullha and S. S. Osman, CR incorporation in mesoporous silica matrix for fiber optic pH sensing, *Sens. Actuators, A*, 2018, **280**, 429–436.
27. W. K. Setiawan and K. Y. Chiang, Crop residues as potential sustainable precursors for developing silica materials: a review, *Waste Biomass Valorization*, 2021, **12**, 2207–2236.
28. S. R. Priyan, G. S. Kumar, K. Lalithambigai, M. Shkir, A. Khan, R. Rajendran and G. Arumugam, Microwave-assisted sol-gel synthesis of mesoporous NiO-decorated silica nanostructures utilizing biogenic silica source for supercapacitor applications, *J. Alloys Compd.*, 2024, **976**, 173206.
29. R. J. Haynes, The nature of biogenic Si and its potential role in Si supply in agricultural soils, *Agric., Ecosyst. Environ.*, 2017, **245**, 100–111.
30. N. Borrelli, M. F. Alvarez, M. L. Osterrieth and J. E. Marcovecchio, Silica content in soil solution and its relation with phytolith weathering and silica biogeochemical cycle in Typical Argiudolls of the Pampean Plain, Argentina—a preliminary study, *J. Soils Sediments*, 2010, **10**, 983–994.
31. I. Weir, S. Fawcett, S. Smith, D. Walker, T. Bornman and S. Fietz, Winter biogenic silica and diatom distributions in the Indian sector of the Southern Ocean, *Deep Sea Res., Part I*, 2020, **166**, 103421.
32. M. Yadav, V. Dwibedi, S. Sharma and N. George, Biogenic silica nanoparticles from agro-waste: Properties, mechanism of extraction and applications in environmental sustainability, *J. Environ. Chem. Eng.*, 2022, **10**(6), 108550.
33. H. Ahmad-Alyosef, H. Uhlig, T. Münster, G. Kloess, W. D. Einicke, R. Gläser and D. Enke, Biogenic silica from rice husk ash–Sustainable sources for the synthesis of value added silica, *Chem. Eng.*, 2014, **37**, 667–672.
34. D. J. DeMaster, The accumulation and cycling of biogenic silica in the Southern Ocean: revisiting the marine silica budget, *Deep Sea Res., Part II*, 2002, **49**(16), 3155–3167.
35. D. J. Conley, Terrestrial ecosystems and the global biogeochemical silica cycle, *Global Biogeochem. Cycles*, 2002, **16**(4), 68-1–68-8.
36. P. B. Kaufman, P. Dayanandan, Y. Takeoka, W. C. Bigelow, J. D. Jones and R. Iler, Silica in shoots of higher plants, in *Silicon and siliceous structures in biological systems*, Springer New York, New York, NY, 1981, pp. 409–449.
37. L. H. Jones and K. A. Handreck, Silica in soils, plants, and animals, *Adv. Agron.*, 1967, **19**, 107–149.
38. H. A. Currie and C. C. Perry, Silica in plants: biological, biochemical and chemical studies, *Ann. Bot.*, 2007, **100**(7), 1383–1389.

39. E. Trembath-Reichert, J. P. Wilson, S. E. McGlynn and W. W. Fischer, Four hundred million years of silica biomineralization in land plants, *Proc. Natl. Acad. Sci. U. S. A.*, 2015, **112**(17), 5449–5454.
40. J. D. Meunier, F. Guntzer, S. Kirman and C. Keller, Terrestrial plant-Si and environmental changes, *Mineral. Mag.*, 2008, **72**(1), 263–267.
41. L. Cary, A. Alexandre, J. D. Meunier, J. L. Boeglin and J. J. Braun, Contribution of phytoliths to the suspended load of biogenic silica in the Nyong basin rivers (Cameroon), *Biogeochemistry*, 2005, **74**, 101–114.
42. D. K. Chauhan, D. K. Tripathi, N. K. Rai and A. K. Rai, Detection of biogenic silica in leaf blade, leaf sheath, and stem of bermuda grass (Cynodon dactylon) using LIBS and phytolith analysis, *Food Biophys.*, 2011, **6**, 416–423.
43. T. Tkachenko, V. Yevdokymenko, D. Kamenskyh, Y. Sheludko, V. Povazhnyi and V. Kashkovskyi, Physico-chemical properties of biogenic SiO_2 nanoparticles obtained from agriculture residue, *Appl. Nanosci.*, 2020, **10**(12), 4617–4623.
44. S. Porrang, N. Rahemi, S. Davaran, M. Mahdavi and B. Hassanzadeh, Preparation and *in-vitro* evaluation of mesoporous biogenic silica nanoparticles obtained from rice and wheat husk as a biocompatible carrier for anti-cancer drug delivery, *Eur. J. Pharm. Sci.*, 2021, **163**, 105866.
45. Y. Basaran Elalmis, S. Yucel and I. Aydin, Amorphous biogenic silica production and utilization in experimental dental composites: Effect of silica gel formation pH on silica and composite properties, *Polym. Compos.*, 2021, **42**(10), 5111–5124.
46. A. Zareihassangheshlaghi, H. Beidaghy Dizaji, T. Zeng, P. Huth, T. Ruf, R. Denecke and D. Enke, Behavior of metal impurities on surface and bulk of biogenic silica from rice husk combustion and the impact on ash-melting tendency, *ACS Sustainable Chem. Eng.*, 2020, **8**(28), 10369–10379.
47. N. Setyawan, H. Purwaningsih and N. Apriliani, Morphology, extraction yield, and properties of biogenic silica nanoparticles from Indonesian rice husk as influenced by solvent type and aging time, *IOP Conf. Ser.: Earth Environ. Sci.*, 2022, **1024**(1), 012076.
48. J. C. Marchal, D. J. Krug III, P. McDonnell, K. Sun and R. M. Laine, A low cost, low energy route to solar grade silicon from rice hull ash (RHA), a sustainable source, *Green Chem.*, 2015, **17**(7), 3931–3940.
49. W. K. Setiawan and K. Y. Chiang, Amine-functionalized biogenic silica incorporation effect on poly (ether-block-amide) membrane CO_2/N_2 separation performance, *J. Membr. Sci.*, 2023, **680**, 121732.
50. A. M. Grimm, L. Y. Dorsch, G. H. Kloess, D. Enke and A. Roppertz, Transition metal promoted combustion of rice husk and rice straw towards an energy optimized synthesis of biogenic silica, *Biomass Bioenergy*, 2021, **155**, 106282.
51. R. A. Huang, X. Hu, Y. Guo, J. Wang and B. Yang, Highly hierarchical fibrillar biogenic silica with mesoporous structure derived from the

perennial plant equisetum fluviatile, *ACS Appl. Mater. Interfaces*, 2020, **12**(31), 35259–35265.

52. Y. R. Corrales-Ureña, C. Villalobos-Bermúdez, R. Pereira, M. Camacho, E. Estrada, O. Argüello-Miranda and J. R. Vega-Baudrit, Biogenic silica-based microparticles obtained as a sub-product of the nanocellulose extraction process from pineapple peels, *Sci. Rep.*, 2018, **8**(1), 10417.
53. D. Schneider, A. G. Attallah, S. Wassersleben, M. Wenzel, J. Matysik, R. Krause-Rehberg and D. Enke, Advanced textural characterization of biogenic silica by nitrogen physisorption, positron annihilation lifetime spectroscopy and hyperpolarized ^{129}Xe NMR spectroscopy, *Microporous Mesoporous Mater.*, 2020, **307**, 110515.
54. J. Athinarayanan, V. S. Periasamy, M. Alhazmi and A. A. Alshatwi, Synthesis and biocompatibility assessment of sugarcane bagasse-derived biogenic silica nanoparticles for biomedical applications, *J. Biomed. Mater. Res., Part B*, 2017, **105**(2), 340–349.
55. A. L. Carnelli, M. Madella and J. P. Theurillat, Biogenic silica production in selected alpine plant species and plant communities, *Ann. Bot.*, 2001, **87**(4), 425–434.
56. D. Puppe, Review on protozoic silica and its role in silicon cycling, *Geoderma*, 2020, **365**, 114224.
57. Y. Aoki, M. Hoshino and T. Matsubara, Silica and testate amoebae in a soil under pine–oak forest, *Geoderma*, 2007, **142**(1–2), 29–35.
58. M. Sommer, H. Jochheim, A. Höhn, J. Breuer, Z. Zagorski, J. Busse, D. Barkusky, K. Meier, D. Puppe, M. Wanner and D. Kaczorek, Si cycling in a forest biogeosystem–the importance of transient state biogenic Si pools, *Biogeosciences*, 2013, **10**(7), 4991–5007.
59. D. Puppe, D. Kaczorek and J. Schaller, Biological impacts on silicon availability and cycling in agricultural plant–soil systems, in *Silicon and nano-silicon in environmental stress management and crop quality improvement*, Academic Press, 2022, pp. 309–324.
60. D. J. Lahr, T. Bosak, E. Lara and E. A. Mitchell, The Phanerozoic diversification of silica-cycling testate amoebae and its possible links to changes in terrestrial ecosystems, *PeerJ*, 2015, **3**, e1234.
61. Z. V. Finkel and B. Kotrc, Silica use through time: macroevolutionary change in the morphology of the diatom fustule, *Geomicrobiol. J.*, 2010, **27**(6–7), 596–608.
62. D. A. Caron, P. D. Countway, A. C. Jones, D. Y. Kim and A. Schnetzer, Marine protistan diversity, *Annu. Rev. Mar. Sci.*, 2012, **4**(1), 467–493.
63. A. Leynaert, C. Fardel, B. Beker, C. Soler, G. Delebecq, A. Lemercier, P. Pondaven, P. E. Durand and K. Heggarty, Diatom frustules nanostructure in pelagic and benthic environments, *Silicon*, 2018, **10**, 2701–2709.
64. P. J. Tréguer and C. L. De La Rocha, The world ocean silica cycle, *Annu. Rev. Mar. Sci.*, 2013, **5**(1), 477–501.
65. N. Llopis Monferrer, D. Boltovskoy, P. Tréguer, M. M. Sandin, F. Not and A. Leynaert, Estimating biogenic silica production of Rhizaria

in the global ocean, *Global Biogeochem. Cycles*, 2020, **34**(3), e2019GB006286.

66. J. Livage and P. J. Lopez, Biogenic silica glasses, in *Encyclopedia of Glass Science, Technology, History, and Culture*, 2021, vol. 2, pp. 955–967.
67. J. Zang, S. Liu, Y. Liu, Y. Ma and X. Ran, Contribution of phytoliths to total biogenic silica volumes in the tropical rivers of Malaysia and associated implications for the marine biogeochemical cycle, *Chin. J. Oceanol. Limnol.*, 2016, **34**(5), 1076–1084.
68. X. Ran, B. Xu, J. Liu, C. Zhao, S. Liu and J. Zang, Biogenic silica composition and $\delta^{13}C$ abundance in the Changjiang (Yangtze) and Huanghe (Yellow) Rivers with implications for the silicon cycle, *Sci. Total Environ.*, 2017, **579**, 1541–1549.
69. J. T. Cornelis, B. Delvaux, R. B. Georg, Y. Lucas, J. Ranger and S. Opfergelt, Tracing the origin of dissolved silicon transferred from various soil-plant systems towards rivers: a review, *Biogeosciences*, 2011, **8**(1), 89–112.
70. E. Struyf, S. Van Damme, B. Gribsholt, J. J. Middelburg and P. Meire, Biogenic silica in tidal freshwater marsh sediments and vegetation (Schelde estuary, Belgium), *Mar. Ecol.: Prog. Ser.*, 2005, **303**, 51–60.
71. W. E. Müller, A. Boreiko, U. Schloßmacher, X. Wang, M. N. Tahir, W. Tremel, D. Brandt, J. A. Kaandorp and H. C. Schröder, Fractal-related assembly of the axial filament in the demosponge Suberites domuncula: relevance to biomineralization and the formation of biogenic silica, *Biomaterials*, 2007, **28**(30), 4501–4511.
72. X. Wang, H. C. Schröder, Q. Feng, F. Draenert and W. E. Müller, The deep-sea natural products, biogenic polyphosphate (Bio-PolyP) and biogenic silica (Bio-Silica), as biomimetic scaffolds for bone tissue engineering: fabrication of a morphogenetically-active polymer, *Mar. Drugs*, 2013, **11**(3), 718–746.
73. A. Woesz, J. C. Weaver, M. Kazanci, Y. Dauphin, J. Aizenberg, D. E. Morse and P. Fratzl, Micromechanical properties of biological silica in skeletons of deep-sea sponges, *J. Mater. Res.*, 2006, **21**(8), 2068–2078.
74. D. J. DeMaster, The supply and accumulation of silica in the marine environment, *Geochim. Cosmochim. Acta*, 1981, **45**(10), 1715–1732.
75. E. Sañé, E. Isla, M. Á. Bárcena and D. J. DeMaster, A shift in the biogenic silica of sediment in the Larsen B continental shelf, off the eastern Antarctic Peninsula, resulting from climate change, *PLoS One*, 2013, **8**(1), e52632.
76. V. Panwar and T. Dutta, Diatom biogenic silica as a felicitous platform for biochemical engineering: Expanding frontiers, *ACS Appl. Bio Mater.*, 2019, **2**(6), 2295–2316.
77. S. R. Cicco, D. Vona, R. Gristina, E. Sardella, R. Ragni, M. Lo Presti and G. M. Farinola, Biosilica from living diatoms: investigations on biocompatibility of bare and chemically modified Thalassiosira weissflogii silica shells, *Bioengineering*, 2016, **3**(4), 35.

78. M. Rembauville, S. Blain, C. Manno, G. Tarling, A. Thompson, G. Wolff and I. Salter, The role of diatom resting spores in pelagic–benthic coupling in the Southern Ocean, *Biogeosciences*, 2018, **15**(10), 3071–3084.
79. S. Dhali, R. Jain, A. Malik, S. Sharma and R. Raliya, Cultivation of Navicula sp. on rice straw hydrolysate for the production of biogenic silica, *Bioresour. Technol.*, 2022, **360**, 127577.
80. Y. Churakova, A. Aguilera, E. Charalampous, D. J. Conley, D. Lundin, J. Pinhassi and H. Farnelid, Biogenic silica accumulation in picoeukaryotes: Novel players in the marine silica cycle, *Environ. Microbiol. Rep.*, 2023, **15**(4), 282–290.
81. L. Zhang, R. Wang, M. Chen, J. Liu, L. Zeng, R. Xiang and Q. Zhang, Biogenic silica in surface sediments of the South China Sea: Controlling factors and paleoenvironmental implications, *Deep Sea Res., Part II*, 2015, **122**, 142–152.
82. V. P. Silva, M. C. Gonçalves and I. V. Yoshida, Biogenic silica short fibers as alternative reinforcing fillers of silicone rubbers, *J. Appl. Polym. Sci.*, 2006, **101**(1), 290–299.
83. C. O. Prempeh, S. Formann, T. Schliermann, H. B. Dizaji and M. Nelles, Extraction and characterization of biogenic silica obtained from selected agro-waste in Africa, *Appl. Sci.*, 2021, **11**(21), 10363.
84. M. Wysokowski, T. Jesionowski and H. Ehrlich, Biosilica as a source for inspiration in biological materials science, *Am. Mineral.*, 2018, **103**(5), 665–691.
85. S. Mann and C. C. Perry, Structural aspects of biogenic silica, in *InCiba Foundation Symposium 121-Silicon Biochemistry: Silicon Biochemistry: Ciba Foundation Symposium 121*, John Wiley & Sons, Ltd, Chichester, UK, 2007, pp. 40–58.
86. T. A. Hamza, A. H. Sherif and E. A. Abdalla, A novel approach to reinforce provisional material using silica gel powder, *Stomatol. Dis. Sci.*, 2017, **1**, 3–7.
87. C. C. Perry, Biogenic silica: a model of amorphous structure control, in *Growth, Dissolution and Pattern Formation in Geosystems*, Springer Netherlands, Dordrecht, 1999, pp. 237–251.
88. X. Liu, S. M. Colman, E. T. Brown, E. C. Minor and H. Li, Estimation of carbonate, total organic carbon, and biogenic silica content by FTIR and XRF techniques in lacustrine sediments,*J. Paleolimnol.*, 2013, **50**, 387–398.
89. M. Sumper and N. Kröger, Silica formation in diatoms: the function of long-chain polyamines and silaffins, *J. Mater. Chem.*, 2004, **14**(14), 2059–2065.
90. D. Puppe and M. Leue, Physicochemical surface properties of different biogenic silicon structures: Results from spectroscopic and microscopic analyses of protistic and phytogenic silica, *Geoderma*, 2018, **330**, 212–220.
91. A. Mcheik, S. Cassaignon, J. Livage, A. Gibaud, S. Berthier and P. J. Lopez, Optical properties of nanostructured silica structures from marine organisms, *Front. Mar. Sci.*, 2018, **5**, 123.

92. U. T. Uthappa, V. Brahmkhatri, G. Sriram, H. Y. Jung, J. Yu, N. Kurkuri, T. M. Aminabhavi, T. Altalhi, G. M. Neelgund and M. D. Kurkuri, Nature engineered diatom biosilica as drug delivery systems, *J. Controlled Release*, 2018, **281**, 70–83.
93. C. O. Prempeh, S. Formann, I. Hartmann and M. Nelles, An improved method for the production of biogenic silica from cornhusk using sol–gel polymeric route, *Biomass Convers. Biorefin.*, 2022, 1.
94. R. V. Piloni, V. Brunetti, R. C. Urcelay, I. C. Daga and E. L. Moyano, Chemical properties of biosilica and bio-oil derived from fast pyrolysis of Melosira varians, *J. Anal. Appl. Pyrolysis*, 2017, **127**, 402–410.
95. H. K. Tchakouté, C. H. Rüscher, S. Kong and N. Ranjbar, Synthesis of sodium waterglass from white rice husk ash as an activator to produce metakaolin-based geopolymer cements, *J. Build. Eng.*, 2016, **6**, 252–261.
96. A. A. Nayl, A. I. Abd-Elhamid, A. A. Aly and S. Bräse, Recent progress in the applications of silica-based nanoparticles, *RSC Adv.*, 2022, **12**(22), 13706–13726.
97. A. Luthfiah, Y. Deawati, M. L. Firdaus, I. Rahayu and D. R. Eddy, Silica from natural sources: a review on the extraction and potential application as a supporting photocatalytic material for antibacterial activity, *Sci. Technol. Indones.*, 2021, **6**(3), 144–155.
98. H. A. Alyosef, A. Eilert, J. Welscher, S. S. Ibrahim, R. Denecke, W. Schwieger and D. Enke, Characterization of biogenic silica generated by thermo chemical treatment of rice husk, *Part. Sci. Technol.*, 2013, **31**(5), 524–532.
99. V. Vaibhav, U. Vijayalakshmi and S. M. Roopan, Agricultural waste as a source for the production of silica nanoparticles, *Spectrochim. Acta, Part A*, 2015, **139**, 515–520.
100. E. B. Ekwenna, Y. Wang and A. Roskilly, The production of bio-silica from agro-industrial wastes leached and anaerobically digested rice straws, *Bioresour. Technol. Rep.*, 2023, **22**, 101452.
101. W. K. Setiawan and K. Y. Chiang, Eco-friendly rice husk pre-treatment for preparing biogenic silica: Gluconic acid and citric acid comparative study, *Chemosphere*, 2021, **279**, 130541.
102. J. Y. Park, Y. M. Gu, J. Chun, B. I. Sang and J. H. Lee, Pilot-scale continuous biogenic silica extraction from rice husk by one-pot alkali hydrothermal treatment and ball milling, *Chem. Biol. Technol. Agric.*, 2023, **10**(1), 102.
103. S. Venkateswaran, R. Yuvakkumar and V. Rajendran, Nano silicon from nano silica using natural resource (RHA) for solar cell fabrication, *Phosphorus, Sulfur Silicon Relat. Elem.*, 2013, **188**(9), 1178–1193.
104. S. Steven, E. Restiawaty and Y. Bindar, Routes for energy and bio-silica production from rice husk: A comprehensive review and emerging prospect, *Renewable Sustainable Energy Rev.*, 2021, **149**, 111329.
105. N. N. Maseko, D. Schneider, S. Wassersleben, D. Enke, S. A. Iwarere, J. Pocock and A. Stark, The production of biogenic silica from different

South African agricultural residues through a thermo-chemical treatment method, *Sustainability*, 2021, **13**(2), 577.
106. M. Estevez, S. Vargas, V. M. Castano and R. Rodriguez, Silica nanoparticles produced by worms through a bio-digestion process of rice husk, *J. Non-Cryst. Solids*, 2009, **355**(14–15), 844–850.
107. K. Rohatgi, S. V. Prasad and P. K. Rohatgi, Release of silica-rich particles from rice husk by microbial fermentation, *J. Mater. Sci. Lett.*, 1987, **6**(7), 829–831.
108. L. A. Zemnukhova, G. A. Fedorishcheva, A. G. Egorov and V. I. Sergienko, Recovery conditions, impurity composition, and characteristics of amorphous silicon dioxide from wastes formed in rice production, *Russ. J. Appl. Chem.*, 2005, **78**, 319–323.
109. A. Mohamed, M. M. Ghobara, M. K. Abdelmaksoud and G. G. Mohamed, A novel and highly efficient photocatalytic degradation of malachite green dye *via* surface modified polyacrylonitrile nanofibers/biogenic silica composite nanofibers, *Sep. Purif. Technol.*, 2019, **210**, 935–942.
110. M. C. Pereira, L. C. Oliveira and E. Murad, Iron oxide catalysts: Fenton and Fentonlike reactions–a review, *Clay Miner.*, 2012, **47**(3), 285–302.
111. G. Purwiandono, I. Fatimah, I. Sahroni, P. W. Citradewi, A. Kamari, S. Sagadevan, W. C. Oh and R. A. Doong, Fe_3O_4@SiO_2 nanoflakes synthesized using biogenic silica from Salacca zalacca leaf ash and the mechanistic insight into adsorption and photocatalytic wet peroxidation of dye, *Green Process Synth.*, 2022, **11**(1), 345–360.
112. M. Sprynskyy, P. Szczyglewska, I. Wojtczak, I. Nowak, A. Witkowski, B. Buszewski and A. Feliczak-Guzik, Diatom biosilica doped with palladium (II) chloride nanoparticles as new efficient photocatalysts for methyl orange degradation, *Int. J. Mol. Sci.*, 2021, **22**(13), 6734.
113. E. Van Eynde, Z. Y. Hu, T. Tytgat, S. W. Verbruggen, J. Watté, G. Van Tendeloo, I. Van Driessche, R. Blust and S. Lenaerts, Diatom silica–titania photocatalysts for air purification by bio-accumulation of different titanium sources, *Environ. Sci.: Nano*, 2016, **3**(5), 1052–1061.
114. I. Fatimah, N. I. Prakoso, I. Sahroni, M. M. Musawwa, Y. L. Sim, F. Kooli and O. Muraza, Physicochemical characteristics and photocatalytic performance of TiO_2/SiO_2 catalyst synthesized using biogenic silica from bamboo leaves, *Heliyon*, 2019, **5**(11), e02766.
115. G. Singh, H. B. Dizaji, H. Puttuswamy and S. Sharma, Biogenic nanosilica synthesis employing agro-waste rice straw and its application study in photocatalytic degradation of cationic dye, *Sustainability*, 2022, **14**(1), 539.
116. D. Yang, T. Fan, H. Zhou, J. Ding and D. Zhang, Biogenic hierarchical TiO_2/SiO_2 derived from rice husk and enhanced photocatalytic properties for dye degradation, *PLoS One*, 2011, **6**(9), e24788.
117. Y. S. Lin and C. L. Haynes, Synthesis and characterization of biocompatible and size-tunable multifunctional porous silica nanoparticles, *Chem. Mater.*, 2009, **21**(17), 3979–3986.

118. R. Jugdaohsingh, K. L. Tucker, N. Qiao, L. A. Cupples, D. P. Kiel and J. J. Powell, Dietary silicon intake is positively associated with bone mineral density in men and premenopausal women of the Framingham Offspring cohort, *J. Bone Miner. Res.*, 2004, **19**(2), 297–307.
119. J. Athinarayanan, V. S. Periasamy, M. Alhazmi and A. A. Alshatwi, Synthesis and biocompatibility assessment of sugarcane bagasse-derived biogenic silica nanoparticles for biomedical applications, *J. Biomed. Mater. Res., Part B*, 2017, **105**(2), 340–349.
120. A. A. Alshatwi, J. Athinarayanan and V. S. Periasamy, Biocompatibility assessment of rice husk-derived biogenic silica nanoparticles for biomedical applications, *Mater. Sci. Eng., C*, 2015, **47**, 8–16.
121. J. Athinarayanan, V. S. Periasamy, M. Alhazmi, K. A. Alatiah and A. A. Alshatwi, Synthesis of biogenic silica nanoparticles from rice husks for biomedical applications, *Ceram. Int.*, 2015, **41**(1), 275–281.
122. S. A. Jaafari, J. Athinarayanan, V. S. Periasamy and A. A. Alshatwi, Biogenic silica nanostructures derived from Sorghum bicolor induced osteogenic differentiation through BSP, BMP-2 and BMP-4 gene expression, *Process Biochem.*, 2020, **91**, 231–240.
123. J. Athinarayanan, V. S. Periasamy and A. A. Alshatwi, Biogenic silica–metal phosphate (metal = Ca, Fe or Zn) nanocomposites: fabrication from rice husk and their biomedical applications, *J. Mater. Sci.: Mater. Med.*, 2014, **25**, 1637–1644.
124. R. Periakaruppan, S. P. Samadh, S. A. Abed, N. Al-Dayan, S. Dhanasekaran and S. H. Aldhayan, Allium cepa-assisted biogenic silica dioxide nanoparticles: their production, structural characterization, and antioxidant activity, *Biomass Convers. Biorefin.*, 2023, 1–9.
125. L. B. Capeletti, L. F. de Oliveira, K. D. Goncalves, J. F. de Oliveira, A. Saito, J. Kobarg, J. H. Santos and M. B. Cardoso, Tailored silica–antibiotic nanoparticles: overcoming bacterial resistance with low cytotoxicity, *Langmuir*, 2014, **30**(25), 7456–7464.
126. S. K. Sharma, A. R. Sharma, S. D. Pamidimarri, J. Gaur, B. P. Singh, S. Sekar, D. Y. Kim and S. S. Lee, Bacterial compatibility/toxicity of biogenic silica (b-SiO_2) nanoparticles synthesized from biomass rice husk ash, *Nanomaterials*, 2019, **9**(10), 1440.
127. M. Castro-Mora, M. Vásquez-González, J. Cordero-Guerrero, M. Benavides-Acevedo, J. González, M. J. López-Brenes, J. Vega-Baudrit and Y. Corrales-Ureña, Bacterial anti-adhesive films of PDMS coated with microstructures of biogenic silica rosettes extracted from pineapple peels residues, *Surf. Interfaces*, 2022, **30**, 101881.
128. R. Periakaruppan, P. Ranjitha and J. Danaraj, Biosynthesis of silica nanoparticles using the leaf extract of Punica granatum and assessment of its antibacterial activities against human pathogens, *Appl. Biochem. Biotechnol.*, 2022, **194**(11), 5594–5605.
129. A. Sola-Rabada, P. Sahare, G. J. Hickman, M. Vasquez, L. T. Canham, C. C. Perry and V. Agarwal, Biogenic porous silica and silicon sourced from Mexican Giant Horsetail (Equisetum myriochaetum) and their

application as supports for enzyme immobilization, *Colloids Surf., B*, 2018, **166**, 195–202.
130. M. Verma, A. I. Dar and A. Acharya, Facile synthesis of biogenic silica nanomaterial loaded transparent tragacanth gum hydrogels with improved physicochemical properties and inherent anti-bacterial activity, *Nanoscale*, 2022, **14**(32), 11635–11654.
131. S. Ariaeenejad, F. Jokar, P. Hadian, L. Ma'mani, S. Gharaghani, M. Fereidoonnezhad and G. H. Salekdeh, An efficient nano-biocatalyst for lignocellulosic biomass hydrolysis: Xylanase immobilization on organically modified biogenic mesoporous silica nanoparticles, *Int. J. Biol. Macromol.*, 2020, **164**, 3462–3473.
132. E. Onoja, S. Chandren, F. I. Razak and R. A. Wahab, Extraction of nanosilica from oil palm leaves and its application as support for lipase immobilization, *J. Biotechnol.*, 2018, **283**, 81–96.
133. N. Elias, R. A. Wahab, S. Chandren, J. Jamalis, N. A. Mahat and L. W. Jye, Structure and properties of lipase activated by cellulose-silica polyethersulfone membrane for production of pentyl valerate, *Carbohydr. Polym.*, 2020, **245**, 116549.
134. A. G. Jacob, R. A. Wahab and N. A. Mahat, Ternary biogenic silica/magnetite/graphene oxide composite for the hyperactivation of Candida rugosa lipase in the esterification production of ethyl valerate, *Enzyme Microb. Technol.*, 2021, **148**, 109807.
135. Y. C. Zhang, Y. You, S. Xin, Y. X. Yin, J. Zhang, P. Wang, X. S. Zheng, F. F. Cao and Y. G. Guo, Rice husk-derived hierarchical silicon/nitrogen-doped carbon/carbon nanotube spheres as low-cost and high-capacity anodes for lithium-ion batteries, *Nano Energy*, 2016, **25**, 120–127.
136. N. B. Machado, J. P. Miguez, I. C. Bolina, A. B. Salviano, R. A. Gomes, O. L. Tavano, J. H. Luiz, P. W. Tardioli, É. C. Cren and A. A. Mendes, Preparation, functionalization and characterization of rice husk silica for lipase immobilization via adsorption, *Enzyme Microb. Technol.*, 2019, **128**, 9–21.
137. D. P. Wong, R. Suriyaprabha, R. Yuvakumar, V. Rajendran, Y. T. Chen, B. J. Hwang, L. C. Chen and K. H. Chen, Binder-free rice husk-based silicon–graphene composite as energy efficient Li-ion battery anodes, *J. Mater. Chem. A*, 2014, **2**(33), 13437–13441.
138. J. Cui, F. Cheng, J. Lin, J. Yang, K. Jiang, Z. Wen and J. Sun, High surface area C/SiO_2 composites from rice husks as a high-performance anode for lithium ion batteries, *Powder Technol.*, 2017, **311**, 1–8.
139. L. S. Jiao, J. Y. Liu, H. Y. Li, T. S. Wu, F. Li, H. Y. Wang and L. Niu, Facile synthesis of reduced graphene oxide-porous silicon composite as superior anode material for lithium-ion battery anodes, *J. Power Sources*, 2016, **315**, 9–15.
140. Y. Ju, J. A. Tang, K. Zhu, Y. Meng, C. Wang, G. Chen, Y. Wei and Y. Gao, SiOx/C composite from rice husks as an anode material for lithium-ion batteries, *Electrochim. Acta*, 2016, **191**, 411–416.

141. D. Schneider, R. Kircheis, S. Wassersleben, W. D. Einicke, R. Gläser and D. Enke, Low-cost Microwave-Assisted partial Pseudomorphic transformation of biogenic silica, *Front. Chem.*, 2019, **7**, 575.
142. D. Schneider, S. Wassersleben, M. Weiß, R. Denecke, A. Stark and D. Enke, A generalized procedure for the production of high-grade, porous biogenic silica, *Waste Biomass Valorization*, 2020, **11**, 1–5.
143. T. H. Liou, F. W. Chang and J. J. Lo, Pyrolysis kinetics of acid-leached rice husk, *Ind. Eng. Chem. Res.*, 1997, **36**(3), 568–573.
144. L. Wang, D. Fan, W. Li, Y. Liao, X. Zhang, M. Liu and Z. Yang, Grain-size effect of biogenic silica in the surface sediments of the East China Sea, *Cont. Shelf Res.*, 2014, **81**, 29–37.
145. D. Merino-Maldonado, A. Antolín-Rodríguez, L. Serrano-González, S. Blanco, A. Juan-Valdés, J. M. Morán-del Pozo and J. García-González, Innovative approach for the protection of recycled concrete by biogenic silica biodeposition, *Constr. Build. Mater.*, 2023, **368**, 130475.
146. O. Ragueneau, M. Gallinari, L. Corrin, S. Grandel, P. Hall, A. Hauvespre, R. S. Lampitt, D. Rickert, H. Stahl, A. Tengberg and R. Witbaard, The benthic silica cycle in the Northeast Atlantic: annual mass balance, seasonality, and importance of non-steady-state processes for the early diagenesis of biogenic opal in deep-sea sediments, *Prog. Oceanogr.*, 2001, **50**(1–4), 171–200.
147. D. Shrestha, T. Nayaju, M. R. Kandel, R. R. Pradhananga, C. H. Park and C. S. Kim, Rice husk-derived mesoporous biogenic silica nanoparticles for gravity chromatography, *Heliyon*, 2023, **9**(4), e15142.
148. A. E. Karaca, C. Özel, A. C. Özarslan and S. Yücel, The simultaneous extraction of cellulose fiber and crystal biogenic silica from the same rice husk and evaluation in cellulose-based composite bioplastic films, *Polym. Compos.*, 2022, **43**(10), 6838–6853.
149. W. Yang, P. J. Lopez and G. Rosengarten, Diatoms: self assembled silica nanostructures, and templates for bio/chemical sensors and biomimetic membranes, *Analyst*, 2011, **136**(1), 42–53.

CHAPTER 2

Biogenic Silica in Ocean and Terrestrial Environments

DIKSHA SUMAN, SOURAV KUMAR, ASHOK KUMAR NADDA AND RAHUL SHRIVASTAVA*

Department of Biotechnology and Bioinformatics, Jaypee University of Information Technology, Waknaghat, Solan, Himachal Pradesh 173234, India.
*Email: juit.rahul@gmail.com

2.1 Introduction

The 2nd most common element in earth is silica, which ranks seventh within the cosmos in terms of abundance. Dissolved silicic acid (DSi; $Si(OH)_4$) is produced within water because of the weathering of the Earth's crust by carbon dioxide rich rainwater, a crucial step in the regulation of atmospheric carbon dioxide. Biogenic silica (BSi; SiO_2) is precipitated as internal or exterior structures by silicifiers using DSi. Microorganisms like diatoms, rhizarians, silicoflagellates, and several types of choanoflagellates are among the most significant aquatic organisms, as are macroorganisms like siliceous sponges. Diatoms and other phototrophic silicifiers absorb enormous amounts of silicon along with nitrogen, phosphorus, and inorganic carbon. This connects their biogeochemistry and supports the ocean's capacity to absorb CO_2. The BSi that has been dubbed "dark silica" is produced by heterotrophic organisms like rhizarians, sponges and choanoflagellates without the aid of photoautrophic digestion of C and N.[1]

Diatoms are the group of marine algae with the greatest diversity of species, and they usually predominate in eutrophic habitats.[2] According to Milligan and Morel (2002), these photosynthetic protists precipitate opaline

Biogenic Silica: Fundamentals and Applications
Edited by Ashok Kumar Nadda and Vijai Kumar Gupta

Published by the Royal Society of Chemistry, www.rsc.org

silica to create their cell walls after absorbing dissolved silicic acid from the water. Diatoms' great diversity along with extensive dispersion allow them to provide about one-fourth of the world's primary output.[3] Diatoms are thought to be one of the major organisms responsible for the transport of silicon and carbon to depths as well as the minimal amount of dissolved form orthosilicic acid observed on the surface of the ocean.[4] Therefore, it is believed that the marine Si cycle influences the growth of diatoms, which in turn links it mechanistically to the C cycle. Diatoms dominate BSi stocks by a large margin, despite a number of siliceous taxa, such as Rhizarians and silicoflagellates, which may decrease the relative contribution of diatoms to the standing stocks of water column biogenic silica. This is a basic comprehension of the biome function in the marine silica cycle.[1,5]

Diatom production requires the presence of dissolved silicic acid, which diatoms used to generate their biogenic silica shells. It has been discovered that silicic acid within the Norwegian Seas is low (5 M) and that it is dropping with time.[6] A more recent study found that pre-bloom orthosilicic acid concentrations in the North Atlantic subpolar and polar areas have dropped by 1–2 M during the previous 25 years. The 10 to 60 M orthosilicic acid found in the Southern Ocean's surface waters and the peripheral ice zone surrounding Antarctica stands in sharp contrast to this.[7,8] It is unlikely that orthosilicic acid would inhibit diatom development unless iron is plentiful and promotes extraordinary blooms that eat Si, or unless assemblages are very poor at absorbing Si. Furthermore, the stoichiometry of orthosilicic acid availability in relation to NO_3^- ($Si:N<1$) in the source waters, which support the spring bloom in the majority of the North Atlantic and European polar seas, implies that, in the course of a bloom cycle, diatoms may encounter silica limitation before nitrogen limitation, particularly if they consumed silica along with nitrogen in nearly equal amounts as in other diatom bloom regions.[9–11]

Natural silicon is known as silica (SiO_2). The quantities of silicic acid within contemporary freshwaters and ocean's range from 10 to 180 M; however, at concentrations higher than 2 mM, silicic acid auto-polymerizes to silica, with a strong pH dependency.[12] Due to the necessity for highly specialized biochemical systems, which are in charge of producing a variety of siliceous structures, silicon uptake and the development of biosilica within live cells are both reliant on these mechanisms.[13] Numerous organisms, including bacteria, plants, algae, yeast, fungus, sponges, protists, ascidians, mollusks, brachiopods and crustaceans have been characterized as having these structures.[14,15] As a result of pathological biomineralization, fish and animals, including humans, can also develop biosilica.[16] In addition to showing variance in structure, density, and content, biogenic silica exhibits a range of structural patterns, including sheet-like, a rounded form, fibrillar, spherical, tube shaped, and foldable sheets. In nature, there are many levels of biosilica structural organization.[13] These can be as little as nanogranules in bacteria or mitochondria or as large as the three-dimensional siliceous skeletons of two-meter-tall oceanic glass sponges.[17] Vital activity has deposited the amorphous silica structures in diatoms, radiolarians, sponges, and silico-flagellates; these structures seem to be kept intact and linked to the live organisms in their aquatic habitat, preventing them from disintegrating.[18]

At temperatures between −2 and 4 °C, the psychrophilic organisms known as glass sponges may create enormous biosilica-based skeletons.[17] For instance, a group of *Aphrocallistes vastus* discovered in cold, deep seas near western Canada had some of the greatest diameters ever recorded for any sponge. According to Austin and Conway (2007), the colony of hexactinellid was 3.4 meter long, 1.1 meter tall, and 0.5 meter wide.[19] It is reported, one type of hexactinellid glass sponge which belongs subfamily Lanuginellinae and family Rossellidae has reportedly been measured to be over 3.5 meters long, 2 meters tall, and 1.5 meters wide in recent years.[17,19]

It has been widely documented that Si has the ability to suppress both chemical and physical stresses, such as salinity, nutritional imbalance, and metal toxicity. Examples of physical stress include drought, high temperatures, UV, loading, and freezing.[20,21] The beneficial physiological effects of silicon on plant improvement work hand in hand with plants' endogenous stress responses under various environmental conditions.[20] By replenishing the soluble protein content that is lost when plants are exposed to salinity stress, silicon has the ability to boost the soluble protein content of plant leaves. In plants under salt stress, silicon can also boost the activity of the antioxidant enzymes superoxide dismutase, guaiacol peroxidase, ascorbate peroxidase, dehydroascorbate reductase, and glutathione reductase.[20,22,23]

Since the late 1980s, the relevance of BSi in the terrestrial ecosystem has been acknowledged; nonetheless, it has only lately become a focus of attention for scientists working to understand the biogeochemical-climate feedback processes. The need for more research is greatest in ecosystems that are affected by the terrestrial biogeochemical cycle of silica, particularly soils.[24] The contribution of terrestrial biogenic silica (TBSi) to continental- and global-scale carbon and silicon cycles has been largely ignored since the 1980s and continues to this day. As a result, we currently have little knowledge about the terrestrial biogenic silica cycle (TBSC). Physical mechanisms and equations involved in the movement of silicon from land to river are described in recent landscape studies.[25–27] These investigations don't offer a way to measure forecast dissolved biogenic silica (DSi) flows from field to rivers, even though they offer useful details for terrestrial settings and silica cycling. The distinction between DSi and BSi is hazy in innovative framework that describes a mechanistic model for dissolved silica conveyance with an emphasis on soils.[28] The amount of silica entering river system and aquatic systems could be significantly impacted by riverine transport of silica and biogenic contributions across various land cover categories. Living vegetation, soils, and rivers are examples of biogenic silica reservoirs at landscape scales. The accumulation, dissolution, development, and leaching processes of soil have an impact on BSi in these pools. Amorphous silica, which is made up of biogenic silica and inorganic silica, and silica minerals

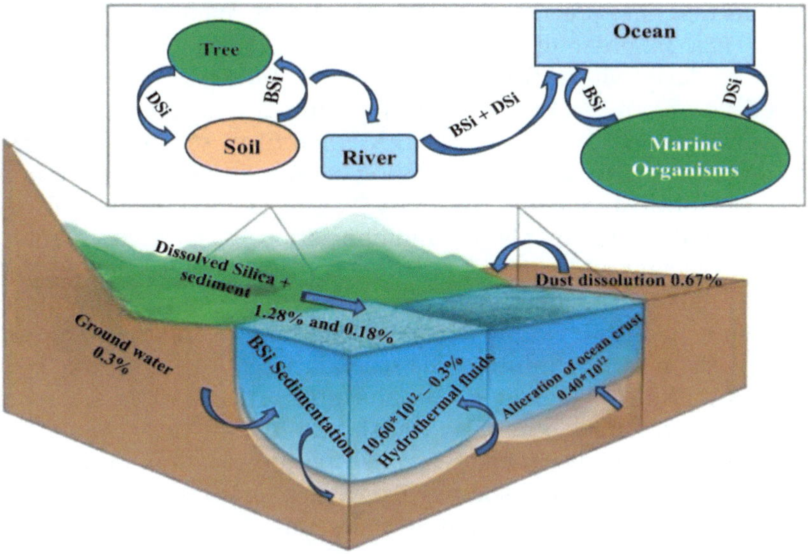

Figure 2.1 Cycle of biosilica production.

such as primary and secondary silicates make up the particulate silica portion from which break down occurs in soils. In soils, the BSi part, which is primarily made up of phytoliths produced by plants,[29] contributes significantly to the Si pool. Silica first goes in the biogenic terrestrial cycle as lithogenic silica, and plants transform it into biogenic structures. Following litterfall and decomposition, soil Si is recharged. As plants absorb soluble Si during growth, deposition rates have an impact on ecophysiological processes that lead to Si buildup. Land cover classes may be connected to particular silica production rates and result changes in the biogenic silica cycle at the landscape scale since flora peculiar with relation to land cover type exhibits characteristic biogenic silica values.[30,31]

A depiction of the current global silicon (Si) cycle is shown in Figure 2.1. The flux magnitudes in 1012 mol year^{-1} and the corresponding δ^{30}Si values in ‰ are shown in the diagram. It illustrates common fractionations (ε, ‰) associated with the synthesis of clay minerals and biogenic silica (BSi). The figure gives a thorough summary of the Si cycle dynamics by differentiating between particulate fluxes, solute fluxes, or transformations.

2.2 Biogenic Silica (BSi) in the Terrestrial Environment

Plants have been found to benefit from silica in overcoming a range of environmental obstacles, such as predators and preserving stem rigidity.[32,33] Silica occurs in two different forms in vegetation: silica bodies and silica cells. Silica bodies are made up of phytoliths and potch, which are silica cell formations that have precipitated, while Si cells are made up of free-floating, unbound BSi. In order to create a cuticular Si double layer, Si cells are formed as a layer within the area underneath the leaf's cuticle layer.[33] Si dissolved as organic matter is deposited annually, and biogenic silica is subsequently added to the soil reserve.[34] Based on soil temperature, soil acidity, specific surface area, and sorption of aluminium (Al), different plant species' phytolith dissolving rates vary.[35–37] Si is recycled in soils by re-absorption and release by plants, immobilization by net accumulation, development of plant and soil preservation and finally soil leaching.[38] Land cover classes affect DSi dissolution and distribution and have an impact on soil attributes like soil hydrology and soil chemistry. The discharge of biogenic silica from soils is inhibited by secondary binding of soluble silicates to aluminium silicates, plant absorption and export to rivers, as well as precipitation on other minerals, translocation, and desiccation.[39] Si is moved through soils, rivers, and ultimately oceans from both the lithogenic (LSiD) and biogenic (BSiD) sources. However, significant quantity of disintegrated silica can be preserved in river systems and lake sediments during river transport processes.[40,41]

According to previous studies, the flow of biogenic silica to the oceans is less than that of lithogenic silica but is still important for determining the overall silica content in the seas.[42,43] Understanding how this value fluctuates with changes in the terrestrial environment might have a big impact on how much Si gets into the ocean, moves around in there, and is eventually deposited there. This relation alerts that the possible modifications to the silica biological, geological and chemical aspects is caused by changes in the type of land cover. Additionally, significantly more silica is held in plants and soil than is carried away from the continents, allowing silica to be maintained within terrestrial ecosystems and delaying the movement of silica to the oceans.[29,44,45] The retention of silica in soil and plants in far huge amounts than what is carried far from the continents causes a delay in the movement of silica to the oceans. This is especially true for swamplands, where sediment storage is both spatially and momentarily vast. Si buildup and the pulse events that release it might constitute a key earth system process in stable settings.[46]

Through their roots, terrestrial plants absorb silicic acid from the matrix of soil. After that, it is transported by the transpiration stream, where it accumulates permanently as hydrated silica that takes the form of an amorphous substance found in leaves, fruits, roots and stems. With n being less than two and m being very big, the typical formula for biogenic silica is $[\mathrm{SiO}n{\cdot}\mathrm{OH}(4-2n)]m$, indicating a variety of composition between $m\mathrm{SiO_2}$ and $m\mathrm{Si(OH)_4}$ as the end components. Discrete silica bodies, also referred to as opal phytoliths, are found in plant tissues. Leaf SiO_2 levels have been reported to vary between ~0.5% and ~19% dry weight; nevertheless, significant variances are typically seen between various sections of the same plant. Significant phylogenetic heterogeneity is also seen in the silica contents, with shoot silica concentrations declining in the following order: ferns > liverworts > horsetails > clubmosses > mosses > angiosperms > gymnosperms.[47] They found that monocots (such as grasses, particularly bamboo and wetland species like rice and Phragmites), large quantity of angiosperm were found in palms, bananas, sedges, cannabis and several composite like sunflower. A few dicots including some tropical forestries like maple, oak, elm, beech, sugar were also present. Some conifers, or gymnosperms, are notable Si accumulators.[48] Examples of these include larch and spruce.[49] There have also been reports of silica within the shade of several hundred rainforest trees with sturdy wood. Three types of higher plants exist: non-Si accumulators, intermediate, and high. SiO_2 concentrations in high accumulators are usually greater than 1% dry weight, and they absorb DSi more quickly than water, indicating the presence of an active-uptake mechanism. In rice, a Si-transporter gene was discovered by researchers. Plants falling in the intermediate Si accumulator group have less than 1% SiO_2 and passively absorb DSi through water, whereas non-depositor plants have significantly lower biogenic silica concentrations.[50] The quantity of dissolved silica (BSi) that accumulates within a given group varies depending on several factors, including type of soil, pH, dissolved silica concentration, plant maturity, the rate of transpiration, the prevalence of ectomycorrhizal fungus on roots of plants, and the presence of nutrients like nitrogen and phosphorus.[51]

2.2.1 Biogenic Silica in Rivers

When DSi leaks from terrestrial habitats, freshwater organisms aggressively cycle it. While not many vascular plant species that are submerged or float have been observed to have appreciable levels of biogenic silica, the quantity is growing. Notable examples are the water chestnut *Trapa natans* and the Salt Lake colonizing *Ruppia maritima*. Nevertheless, a tiny but important group of aquatics "silica bioengineers," such as diatoms, sponges, and chrysophytes (flagellates), depend on Si as a necessary nutrient. Chrysophytes and diatoms both require a dissolved form of inorganic carbon for photosynthesis.[52] Diatoms are generally regarded as the most significant silica drain in lakes and rivers because of their prevalence throughout a range of depth habitats, water chemistries, and flow regimes. They produce DSi into nanospheres when proteins called silaffins are present.[53] Then, these spheres serve as the building blocks for the development of frustules. The frustule is shielded from disintegration by an organic membrane; thick polysaccharide gels linked to specific species enhance this layer. The membrane behaves as a physical barrier along with facilitating the adsorption of cations that impede dissolution, such Fe and Al. After dying, it breaks down quickly, in part because of bacterial activity that speeds up the disintegration process and suggests the presence of a bacterial loop within the silica cycle.[54]

Siliceous sponges are found in freshwater and marine environments all over the world. Approximately 40% of the biogenic silica in the surface sediment of Florida lakes is made up of them. They are common in many shallower waterbodies as well as plentiful in deep lakes with hydrothermal silica inputs (like Tanganyika, Taupo and Baikal). Sponge biosilicification is comparable to diatoms. They release spicules that resemble needles, which serve as a protection against predators and to give structural support.[55] The protein silicatein, which is present in an axial organic filament encased by the spicules, encourages the condensation of amorphous silica. Researchers Maldonado and Ferreira contended that sponges have a far smaller part in the global biogeochemical Si cycle than previously thought, primarily because sponge spicules are much bigger than diatom frustules and have a smaller specific surface area.[56] They might also be noticeably more resistant to disintegration if they have a higher Al concentration.

Compared to diatoms or sponges, chysophyte microalgae have a more limited distribution, despite being effective competitors for resources.[54] They are widespread in certain tropical swamps and the chilly, oligotrophic waters of lakes in the Arctic and Antarctic. Chrysophytes, unlike diatoms, are generally uncommon in alkaline, eutrophic waters, however there are some exceptions.[57] They take on several forms of life and are particularly prevalent in low-trophic lake plankton, where they compete for nutrients by using their flagella to remain in the pelagic zone. Scales, cysts are among its silicates portion, which are synthesized similarly to sponge and diatom components.[58]

Lakes work well as silica traps. In the arid season of Lake Malawi, when upwelling takes nutrients from anoxic hypolimnion to the surface, diatoms predominate over phytoplankton. The outflow loses just around 1 percent of the 600–1100 mmol silica m^{-2} $year^{-1}$ external Si input that entered the lake.[59] Conversely, the diatom production cycle is around 9000 mmol silica m^{-2} $year^{-1}$, with 7–11% of that amount being buried in the sediments. The majority of DSi's fluvial transfer to the oceans occurs *via* tropical rivers. Large amounts of DSi are naturally stored by siliceous organisms; however, siliceous recycling may occur more quickly and BSi storage is less persistent than in water bodies in flood plains and river channels.[57,60]

Globally, riverine ecosystems reduce the net export of dissolved silica to the seas by converting approximately 16 percent of the total riverine dissolved silica load to suspended biogenic silica, primarily diatoms. Building reservoirs is a severe illustration of how human behaviour affects riverine filtration of silica on the continents.[61] Other human influences include eutrophication, altering river channels, and manipulating macrophyte ecosystems. Several lakes and coastal waterways, including the Black and Baltic Seas, are being negatively impacted by diminishing river flow and dissolved silica availability, which are encouraging hazardous blooms of flagellates or N-fixing cyanobacteria.[51]

2.2.2 Biogenic Silica in Ocean Environment

Through chemical weathering processes of silicate minerals, which move CO_2 from the atmosphere to the lithosphere, the Si biogeochemical cycle influences the global CO_2 concentrations.[62] Due to the substantial contribution made by the siliceous phytoplankton group called diatoms to atmospheric CO_2 drawdown, silica also influences carbon cycling through a biological mechanism.[63] According to a study, as diatoms are the only environmentally significant phytoplankton species that require orthosilicic acid [$Si(OH_4)$] for their development, they contribute between 25% and 75% of the ocean's primary productivity.[64] Despite being the 3rd most abundant element in the universe, silica is one of the limiting nutrients in many regions of the world's seas.[10] The main cause of this is biological uptake by diatoms, which export particulate biogenic silica to depth by removing silicic acid from surface waters. The capacity of diatoms to produce substantial blooms, which then sediment in bulk at rapid declining rates may push more organic carbon towards the deep sea than the majority of different phytoplankton groups, further ties the Si and C cycles together. The preeminence of diatoms over other extensive groups like coccolithophorids (calcifying organisms), and the variations in atmospheric CO_2 associated with glacial/interglacial transitions may be affected by change in the supply of silica along with remnant metals (primarily Fe) to the surface of oceans through aeolian along with river inputs.[65,66] A deeper comprehension of the regulation of diatom growth by silicic acid availability is therefore necessary to comprehend notable shifts in the composition of phytoplankton communities, their implications on CO_2 uptake from the atmosphere, and, eventually, to develop practical models that can forecast future climate changes.[67]

In order to study the silicon biology and geochemistry along with requirements of diatoms, large specifications like BSi, orthosilicic acid concentrations, along with biological silicic acid ingestion rates are also determined using chemical and isotopic techniques. The advancement of multi elemental, multifunctional group biological, geological and chemical models recently emphasizes the significance of the flux's attribution of the three major nutrients, silica, nitrogen, and phosphorus, to specific phytoplankton groups, sometimes downward to the level of ecological significant species.[68–70] In order to improve these biological, geological and chemical models, one latest orientation in oceanography is the investigation for new instruments that can be used to trace the activity of key functional groups, such as N_2 fixation by diazotrophs, calcification by silicification by diatoms, or coccolithophorids.[71]

2.2.3 River Contribution of Biogenic Silica to Oceans

The rivers carrying dissolved silicate (DSi) inputs from continental weathering are the main sources of Si into the world's ocean. Since it is commonly accepted that the amount of suspended minerals is not mix up and contribute to dissolved silicate input, particle inputs, which are predominantly composed of rock, minerals, and soil, fragments from weathered rocks have not been taken into account in global budgets up to this point.[72]

Diatoms provide almost 90% of the main content of biogenic silica (BSi), with an estimated 240 Tmol Si $year^{-1}$ of BSi produced annually in marine environments.[73] When it comes to biological processes like DSi absorption and biogenic silica (BSi) dissolution, mineral silicate weathers and dissolves slowly enough that Si found in suspended particle matter in rivers can be regarded as inaccessible. This is because mineral silicate dissolves sufficiently slowly and weathers slowly, at -6×10^{-14} mol cm^{-2} s^{-1}.[74] However, at a rate of -2×10^{-9} mol cm^{-2} s^{-1}, the amorphous silica present in diatoms dissolves five times faster than that of crystalline silicates. As a result, it may participate in the biogeochemical cycle at a rate that is pertinent to biological activities.[75]

For the past 15 years, methods for separating further refractory mineral silicates from particulate matter from dissolvable nebulous silica present in diatoms (such as biogenic silica) have been widely available. To date, there has not been enough information to assess how much BSi is added to the global ocean Si budget by rivers.[75–79] See Table 2.1.

The worldwide pattern of net CO_2 drawdown by silicate weathering is depicted (Figure 2.2). The estimated CO_2 drawdown by silicate weathering is derived from geochemical analysis of data related to river movement. It illustrates the significance of freshwater and terrestrial ecosystems in the global carbon cycle and sheds information on how the continental biota affects the silicon cycle.[81] Additionally, how silicon moves through plants, surface waters, and soils before ending up in sediments, highlights the significance of Late Quaternary fluctuations in the biocycling of silicon and how hydroxy aluminosilicates (HAS) such as allophane function in the silicon cycle, highlighting their importance in the biogeochemical processes. There are close connections between the silicon biogeochemical cycles in marine and terrestrial settings.[80] In this cycle, diatoms—siliceous microalgae—play a major role by controlling the carbon export to the deep ocean. The silicate weathering process was greatly accelerated by the growth of C4 grasslands throughout the Miocene, which resulted in higher diatom accumulation rates and variety.

2.3 Biogenic Silica in Oceanic Diatoms

Up to 50% of the world's primary output is produced by diatoms (Bacillariophyceae), which create exoskeletons made of amorphous silica or opal. Within the euphotic zone, among 10 to 100 percentage of mass of biogenic silica generated re-dissolves.[88–90] The remaining diatomaceous opal quickly settles to the sea floor,[91,92] where continued dissolution causes a build-up of silicic acid in the pore water within the first 5 to 20 centimeters beneath the water-sediment interface.[93–95] Diatoms do not absorb orthosilicic acid at a steady amount or continuously during the course of the cell cycle. For a number of species, Si uptake has been shown just before cell

Table 2.1 Concentrations of biogenic silica (BSi) and dissolved silicate (DSi). It shows the percentage of DSi and BSi in various rivers around the world.

S. no.	River	Country	Flow	*n*	DSi (μmol liter $^{-1}$)	BSi (μmol liter^{-1})	% as BSi	% Sm as BSi	Ref.
1	Amazon River	Brazil	6300	4	140	80	43.6	0.98	80
2	Zaire River	Congo	1300	21	185	41	17.2	10.28	80 and 81
3	Lena River	Russia	525	11	73	8.5	11.6	—	80
4	Mississippi River	USA	400	8	107	15.89	12.6	1.24	82
5	Rhine River	Netherlands	206	62	49.1	24.4	31.5	6.65	83
6	Danube River	Romania	204	1	55.1	18.1	24.7	—	80 and 83
7	Atchafalaya River	USA	175	2	113	25.7	18.9	2.72	84
8	Columbia River	USA	112	51	116	39.4	35.5	19.1	85
9	Susquehanna River	USA	36	17	52.3	24.7	40.3	—	86
10	Connecticut River	USA	18	4	121	32.1	23.2	—	87

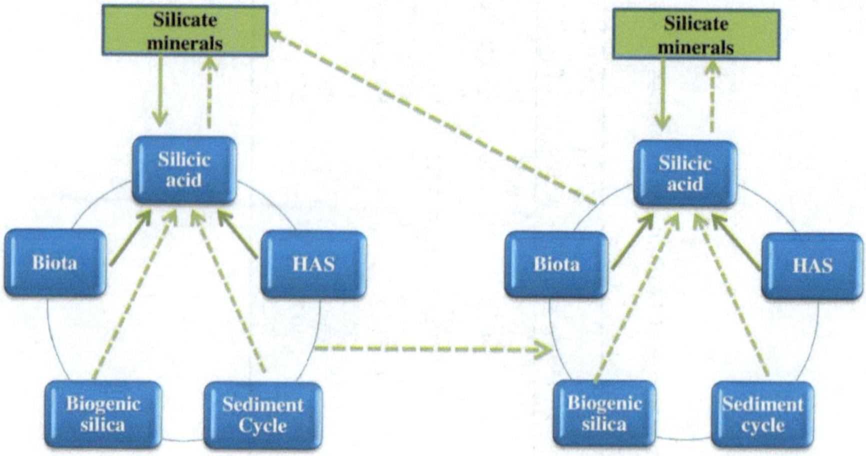

Figure 2.2 The terrestrial and marine biological, geological and chemical cycles of silica.

division and valve development.[96] An important factor in the organic carbon export to the deep sea is diatoms. The comprehension of the process governing the effectiveness of the biological pump of carbon in different locations of the modern ocean depends on our study of the factors influencing the contribution of diatoms to overall production. Therefore, reconstructing the transshipment of BSi is also essential for determining the export flow of organic carbon to the deep sea by settling, which are two critical components for managing surface water alkalinity and total carbon dioxide, and therefore, atmospheric carbon dioxide.[97] Due to the existence of opal-rich sediments at all latitudes, the Pacific, the Southern Ocean, the Northwest Pacific, South African coast, and all oceanic climate zones, the biogenic opal record is significant on a worldwide scale, particularly in high nutrient low chlorophyll regions that are crucial to Earth's climate. Since the Southern Ocean is a system where extremely significant biogenic silica buildup occurs in the current ocean, it serves as the greatest example of this process. The return of silicic acid to the water column is driven by the gradients in pore water concentration.[98] Only a little portion of aquatic BSi production ultimately gets eliminated by burial in sediments (approximately 3 percent in the current oceans). This removal is made up for by rivers and interactions between seawater and basalt supplying reactive silica to the oceans. The effectiveness of nutritional silicon recycling in the oceans is controlled by biogenic silica dissolution in the water column and sediments. Using silica isotopic tracers, the rates of silica formation in surface water of the ocean have been directly monitored.[99] The accessibility of DSi, which diatoms employ to create their biogenic silica ($bSiO_2$) shells, is necessary for diatom formation. In the Norwegian Seas, silicic acid is declining (5 M) and is decreasing with time.[100,101]

2.4 Biogenic Silica in Soil

Biogenic Si is a common and important element found in soils that are formed in a variety of environmental settings. It has been observed in the soils of marsh regions, temperate farms, tropical rain forests, woods, and grassland, both deciduous and evergreen.[102–104] It is possible to divide the biogenic reservoir of silica into components that are phytogenic, zoogenic, microbial, and protistic. Although studies on soils have looked at the magnitude and change in value of the protozoic silica reservoir, the focus had been on phytogenic Si. There is only qualitative proof that the other Si pools—such as microbiological, zoogenic, and protophytic—exist in soils.[105] For instance, nothing is called to the protophytic silica reservoir in soils, even though diatoms have long been recognised to live in soil conditions. Even though sponges are marine, the soil fauna will contain silica, however it has not yet been identified. The biomass of soil microorganisms has not been increased. Nevertheless, the majority of the biogenic soil pool is unquestionably accounted for by the phytogenic pool of Si.[106] All of the silica formed from plants that is present in the soil is commonly referred to as the phytogenic pool of Si in soils, whereas phytolith silica typically refers to that portion of soil that can be recovered that is larger than 2 μm. Soils can have a phytolith content of 0.1 to 50 percent w/w (0.004 to 21 percent silica), however it often varies from 0.1 to 4 percent w/w (0.04 to 1.68 percent silica). Total phytogenic silica in soils is even more elevated since phytolith evaluation is often limited to small sized particles and discards the percentage less than 5 μm. Plant litter is a source of phytogenic silica for the soil.[102,107] Thus, the type of phytogenic silica found in the surrounding soil is determined by the amount and pattern of silica deposition in plants.

The ability of plants to accumulate silicon varies widely, and concentrations in over-ground sections can range from 1–100 $g\,kg^{-1}$. Typically, these concentrations are similar to those of macronutrients like nitrogen and potassium.[108] According to the classification of the Si uptake process as active, passive, or rejective plants can be categorised as intermediates, compiler, or eliminator. When the concentration of silica in the tissue is greater than 10 $g\,kg^{-1}$, plants are classified as collectors of Si, and when it is less than 5 $g\,kg^{-1}$, they are classified as excluders.[109] Ten major crops are listed by global production, and seven of them—sugarcane, rice, wheat, barley, sugar beetroot, soybeans, and tomatoes—are considered accumulators. The rate at which an organ transpires determines the distribution of silicon inside its tissue and organs.[110] The majority of silica is accumulated at the ends of the transpiration flow, primarily in outside walls of the epidermal cells of shoots and stays in the apoplast. Phytogenic Si deposits are created when the deposited silica polymerizes to form amorphous silica. ($SiO_2 \cdot nH_2O$) and is concentrated by transpiration loss of water.[110] Tiny concentrations of contaminants including iron, titanium, aluminium, manganese, copper, nitrogen, phosphorus and carbon may be present in these silica deposits. The vascular system and the root endodermis can also accumulate silica. Since amorphous phytogenic silica is stationary inside the plant, it tends to accumulate in the tissues of older plants.[106]

Si is frequently applied to plants to increase their tolerance to a variety of biotic and abiotic stressors. The primary abiotic stresses that are reduced are salinity, drought, and metal toxicity; the primary biotic stresses that are reduced are pathogen and insect pest damage. Because field-grown plants are always exposed to a variety of challenges, fertiliser boosts output. Applications of silica on silica-deficient soils are not surprising.[111] Slag containing silicon is usually applied to rice at a rate of around 225 to 300 kg silica ha^{-1} per year and the sugarcane at a rate of 600 to 1125 kg silica ha^{-1} every 4 to 5 years (during replantation).[112,113]

It has been demonstrated that silicon shields plants from a variety of diseases. There are two basic ways in which the improved resistance to illness arises. Initially, when epidermal cell walls are impregnated with silica, a strong external barrier is created that physically prevents fungal penetration and blocks the infection process.[114] Second, through physiological and biochemical/molecular pathways, soluble Si can modulate the host's resistance to infections. Applications of Si can also increase a plant's resilience to insect infestations.[113] In addition to providing a physical defense against leaf-eating caterpillars and sucking insects, silica is deposited in epidermal layers and contributes inducing chemical defense mechanisms against herbivores. Si can improve drought resistance in a variety of crops such as wheat, sorghum, sunflower, cucumber, rice, and pepper, according to numerous reports. In a comparable range of crops, silicon supplements have also been demonstrated to boost salinity tolerance.[109,115] A variety of physiological (such as altered transpiration rates, enhanced photosynthesis, and reduced osmotic adjustment with organic solutes and improved antioxidant defenses) as well as biochemical/molecular processes (such as absorption of Na^+ and Cl^-) that are Si-stimulated contribute to this tolerance to salinity and drought.[116]

Applications involving silicon have the potential to reduce the toxicity of metals such as aluminium, magnesium, iron, cadmium, lead, and zinc. The intricate mechanisms that are accountable for this can be categorized as either internal (such as the co-accumulation of metals and silica in soil solution) or external (such as the co-accumulation of metals along with silica within roots, blockage of metal transport from roots to stem.[114]

2.5 Conclusion

This chapter functions as a springboard for future investigation and study into the dynamic interactions between biogenic silica and the natural world, emphasizing the material's applicability in both terrestrial and marine contexts as well as in the larger framework of climate change on a worldwide scale. In the book "*The Biogenic Silica in Ocean and Terrestrial Environments*", the complex role of silica in forming ecosystems and affecting biogeochemical cycles is explored. Biogenic silica is extremely significant as it is sourced from diatoms and other silica-producing organisms and has an impact on carbon sequestration and nutrient cycle.

Diatoms play a crucial role in the biological pump that moves carbon from the ocean's surface to its depths through the silica shells they carry. This mechanism helps to regulate the global temperature and has an impact on marine production. On the other hand, in terrestrial settings, silica, plants, and soils interact in various ways, influencing the availability of nutrients, the structure of the plants, and their capacity to withstand environmental stresses. Our understanding of biogenic silica's many contributions to Earth's systems is growing, and the study of biogenic silica provides doors to creative methods in environmental science and conservation, from presenting options for sustainable agriculture and carbon sequestration to revealing information about former climates through sedimentary records.

2.6 Future Perspective

There is great potential for future progress in the research of biogenic silica in terrestrial and maritime ecosystems. Understanding the complex mechanisms behind the biomineralization of silica in a variety of organisms—from diatoms and sponges in marine ecosystems to grasses and trees in terrestrial habitats—is an important field of research. This information will not only help us understand ecological interactions better, but it will also open up new possibilities for biotechnology and materials science applications. Furthermore, the assessment of biogenic silica's involvement in carbon and nutrient cycling is becoming more and more important as climate change continues to affect global ecosystems, particularly in light of changing climatic circumstances. By using sophisticated analytical methods like proteomics, genomics, and high-resolution imaging, scientists will be able to better understand the intricate relationships between silica dynamics at various temporal and geographical scales. The field of biogenic silica study holds great promise for providing novel insights into how Earth's ecosystems function and adapt to changing environmental conditions, particularly *via* the promotion of multidisciplinary cooperation and the utilization of developing technology.

Acknowledgements

We are thankful to Jaypee University of Information Technology, Waknaghat, Solan for providing infrastructure and required facilities to write this book chapter. We extend our thanks to the Royal Society of Chemistry for providing us with the opportunity to contribute to this book chapter. Their support and encouragement have been instrumental in making this endeavor possible.

References

1. P. J. Tréguer, J. N. Sutton and M. Brzezinski, *et al.*, Reviews and syntheses: the biogeochemical cycle of silicon in the modern ocean, *Biogeosciences*, 2021, **18**(4), 1269–1289.
2. D. G. Mann, The species concept in diatoms, *Phycologia*, 1999, **38**(6), 437–495.
3. C. E. Hamm, R. Merkel and O. Springer, *et al.*, Architecture and material properties of diatom shells provide effective mechanical protection, *Nature*, 2003, **421**(6925), 841–843.
4. S. B. Baines, B. S. Twining and M. A. Brzezinski, *et al.*, Significant silicon accumulation by marine picocyanobacteria, *Nat. Geosci.*, 2012, **5**(12), 886–891.
5. Y. Wei, Z. Zhang, Z. Cui and J. Sun, Size-fractionated biogenic silica standing stocks and carbon biomass in the western tropical North Pacific: evidence for the ecological importance of pico-sized plankton in oligotrophic gyres, *Front. Mar. Sci.*, 2021, **8**, DOI: 10.3389/fmars.2021.691367.
6. F. Rey, Declining silicate concentrations in the Norwegian and Barents Seas, *ICES J. Mar. Sci.*, 2012, **69**(2), 208–212.
7. H. Hátún, K. Azetsu-Scott and R. Somavilla, *et al.*, The subpolar gyre regulates silicate concentrations in the North Atlantic, *Sci. Rep.*, 2017, 7(1), 14576.
8. M. A. Brzezinski, D. M. Nelson, V. M. Franck and D. E. Sigmon, Silicon dynamics within an intense open-ocean diatom bloom in the Pacific sector of the Southern Ocean, *Deep Sea Res., Part II*, 2001, **48**(19–20), 3997–4018.
9. M. A. Brzezinski, D. R. Phillips, F. P. Chavez, G. E. Friederich and R. C. Dugdale, Silica production in the Monterey, California, upwelling system, *Limnol. Oceanogr.*, 1997, **42**(8), 1694–1705.
10. R. C. Dugdale, F. P. Wilkerson and H. J. Minas, The role of a silicate pump in driving new production, *Deep Sea Res., Part I*, 1995, **42**(5), 697–719.
11. M. A. Brzezinski, The Si : C : N ratio of marine diatoms: interspecific variability and the effect of some environmental variables, *J. Phycol.*, 1985, **21**(3), 347–357.
12. A. O. Marron, M. J. Alston and D. Heavens, *et al.*, A family of diatom-like silicon transporters in the siliceous loricate choanoflagellates, *Proc. Biol. Sci.*, 2013, **280**(1756), 20122543.
13. S. Görlich, A. J. Samuel and R. J. Best, *et al.*, Natural hybrid silica/protein superstructure at atomic resolution, *Proc. Natl. Acad. Sci. U. S. A.*, 2020, **117**(49), 31088–31093.
14. M. Preari, K. Spinde, J. Lazic, E. Brunner and K. D. Demadis, Bioinspired insights into silicic acid stabilization mechanisms: the dominant role of polyethylene glycol-induced hydrogen bonding, *J. Am. Chem. Soc.*, 2014, **136**(11), 4236–4244.

15. D. J. Belton, O. Deschaume and C. C. Perry, An overview of the fundamentals of the chemistry of silica with relevance to biosilicification and technological advances, *FEBS J.*, 2012, **279**(10), 1710–1720.
16. H. Ehrlich, R. Deutzmann and E. Brunner, *et al.*, Mineralization of the metre-long biosilica structures of glass sponges is templated on hydroxylated collagen, *Nat. Chem.*, 2010, **2**(12), 1084–1088.
17. M. Wysokowski, T. Jesionowski and H. Ehrlich, Biosilica as a source for inspiration in biological materials science, *Am. Mineral.*, 2018, **103**(5), 665–691.
18. J. C. Lewin, The dissolution of silica from diatom walls, *Geochim. Cosmochim. Acta*, 1961, **21**(3–4), 182–198.
19. R. M. Custódio and M. Nacional, *Porifera Research: Biodiversity, Innovation and Sustainability*, Museu Nac, 2007.
20. M. Sahebi, M. M. Hanafi and A. Siti Nor Akmar, *et al.*, Importance of silicon and mechanisms of biosilica formation in plants, *BioMed Res. Int.*, 2015, **2015**, 396010.
21. J. F. Ma, Role of silicon in enhancing the resistance of plants to biotic and abiotic stresses, *Soil Sci. Plant Nutr.*, 2004, **50**(1), 11–18.
22. Y. Liang, Effects of silicon on enzyme activity and sodium, potassium and calcium concentration in barley under salt stress, *Plant Soil*, 1999, **209**(2), 217–224.
23. Y. Liang, Q. Chen, Q. Liu, W. Zhang and R. Ding, Exogenous silicon (Si) increases antioxidant enzyme activity and reduces lipid peroxidation in roots of salt-stressed barley (Hordeum vulgare L.), *J. Plant Physiol.*, 2003, **160**(10), 1157–1164.
24. E. Struyf and D. J. Conley, Silica: an essential nutrient in wetland biogeochemistry, *Front. Ecol. Environ.*, 2009, 7(2), 88–94.
25. S. W. Blecker, R. L. McCulley, O. A. Chadwick and E. F. Kelly, Biologic cycling of silica across a grassland bioclimosequence, *Global Biogeochem. Cycles*, 2006, **20**(3), DOI: 10.1029/2006GB002690.
26. J.-T. Cornelis, B. Delvaux, R. B. Georg, Y. Lucas, J. Ranger and S. Opfergelt, Tracing the origin of dissolved silicon transferred from various soil-plant systems towards rivers: a review, *Biogeosciences*, 2011, **8**(1), 89–112.
27. A. Alexandre, M. Bouvet and L. Abbadie, The role of savannas in the terrestrial Si cycle: A case-study from Lamto, Ivory Coast, *Glob Planet Change*, 2011, **78**(3–4), 162–169.
28. B. Ronchi, W. Clymans and A. L. P. Barão, *et al.*, Transport of dissolved Si from soil to river: a conceptual mechanistic model, *Silicon*, 2013, **5**(1), 115–133.
29. L. A. Derry, A. C. Kurtz, K. Ziegler and O. A. Chadwick, Biological control of terrestrial silica cycling and export fluxes to watersheds, *Nature*, 2005, **433**(7027), 728–731.
30. J. C. Carey and R. W. Fulweiler, The terrestrial silica pump, *PLoS One*, 2012, 7(12), e52932.

31. P. J. Frings, W. Clymans, G. Fontorbe, C. L. De La Rocha and D. J. Conley, The continental Si cycle and its impact on the ocean Si isotope budget, *Chem. Geol.*, 2016, **425**, 12–36.
32. N. K. Savant, G. H. Snyder and L. E. Datnoff, Silicon management and sustainable rice production, *Adv. Agron.*, 1996, **58**, 151–199.
33. J. F. Ma and N. Yamaji, Silicon uptake and accumulation in higher plants, *Trends Plant Sci.*, 2006, **11**(8), 392–397.
34. B. Collin, E. Doelsch, C. Keller, F. Panfili and J.-D. Meunier, Distribution and variability of silicon, copper and zinc in different bamboo species, *Plant Soil*, 2012, **351**(1–2), 377–387.
35. M. J. Hodson and D. E. Evans, Aluminium/silicon interactions in higher plants, *J. Exp. Bot.*, 1995, **46**(2), 161–171.
36. F. Fraysse, O. S. Pokrovsky, J. Schott and J.-D. Meunier, Surface properties, solubility and dissolution kinetics of bamboo phytoliths, *Geochim. Cosmochim. Acta*, 2006, **70**(8), 1939–1951.
37. J.-C. Huang, T.-Y. Lee and T.-C. Lin, *et al.*, Effects of different N sources on riverine DIN export and retention in a subtropical high-standing island, Taiwan, *Biogeosciences*, 2016, **13**(6), 1787–1800.
38. A. F. White, D. V. Vivit, M. S. Schulz, T. D. Bullen, R. R. Evett and J. Aagarwal, Biogenic and pedogenic controls on Si distributions and cycling in grasslands of the Santa Cruz soil chronosequence, California, *Geochim. Cosmochim. Acta*, 2012, **94**, 72–94.
39. M. Sommer, D. Kaczorek, Y. Kuzyakov and J. Breuer, Silicon pools and fluxes in soils and landscapes – a review, *J. Plant Nutr. Soil Sci.*, 2006, **169**(3), 310–329.
40. T. N. M. Luu, J. Garnier and G. Billen, *et al.*, N, P, Si budgets for the Red River Delta (northern Vietnam): how the delta affects river nutrient delivery to the sea, *Biogeochemistry*, 2012, **107**(1–3), 241–259.
41. R. Lauerwald, J. Hartmann, N. Moosdorf, H. H. Dürr and S. Kempe, Retention of dissolved silica within the fluvial system of the conterminous USA, *Biogeochemistry*, 2013, **112**(1–3), 637–659.
42. C. Y. Bernard, H. H. Dürr, C. Heinze, J. Segschneider and E. Maier-Reimer, Contribution of riverine nutrients to the silicon biogeochemistry of the global ocean – A model study, *Biogeosciences*, 2011, **8**(3), 551–564.
43. P. J. Tréguer and C. L. De La Rocha, The world ocean silica cycle, *Annu. Rev. Mar. Sci.*, 2013, **5**, 477–501.
44. D. J. Conley, Riverine contribution of biogenic silica to the oceanic silica budget, *Limnol. Oceanogr.*, 1997, **42**(4), 774–777.
45. S. L. Loucaides, P. V. Cappelle and T. Behrends, Dissolution of biogenic silica from land to ocean: role of salinity and pH, *Limnol. Oceanogr.*, 2008, **53**(4), 1614–1621.
46. B. Opalinska and S. A. Cowling, Modelling the movement of biogenic silica from terrestrial vegetation to riverine systems within the continental USA, *Ecol. Modell.*, 2015, **312**, 104–113.
47. D. R. Piperno, The status of phytolith analysis in the American tropics, *J. World Prehist.*, 1991, **5**(2), 155–191.

48. M. H. Martin, J. R. Cox and F. F. Ibarra, Climatic effects on buffelgrass productivity in the Sonoran Desert, *J. Range Manage.*, 1995, **48**(1), 60–63.
49. L. H. P. Jones and K. A. Handreck, Silica in soils, plants, and animals, *Adv. Agron.*, 1967, **19**, 107–149.
50. J. A. Raven, The transport and function of silicon in plants, *Biol. Rev.*, 1983, **58**(2), 179–207.
51. F. A. Street-Perrott and P. A. Barker, Biogenic silica: a neglected component of the coupled global continental biogeochemical cycles of carbon and silicon, *Earth Surf. Processes Landforms*, 2008, **33**, 1436–1457.
52. M. Maldonado, M. C. Carmona and Z. Velásquez, *et al.*, Siliceous sponges as a silicon sink: an overlooked aspect of benthopelagic coupling in the marine silicon cycle, *Limnol. Oceanogr.*, 2005, **50**(3), 799–809.
53. D. J. Conley and C. L. Schelske, Potential role of sponge spicules in influencing the silicon biogeochemistry of Florida lakes, *Can. J. Fish. Aquat. Sci.*, 1993, **50**(2), 296–302.
54. K. D. Bidle, M. A. Brzezinski, R. A. Long, J. L. Jones and F. Azam, Diminished efficiency in the oceanic silica pump caused by bacteria-mediated silica dissolution, *Limnol. Oceanogr.*, 2003, **48**(5), 1855–1868.
55. J. C. Lewin, The dissolution of silica from diatom walls, *Geochim. Cosmochim. Acta*, 1961, **21**(3–4), 182–198.
56. *Silicon and Siliceous Structures in Biological Systems*, ed. T. L. Simpson and B. E. Volcani, Springer Science+Business Media, 2012, DOI: 10.1007/978-1-4612-5947-3.
57. M. B. Melia, The distribution and relationship between palynomorphs in aerosols and deep-sea sediments off the coast of northwest Africa, *Mar. Geol.*, 1984, **58**(3–4), 345–371.
58. L. F. Ruppert, S. G. Neuzil, C. B. Cecil and J. S. Kane, Inorganic constituents from samples of a domed and lacustrine peat, Sumatra, Indonesia, *Spec. Pap. – Geol. Soc. Am.*, 1993, **286**, 83–96.
59. M. J. Hodson, P. J. White, A. Mead and M. R. Broadley, Phylogenetic variation in the silicon composition of plants, *Ann. Bot.*, 2005, **96**(6), 1027–1046.
60. T. S. Lovering, Significance of accumulator plants in rock weathering, *Geol. Soc. Am. Bull.*, 1959, **70**(6), 781–800.
61. C. E. J. De Ronde, P. Stoffers and D. Garbe-Schönberg, *et al.*, Discovery of active hydrothermal venting in Lake Taupo, New Zealand, *J. Volcanol. Geotherm. Res.*, 2002, **115**(3–4), 257–275.
62. C. L. Blättler and J. A. Higgins, Testing Urey's carbonate–silicate cycle using the calcium isotopic composition of sedimentary carbonates, *Earth Planet. Sci. Lett.*, 2017, **479**, 241–251.
63. D. J. Conley and C. L. Schelske, Biogenic silica, in *Tracking Environmental Change Using Lake Sediments: Terrestrial, Algal, and Siliceous Indicators*, 2002, pp. 281–293.
64. D. M. Nelson, P. Tréguer, M. A. Brzezinski, A. Leynaert and B. Quéguiner, Production and dissolution of biogenic silica in the

ocean: revised global estimates, comparison with regional data and relationship to biogenic sedimentation, *Global Biogeochem. Cycles*, 1995, **9**(3), 359–372.

65. J. H. Martin, M. Gordon and S. E. Fitzwater, The case for iron, *Limnol. Oceanogr.*, 1991, **36**(8), 1793–1802.
66. K. G. Harrison, Role of increased marine silica input on paleo-pCO_2 levels, *Paleoceanography*, 2000, **15**(3), 292–298.
67. D. K. Tripathi, V. P. Singh, S. Gangwar, S. M. Prasad, J. N. Maurya and D. K. Chauhan, Role of silicon in enrichment of plant nutrients and protection from biotic and abiotic stresses, in *Improvement of Crops in the Era of Climatic Changes*, Springer, 2014, vol. 1, pp. 39–56.
68. O. Aumont, E. Maier-Reimer, S. Blain and P. Monfray, An ecosystem model of the global ocean including Fe, Si, P colimitations, *Global Biogeochem. Cycles*, 2003, **17**(2), DOI: 10.1029/2001GB001745.
69. S. C. Doney, Major challenges confronting marine biogeochemical modeling, *Global Biogeochem. Cycles*, 1999, **13**(3), 705–714.
70. J. K. Moore, S. C. Doney, J. A. Kleypas, D. M. Glover and I. Y. Fung, An intermediate complexity marine ecosystem model for the global domain, *Deep Sea Res., Part II*, 2001, **49**(1–3), 403–462.
71. K. Leblanc and D. A. Hutchins, New applications of a biogenic silica deposition fluorophore in the study of oceanic diatoms, *Limnol. Oceanogr.: Methods*, 2005, **3**(10), 462–476.
72. P. Tréguer and A. J. Van Bennekom, The annual production of biogenic silica in the Antarctic Ocean, *Mar. Chem.*, 1991, **35**(1–4), 477–487.
73. N. Ma, Z. Song and B. Wang, *et al.*, Effects of river damming on biogenic silica turnover: implications for biogeochemical carbon and nutrient cycles, *Acta Geochim.*, 2017, **36**(4), 626–637.
74. A. Lerman, Weathering rates and major transport processes an introduction, in *Physical and Chemical Weathering in Geochemical Cycles*, Springer Netherlands, 1988, pp. 1–10.
75. D. J. Conley, Riverine contribution of biogenic silica to the oceanic silica budget, *Limnol. Oceanogr.*, 1997, **42**(4), 774–777.
76. D. W. Eggimann, F. T. Manheim and P. R. Betzer, Dissolution and analysis of amorphous silica in marine sediments, *J. Sediment. Res.*, 1980, **50**(1), 215–225.
77. R. A. Mortlock and P. N. Froelich, A simple method for the rapid determination of biogenic opal in pelagic marine sediments, *Deep-Sea Res., Part A*, 1989, **36**(9), 1415–1426.
78. D. J. DeMaster, The supply and accumulation of silica in the marine environment, *Geochim. Cosmochim. Acta*, 1981, **45**(10), 1715–1732.
79. P. J. Müller and R. Schneider, An automated leaching method for the determination of opal in sediments and particulate matter, *Deep Sea Res., Part I*, 1993, **40**(3), 425–444.
80. D. J. Conley, Riverine contribution of biogenic silica to the oceanic silica budget, *Limnol. Oceanogr.*, 1997, **42**(4), 774–777.

81. H. H. Dürr, M. Meybeck, J. Hartmann, G. G. Laruelle and V. Roubeix, Global spatial distribution of natural riverine silica inputs to the coastal zone, *Biogeosciences*, 2011, **8**(3), 597–620.
82. D. M. Nelson, P. Tréguer, M. A. Brzezinski, A. Leynaert and B. Quéguiner, Production and dissolution of biogenic silica in the ocean: revised global estimates, comparison with regional data and relationship to biogenic sedimentation, *Global Biogeochem. Cycles*, 1995, **9**(3), 359–372.
83. V. Carbonnel, M. Lionard, K. Muylaert and L. Chou, Dynamics of dissolved and biogenic silica in the freshwater reaches of a macrotidal estuary (The Scheldt, Belgium), *Biogeochemistry*, 2009, **96**(1–3), 49–72.
84. B. Quéguiner, P. Tréguer and R. Corvaisier, Seasonal and interannual evolutions of biogenic silica standing stock and production at the KERFIX time-series station off the Kerguelen Islands, in *The Kerguelen Plateau: Marine Ecosystem and Fisheries*, ed. G. Duhamel and D. Welsford, Proceedings of the 1st International Science Symposium on the Kerguelen Plateau, Société Française d'Ichtyologie, 2011, pp. 175–178.
85. M. A. Brzezinski, D. R. Phillips, F. P. Chavez, G. E. Friederich and R. C. Dugdale, Silica production in the Monterey, California, upwelling system, *Limnol. Oceanogr.*, 1997, **42**(8), 1694–1705.
86. M. A. Brzezinski, J. L. Jones, K. D. Bidle and F. Azam, The balance between silica production and silica dissolution in the sea: insights from Monterey Bay, California, applied to the global data set, *Limnol. Oceanogr.*, 2003, **48**(5), 1846–1854.
87. S. Loucaides, P. Van Cappellen and T. Behrends, Dissolution of biogenic silica from land to ocean: role of salinity and pH, *Limnol. Oceanogr.*, 2008, **53**(4), 1614–1621.
88. D. M. Nelson, P. Tréguer, M. A. Brzezinski, A. Leynaert and B. Quéguiner, Production and dissolution of biogenic silica in the ocean: revised global estimates, comparison with regional data and relationship to biogenic sedimentation, *Global Biogeochem. Cycles*, 1995, **9**(3), 359–372.
89. P. J. Tréguer and C. L. De La Rocha, The world ocean silica cycle, *Annu. Rev. Mar. Sci.*, 2013, **5**, 477–501.
90. D. M. Nelson and M. A. Brzezinski, Diatom growth and productivity in an oligotropic midocean gyre: A 3-yr record from the Sargasso Sea near Bermuda, *Limnol. Oceanogr.*, 1997, **42**(3), 473–486.
91. D. S. M. Billett, R. S. Lampitt, A. L. Rice and R. F. C. Mantoura, Seasonal sedimentation of phytoplankton to the deep-sea benthos, *Nature*, 1983, **302**(5908), 520–522.
92. B. V. Bodungen, V. S. Smetacek and M. M. Tilzer, Primary production and sedimentation during spring in the Antarctic Peninsula region, *Deep-Sea Res., Part A*, 1986, **33**(2), 177–194.
93. D. C. Hurd, Interactions of biogenic opal, sediment and seawater in the central equatorial pacific, *Geochim. Cosmochim. Acta*, 1973, **37**(10), 2257–2282.

94. R. Jahnke, D. Heggie, S. Emerson and V. Grundmanis, Pore waters of the central Pacific Ocean: nutrient results, *Earth Planet. Sci. Lett.*, 1982, **61**(2), 233–256.
95. J. McManus, Early diagenesis of biogenic opal: dissolution rates, kinetics, and paleoceanographic implications, *Oceanogr. Lit. Rev.*, 1996, **1**(43), 34.
96. O. Ragueneau, P. Tréguer and A. Leynaert, *et al.*, A review of the Si cycle in the modern ocean: recent progress and missing gaps in the application of biogenic opal as a paleoproductivity proxy, *Global Planet Chan*, 2000, **26**(4), 317–365.
97. J. C. Goldman, Potential role of large oceanic diatoms in new primary production, *Deep Sea Res., Part I*, 1993, **40**(1), 159–168.
98. D. J. DeMaster, The supply and accumulation of silica in the marine environment, *Geochim. Cosmochim. Acta*, 1981, **45**(10), 1715–1732.
99. P. Van Cappellen, S. Dixit and J. van Beusekom, Biogenic silica dissolution in the oceans: reconciling experimental and field-based dissolution rates, *Global Biogeochem. Cycles*, 2002, **16**(4), 23-1–23-10.
100. F. Rey, Declining silicate concentrations in the Norwegian and Barents Seas, *ICES J. Mar. Sci.*, 2012, **69**(2), 208–212.
101. H. Hátún, K. Azetsu-Scott and R. Somavilla, *et al.*, The subpolar gyre regulates silicate concentrations in the North Atlantic, *Sci. Rep.*, 2017, 7(1), 14576.
102. A. Alexandre, J.-D. Meunier, F. Colin and J.-M. Koud, Plant impact on the biogeochemical cycle of silicon and related weathering processes, *Geochim. Cosmochim. Acta*, 1997, **61**(3), 677–682.
103. E. Struyf and D. J. Conley, Silica: an essential nutrient in wetland biogeochemistry, *Front. Ecol. Environ.*, 2009, 7(2), 88–94.
104. S. W. Blecker, R. L. McCulley, O. A. Chadwick and E. F. Kelly, Biologic cycling of silica across a grassland bioclimosequence, *Global Biogeochem. Cycles*, 2006, **20**(3), DOI: 10.1029/2006GB002690.
105. O. Katz, D. Puppe, D. Kaczorek, N. B. Prakash and J. Schaller, Silicon in the soil–plant continuum: intricate feedback mechanisms within ecosystems, *Plants*, 2021, **10**(4), 652.
106. R. J. Haynes, A contemporary overview of silicon availability in agricultural soils, *J. Plant Nutr. Soil Sci.*, 2014, **177**(6), 831–844.
107. J.-T. Cornelis, J. Ranger, A. Iserentant and B. Delvaux, Tree species impact the terrestrial cycle of silicon through various uptakes, *Biogeochemistry*, 2010, **97**(2–3), 231–245.
108. H.-H. Gu, S.-S. Zhan and S.-Z. Wang, *et al.*, Silicon-mediated amelioration of zinc toxicity in rice (Oryza sativa L.) seedlings, *Plant Soil*, 2012, **350**(1–2), 193–204.
109. J. Clarke, The occurrence and significance of biogenic opal in the regolith, *Earth-Sci. Rev.*, 2003, **60**(3–4), 175–194.
110. N. Mitani and J. F. Ma, Uptake system of silicon in different plant species, *J. Exp. Bot.*, 2005, **56**(414), 1255–1261.

111. Y. Liang, W. Sun, Y.-G. Zhu and P. Christie, Mechanisms of silicon-mediated alleviation of abiotic stresses in higher plants: a review, *Environ. Pollut.*, 2007, **147**(2), 422–428.
112. R. L. Fox, J. A. Silva, O. R. Younge, D. L. Plucknett and G. D. Sherman, Soil and plant silicon and silicate response by sugar cane, *Soil Sci. Soc. Am. J.*, 1967, **31**(6), 775–779.
113. S. A. Bharwana, S. Ali, M. A. Farooq, N. Iqbal, F. Abbas and M. S. A. Ahmad, Alleviation of lead toxicity by silicon is related to elevated photosynthesis, antioxidant enzymes suppressed lead uptake and oxidative stress in cotton, *J. Biorem. Biodegrad.*, 2013, **4**(4), 187.
114. R. J. Haynes, A contemporary overview of silicon availability in agricultural soils, *J. Plant Nutr. Soil Sci.*, 2014, **177**(6), 831–844.
115. A. L. Carnelli, M. Madella and J. P. Theurillat, Biogenic silica production in selected alpine plant species and plant communities, *Ann. Bot.*, 2001, **87**(4), 425–434.
116. V. D. Meena, M. L. Dotaniya, V. Coumar, S. Rajendiran, A. S. Kundu and A. S. Rao, A case for silicon fertilization to improve crop yields in tropical soils, *Proc. Natl. Acad. Sci., India, Sect. B*, 2014, **84**, 505–518.

CHAPTER 3

Production and Characterization of Biogenic Silica

DEEPAK KUMAR,*[a] STEFFY ANGURAL,*[b] SAKSHI,[a] MANISH DHAWAN,[c] MANISHA PARMAR[d] AND SWATI SHARMA[a]

[a] Department of Biotechnology, University Institute of Biotechnology (UIBT), Chandigarh University, India; [b] Department of Microbiology, Panjab University, Chandigarh, India; [c] Department of Microbiology, Punjab Agricultural University, Ludhiana, India; [d] Department of Veterinary Microbiology, Guru Angad Dev Veterinary and Animal Sciences University, Ludhiana, India

*Emails: deepakkdhiman14@gmail.com; steffyangural91@gmail.com

3.1 Introduction

Biogenic silica (bSi) is also known as biogenic opal or amorphous silica. It is widely spread worldwide to form microscopic minerals known as phytoliths grown in plants and lower grasses. Naturally, silica is an amorphous metal-oxide which is formed by complex inorganic polymerization processes. It is made up of carbonate and phosphate as crystalline iono-covalent solids naturally. Chemically, biogenic silica is hydrated silica ($SiO_2 \cdot nH_2O$), and it is very important for plants as well as animals. Diatoms extract silica from fresh and salt water for their cell walls as an important component. However, another form of silica is silicon which is found in sponges, protozoa and leafy plants as a structural material and in chicks and rats for skeletal

Biogenic Silica: Fundamentals and Applications
Edited by Ashok Kumar Nadda and Vijai Kumar Gupta

Published by the Royal Society of Chemistry, www.rsc.org

and growth development. It is also found in the human body in connective tissues such as bones, skin, teeth, eyes, glands and various organs.

Various agricultural wastes such as corncob, cornhusk, yam peelings, coconut husks, rice husk, rice, oat spelt husk, horsetail husk and many more are considered as waste, which is accumulated and in many cases burnt in the close proximity of forests. These all agricultural wastes are the major cause for the production of biogenic silica and an evident problem for the environment. These all agricultural wastes are produced in huge amounts in different areas of the world.[1] Large amounts of these agricultural residues are also disposed of and it leads to various health issues. The production of diatoms is dependent on the accessibility of dissolved silicic acid and their application is to produce biogenic silica shells.[2]

Silicate as silicon is concerned with many significant structures and plays a very important role in a different range of organisms, including human beings. In different cells and tissues, the availability of silicate enhances the metal concentrations but till now the best method of enhancement was revealed only with the sponges and diatoms.[3]

To meet the energy demands of enhancing the production of oils and reducing the cost of production, nanotechnology should be used for the proper recovery of the oil reservoirs. In nanotechnology, nanoparticles made from microorganisms using different methods of biotechnology is the best eco-friendly approach. It is also considered as a substitute for methods like conventional, physical and chemical methods. A study indicated the feasibility of biotechnology approaches in the oil industry.[4]

Recent developments in science and technology has encouraged researchers to use green chemistry while performing their experiments to decrease the level of environmental pollution and achieve an eco-friendly and environmental pollution free ecosystem.[5–8]

In this chapter, we discuss various methods for the large biogenic silica production from residue of agricultural waste. The optimal use of end products in different applications is very important. Therefore, we also discuss various characterization techniques of biogenic silica. In this context, several new techniques for production as well as for characterization were discussed.

3.2 Biogenic Silica

The silica cycle is one of the biogeochemical cycles where biogenic silica is transported among the various systems of Earth. Biogenic silica formation is a natural process in the environment. It exists in all the types of biogeochemical cycles.

Out of all the elements, silicon is widely present on the globe and is considered a bio essential element.[9] The silica cycle has considerable common characteristics with the carbon cycle and plays a significant role in sequestration of carbon by biogenic transportation and continental weathering.[10] Biogenic silica is also known as opal, biogenic opal, and amorphous opaline silica. It is one of the most extensive biogenic minerals. Biogenic silica is formed by the process when organisms decompose, their siliceous skeletal parts dissolving. After this step, material settles by the water columns and enriches the water in the deep region with dissolved silica.[11]

3.3 Production of Biogenic Silica

Biogenic silica can exist in different structural forms such as helical, globular, tubular, sheet-like, folded sheets and fibrillar due to their assortment in density, shape and structure, constituents and composition. A detailed study of reviews and literature revealed a number of organisms have siliceous structures such as in bacteria, fungus, yeast, molluscs, sponges, plants, fishes, brachiopods, mammals and humans, *etc.* (Figure 3.1). Biogenic silica exists in a large number of shapes and sizes due to their wide existence.

3.3.1 Microbial Production of Biogenic Silica

Microorganisms are closely associated with different bio-activities like formation of soil, biodegradation and biodeterioration and bioweathering. Inorganic nutrients will liberate after biological activity on silicate minerals. This is the most important fundamental activity of microbial ecology. The interaction between microorganisms and silicate can be classified through the following reactions:

(a) Deterioration and termination of the silicate crystal lattice
(b) Silicon uptake and its utilization
(c) In the end, release and sedimentation of the silicon minerals in the environment

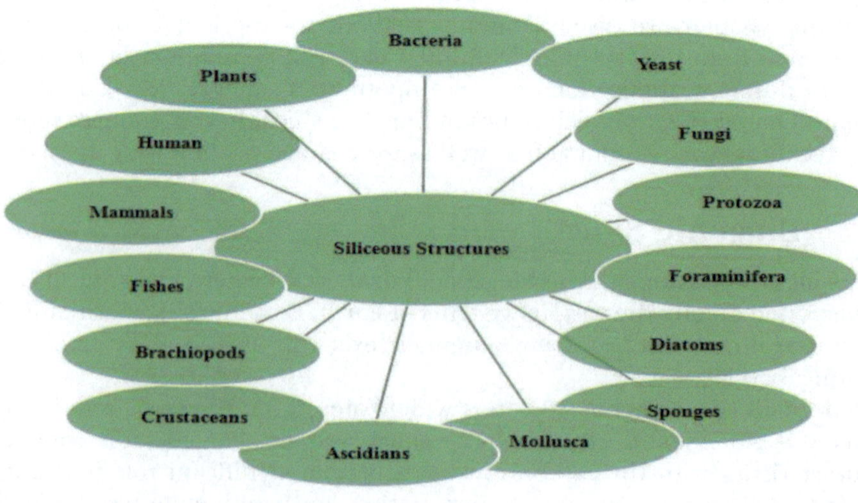

Figure 3.1 Structural diversity of biosilica.

Abiotic polymerization helps the degradation of biological polymeric silicates and later-on its separation process. The communication between silicon and biosphere is done by different methods in its many forms. It is very important to maintain the biological cycle for the ecological balance of lower and higher organisms. In the different phases of biogenic silica, amorphous silica is better than the crystalline silica and it is quite responsible for the cycling of silica in the soil and aquatic environment. Besides all these activities plants also plays a very important part in the silicon biogeochemical cycle. Alexandre *et al.*[12] revealed that more that 90% of biogenic silica is recycled by plants in the soil. Plants acquire silicon minerals from water present in the soil in the outward appearance of water-soluble silicic acid. This soluble form is converted to amorphous silica by polymerization and precipitation.

After this process, later on plant death occurs and generated silica is again solubilized into the soil and the rest of all plant decay produced humic acid which is responsible for weathering activities. Degradation of the final product of silicate minerals is done by metabolic activities of microorganisms. Hence, plants and microorganisms and their metabolic activities with the soil minerals are responsible for the complete silicon cycle. Production and structural diversity of biogenic silica from various microorganisms are discussed in Table 3.1.

3.3.1.1 Production of Biogenic Silica by Bacteria

A number of studies were conducted for the biogenic silica production from bacteria discussed in this chapter. Jones and Renaut[13] observed the water discharge from geysers containing chloride having alkaline pH at 100 °C. He observed that the discharge was coated with amorphous silica in the different regions of the vents termed as "geyser eggs", officially as siliceous "geyser eggs". Detailed study of these eggs revealed their structure has nuclei and laminae. Most of the cortical laminae contain filamentous and spherical microbes. Further detailed study of the organism identified as cyanobacteria. The growth of these eggs was controlled by the temperature of the geyser. In the dormant phase the eggs were exposed sub-aerially and submerged in boiling water, while exposed, the eggs may get dehydrated. Due to their abundance in the laminae the eggs confirmed that they originated in the siliceous coated grains from the environment set up by siliceous thermal spring systems.

An important study was conducted by Chafetz and Guidry[14] in geothermally heated hot-springs where travertine deposition was formed by carbonate minerals precipitation. The travertine deposits contain regular and irregular siliceous shrub-like morphologies and is termed crystal, ray, and bacterial shrubs. All three types of shrubs contain bacterial body fossils and micropores. The morphology of all the shrubs is varied due to the difference in the precipitation induced by bacteria. Therefore, in this environment under unfriendly conditions *Epiphyton* and *Renalcis* were

Table 3.1 Structural diversity and biogenic silica production from different microorganisms.

S. no.	Type/name of the organism	Form and structure of biogenic silica	Conditions	Ref.
Bacteria				
1.	*Cyanobacteria*	Siliceous "geyser eggs"	pH 8.5 to 9.25 at 100 °C	13
2.	*Epiphyton* and *Renalcis*, filamentous microorganisms	Siliceous shrubs (crystal, ray and shrubs)	Precipitation of carbonate minerals, heated hot springs	14
3.	Hyperthermophilic microorganisms	Siliceous and mixed silica-carbonate sinters	High temperature (92–94 °C)	15
Fungi				
4.	*Trentepohlia aurea* green algae	Siliceous coated grains	Rare minerals treated with algae	16
5.	Lab model	Silicic acid and bio chemicals	Rocks treated with mixture of sugar, silicic acid, metals, clays and amino acids in the presence of heat and UV light	17
6.	*Equisetum arvense* silica accumulator	Biogenic silica	Varying the concentrations of N_2 and P, mineral supplements	18
7.	*Fusarium culmorum*	Silica nanoparticles	Corn cobs treated with algae	19
Yeast				
8.	*Saccharomyces cerevisiae*	Silicate polysilicate	Treated with different concentrations of silicate and other metals	3
9.	*Saccharomyces cerevisiae*	Silica nanoparticles	Added sodium silicate as a precursor	4
Diatoms				
10.	Diatoms	Biogenic silica	Euphotic region supply of autotrophic carbon	20
11.	*Thalassiosira weissflogii* and *Thalassiosira pseudonana*	Silica content	Organisms cells treated with heterotrophic dinoflagellates with different concentrations of silica	21
Protists				
12.	*Polycystines* and *Phaeodarians* planktonic protists	Biogenic silica	Marine organic carbon	22

commonly found. Detailed identification concluded that *Epiphyton* and *Renalcis* are filamentous microorganisms and bacteria induced precipitates.

A pilot study was investigated by McKenzie *et al.*,[15] and explained the characterization of deposits and their fluids immobilization of heavy metals with microorganisms at the different sinters (siliceous and mixed silica-carbonate sinters). At high temperature, fluids of hyperthermophilic microorganisms were associated with the sinters deposited. The speed and method of silicification of hyperthermophilic microorganisms can be interrelated with the fluids containing dissolved silica content. It was found that the organic portion of the sinters were linked with high concentrations of heavy metals such as Hg and Tl, but there is no proof that hyperthermophiles were closely related with the heavy metals at high temperature ends of the global thermal hot springs ecosystem.

3.3.1.2 Production of Biogenic Silica by Fungi

According to the study done by Jones *et al.*,[16] there was a strong relationship between microorganisms and rare minerals such as dypingite, monohydrocalcite, and nesquehonite found at Hermitage Castle in the Scottish Borders during the process of biodeterioration. Out of all the minerals, monohydrocalcite has a strong connection with the green algae "*Trentepohlia aurea*", closely related to Chlorophyta division. They have siliceous coated grains. He also explained in detail the applications of monohydrocalcite mineral as a biogenic by-product.

A study illustrated the coating of rocks with silicic acid to form desert varnish. The authors proposed a method where rocks were first treated with amino acids, sugar mixture, silicic acid, metals, and clays in the presence of heat and UV light. A laboratory model having natural conditions was made for the formation of desert varnish by the polymerization of silicic acid. This method is similar to the natural process and also supported the importance of silicic acid for the making of rock coatings.[17]

A very important study was done by Kolb *et al.*,[18] for collecting the biogenic silica ($bSiO_2$) dissolution from the degrading litter. In this study, the colonization of submerged plant litter with bacterial and fungal biomass was investigated using a silica accumulator (*Equisetum arvense*). The effect of microbial colonization was observed using varying concentrations of nitrogen and phosphorus mineral supplements and differences in colonization was observed. It showed microbial litter decomposition influenced the Si remineralization from plants. Therefore, results suggested that the role of microbes in plant litter linked with Si is varied from the usual based on $bSiO_2$ dissolution used in marine ecosystems.

Alfredsson *et al.*,[19] obtained the silica nanoparticles using a fungus *Fusarium culmorum* using corn cob husks. They also studied their shape and morphology using different techniques and suggested that these nanoparticles have an effective role in the medical industry.

3.3.1.3 Production of Biogenic Silica by Yeast

According to Brasser *et al.*,[3] *Saccharomyces cerevisiae* silicate has a varying effect on the intracellular growth of metals. Experimental analysis revealed that growth was normal in the presence of low concentrations of silicate, however it was inhibited in the presence of high concentrations of silicate. The effect of silicate was also observed by various metals with different concentrations and explained the cell binding capacity having polysilicate formation. In this study, toxicity of silicate with respect to other metals was also checked in *Saccharomyces cerevisiae* and found that there was no influence of toxicity for the same.

A study investigated the use of *Saccharomyces cervisiae* for the synthesis of silica nanoparticles. A growth medium containing sodium silicate as a precursor was prepared for cultivation of *Saccharomyces cervisiae*. The role of nanoparticles in oil production was evaluated. It was observed that production efficiency improved with the use of silica nanoparticles synthesized using *Saccharomyces cervisiae* as compared to commercial nanoparticles.[4]

3.3.1.4 Production of Biogenic Silica by Diatoms

Krause *et al.*,[20] studied the contribution of biogenic silica using diatoms. The biogenic silica rate is significant in organic matter, below in mixed layers and low in shallow regions. Diatom carbon contributed more to the production than the autotrophic carbon. Less contribution of diatoms was observed in the export of organic matter. Hence, all these results suggested enriched biogenic silica was observed in the organic matter in the euphotic region.

A study conducted by Zhang *et al.*,[21] examined the silica content in the diatoms cells. To accomplish the low and high content of silica two diatoms species *i.e. Thalassiosira weissflogii* and *Thalassiosira pseudonana* were grown in low and high light parameters. The grown cells were inoculated to single and mixed cultures with heterotrophic dinoflagellates *Noctiluca scintillans* ciliates *Euplotes* sp. and ingested with the diatoms with a low silica content than with a high silica content. Experimental results showed that *Euplotes* sp. seemed to be less influenced by the silica content than was *N. scintillans.* As diatoms are difficult to digest due to their high silica content, *N. scintillans* exhibits low silica prey uptake. This study suggested the digestibility of silica content of diatoms is a result of selectivity and grazing activity and of protozoan grazers.

3.3.1.5 Production of Biogenic Silica by Protists

Llopis *et al.*,[22] studied the planktonic protists that originated from the water column. Siliceous *Polycystines* and *Phaeodarians* contributed to the marine organic carbon and biogenic silica. They first reported the cells of *Polycystine* and *Phaeodarian* capacity to uptake Si. Hence, biogenic silica production was significantly enhanced by these organisms in the marine ocean.

3.4 Methods Used for Biogenic Silica Production from Agricultural Wastes

There are a number of methods reported for biogenic silica production from agricultural wastes. Some of the methods are discussed in this chapter.

3.4.1 Materials Required

Various agricultural wastes such as cornhusk, coconut husks, rice husk, rice, oat spelt husk, horsetail husk, *etc.* are used.

According to Prempeh *et al.*,[23] two steps were performed to synthesize the biogenic silica. In the first step, chemical treatment of agricultural biomass was done using citric acid (1–5% solution) after this a combustion step was performed at 600 °C for 2 h in the oven. This method is considered a good process for biogenic silica synthesis.

Another study was performed by Alyosef *et al.*,[24] and illustrated the method to produce amorphous silica from an Egyptian rice husk as agricultural biomass. He used two steps: (1) Excess water washing of the sample then dry milling. After this, leaching was performed using citric acid. Leaching was performed in two steps *i.e.* for 3 h at 323 K and then for 1 h at 353 K. After this, heat treatment was applied in a muffles furnace at four chronological steps. (2) Same steps were followed except citric acid leaching. These methods were cost effective and had the best environmental impact results.

Basaran *et al.*,[25] reported the synthesis of biogenic silica from rice hull ash (RHA). He explained the method for the biogenic silica production in detail. According to this, RHA was milled at a frequency of 30 s^{-1} for 20 s (Retsch, MM400, Haan, Germany) and then burned in a furnace for 5 h at 450 °C. After this, an alkali extraction method was applied for removing silica[26] and an acid leaching step using HCl for removing metals impurities. Optimization was done by using various extraction methods with microwaving. A product was recovered in the form of sodium silicate solution by separating the residual ash using vacuum filters. This product was subjected to inductively coupled plasma-optical emission spectrometry to determine the $SiO_2 : Na_2O$ molar ratio. Silica gel samples were prepared at different pH and separated by centrifugation technique to form silica xerogel. Xerogels were milled and silica powder was obtained. It is designated as $4SiO_2$, $7SiO_2$, and $9SiO_2$ at the production stage.[12]

3.5 Biogenic Silica Characterization

Biogenic silica characterization is still a problem in spite of its significance and demand worldwide. Characterization is very important for the proper use of agricultural wastes for different industrial processes. Different methods of characterization were used in individual studies. Therefore, in this context various methods of characterization and their uses are discussed collectively.

3.5.1 X-ray Diffraction (XRD)

Materials and thin films have been studied extensively using XRD. Its usefulness is contingent upon possessing a crystalline substance. Although it is primarily an analytical method, the technique can also be used to shed light on changes occurring on surfaces under the optimum conditions. XRD can reveal, for instance, how much a carbon system's surface treatment has changed the substance's bulk. Some surface information can be gleaned from using XRD in a thin-film mode with extremely tiny take-off angles, but in general, it should be considered a bulk structural approach.

3.5.1.1 Biogenic Silica Characterized Using XRD

Prempeh *et al.*,[23] illustrated XRD results of citric acid pre-treated biomass showed arcanite (K_2SO_4), sylvite (KCl) and calcite ($CaCO_3$) dominant crystalline phases in ashes because of the presence of potassium and calcium ions. It was found that the extraction of high quality, amorphous silica form wastes help in the applications as construction material, catalyst support, backing material, and concrete, and act as value added products.

XRD analysis revealed that crystalline silica phases were absent when rice husks were burnt at 873 K. Silica ash that is primarily amorphous was also detected by XRD. Additionally, it is very essential to use enough air to entirely combust the organics of rice husks and to avoid the carbonization of the organics.[24]

Sola-Rabada *et al.*,[27] also checked XRD characterization of plant samples and showed distinctive diffraction patterns of cellulose, that can survive in different polymorphs because of the hydroxyl groups present in the structure. The hydroxyl group in the cellulose can be involved in inter or intra molecular hydrogen bonding showing a different arrangement of atoms.

3.5.2 Fourier-transform Infrared Spectroscopy (FTIR)

The technique of Fourier-transform infrared spectroscopy (FTIR) can be used to investigate the molecular structures and molecular vibrations of various substances. In particular, it makes it possible to research how materials absorb and emit light. The FTIR approach is very helpful for analyzing chemical interactions between atoms due to the flexibility of the IR wavelengths used. As it turns out, infrared-wavelength light is a perfect match for the vibrational frequencies of these connections. Spectroscopy typically operates between 200 and 4000 cm^{-1}, which correspond to the infrared spectrum.

3.5.2.1 *Fourier-transform Infrared Spectroscopy: What Is It Good for?*

Although FTIR spectroscopy has found widespread usage in the examination of carbon-containing organic molecules due to a deep understanding of their unique vibrational frequencies, it may be used to learn about a wide variety of other materials as well. Presence of contaminations in the sample was determined when comparing the different peaks of IR spectrum with frequencies of tabulated characteristics. This is because chemical bonds between atoms vibrate at characteristic frequencies, which have been widely tabulated.

In addition, FTIR may be used to do quality examination of a wide range of substances. Variations in a material's signature spectrum, for instance, can point to oxidation or contamination. Variations from a reference IR spectra can reveal whether a polymer is more or less polymerized (cured) or decomposed. Due to its versatility, FTIR is employed in a broad variety of fields, including those dealing with polymer characterization, forensics, pharmaceuticals, and food testing.

3.5.2.2 *Biogenic Silica Characterized Using FTIR*

Amorphous biogenic silicas were characterized by FTIR and magic angle spinning-nuclear magnetic resonance (MAS-NMR) spectroscopy.[28] It showed the presence of silica in fossil diatoms are in condensed and well organized form as compared with living diatoms after analysis. It also suggests that a few proteins and polysaccharides help in the formation of silica by a biosilicification mechanism.

Ben Yahia *et al.*,[29] illustrated the results of FTIR analysis and showed the predominance of quartz phase by the characteristic spectral bands of SiO_2 crystal. Siliceous rocks have many different clay minerals. They also have high thermal stability because of dehydroxylation of kaolinite.

3.5.3 **Inductively Coupled Plasma-optical Emission Spectrometry (ICP-OES)**

The analytical method of inductively coupled plasma optical emission spectroscopy (ICP-OES) is used to quantify the concentrations of individual elements in a given sample. The principle of ICP-OES relies on the fact that atoms and ions can absorb energy, elevating their electrons from the ground state to an excited state. When using ICP-OES, the energy comes from an argon plasma heated to a temperature of 10 000 K. The ICP-OES method is based on the fact that atoms in an excited state emit light at specified wavelengths as they return to a ground state.

The number of atoms or ions undergoing a transition corresponds to the amount of light emitted at each wavelength. As defined by the Beer Lambert law, the correlation between luminosity and elemental concentration.

3.5.3.1 Biogenic Silica Characterized Using (ICP-OES)

Fischer *et al.*,[30] measured the biogenic silicon content from *C. brevis* and *Thalassiosira* sp. and found a better association between silicon content and number of cells. Silicon content was evaluated as 86 ± 4 fmol cell^{-1}. Besides this observation he also checked the sensitivity effect and found that calibration was more sensitive at low concentrations than the higher concentrations for *Thalassiosira* sp. The reason could be the interference between ICP-OES measurement and the dissolved cell content.

Round *et al.*,[31] showed the effect of biogenic content present in the cell during asexual reproduction. Hence, they concluded that their digestion method and the other methods are effective for the estimation of biogenic silicon content.

A wonderful study also revealed the silica content present in the branches and stems.[27] Besides silica, other elements were also checked in the samples but the element sodium was not found in the plant sample.

3.5.4 Magic Angle Spinning-nuclear Magnetic Resonance (MAS-NMR) Spectroscopy

To get higher resolution NMR spectra, magic-angle spinning (MAS) is a common approach in solid-state NMR spectroscopy. For MAS-NMR, the sample is rotated (often at a frequency of 1 to 130 kHz) at the magic angle *m* (about 54.74°, where $\cos^2 m = 1/3$) with regard to the direction of the magnetic field.

3.5.4.1 Biogenic Silica Characterized Using (MAS-NMR) Spectroscopy

Schneider *et al.*,[32] did a magnificent study that showed the advanced textural characterization of biogenic silica from agricultural wastes. These agricultural wastes showed disordered, non-uniform structure of the pore. They also determined hyperpolarized and physisorption mesoporosity by validation with 129Xenon NMR. Positron annihilation lifetime spectroscopy (PALS) was used to determine the microporosity of biogenic silica and showed micropores in the range of 1.5–1.9 nm. They also concluded that various plant origins biogenic silica have different textural structure and inter-connections among the pores.

Another different study looked at amorphous biogenic silica[28] through MAS-NMR spectroscopy. In this study, sponge spicules and diatoms both exhibited silica but their higher concentration was found in diatoms frustules due to the reason of condensation. Furthermore, in diatoms the silica network appears to be connected with the bio-organic component of the cell. Hence, the biosilification process enhances proteins which favorites the formation of silica.

3.5.5 Scanning Electron Microscopy (SEM)

Scanning electron microscopes are the tools that reveal the composition and topography of a sample. SEM is a versatile and effective technique that may be used in many different fields. It performs well on both synthetic and organic samples.

3.5.5.1 What Happens During SEM?

When using scanning electron microscopy, electron beams are scanned across a sample. These beams are shot from an electron gun and go down the scanning electron microscope column at high speeds. The electron beams are focused by traveling through a set of lenses and openings. This takes place in a vacuum, which prevents the electron beam from interacting with any molecules or atoms already present in the microscope column. As a result, the image quality is guaranteed to be very high. The electron source is shielded from unwanted vibrations and noise by the vacuum environment. In a raster scan, the electron beams move in straight lines from left to right and up and down the sample's surface. Atoms on the sample's surface have reactions to the electrons. This interaction generates signature sample-specific signals in the form of secondary electrons, backscattered electrons, and rays. Microscope detectors pick up these signals, allowing for the construction of detailed images for viewing on a computer monitor.

3.5.5.2 Biogenic Silica Characterized Using SEM

Ben Yahia *et al.*,[29] studied babouchite siliceous rocks using SEM for evaluation of physiochemical characteristics like functional groups, surface type and thermal behavior. Properties of biogenic silica were calculated using numerous techniques such as SEM, FTIR, N_2-sorption, and TGA-DSC. SEM results also revealed the presence of mesopores with different size and distribution in the rocks. The study also revealed a finely microcrystalline quartz structure with parallel-layered surfaces. Adsorbents (babouchite samples) were characterized by XRD, SEM, ICP-ES and oxygen isotopic measurement, and the specific surface area and average pore size were calculated.

Different characterization techniques were used in another study by Alyosef *et al.*,[24] in detail. Firstly, N_2 adsorption-SEM analysis was also used to calculate the textural properties of the samples. A pre-treatment step is crucial for the textural parameter as it may generate intraparticle mesopores and macropores in the sample. It can be revealed by N_2 adsorption-SEM analysis.

3.6 Conclusion

Agricultural residues can be considered as a good resource for producing biogenic silica with a wide range of applications. Microorganisms are closely associated with different bio-activities like the formation of soil, biodegradation and biodeterioration and bioweathering. Biologically associated activities of silicate minerals liberate inorganic nutrients. This is the most important fundamental scrutiny of microbial ecology. Biogenic silica can be produced with the help of different types of organisms using siliceous minerals and compounds. Different methods were employed for the better production of biogenic silica. Production of biogenic silica leads to the formation of various end products. There is a need for characterization techniques on these end products for their proper use in different applications and also to decrease the level of environmental pollution and to achieve an eco-friendly and environmental pollution free ecosystem.

References

1. Y. R. Corrales-Ureña, C. Villalobos-Bermúdez, R. Pereira, M. Camacho, E. Estrada, O. Argüello-Miranda and J. R. Vega-Baudrit, Biogenic silica-based microparticles obtained as a sub-product of the nanocellulose extraction process from pineapple peels, *Sci. Rep.*, 2018, **8**(1), 10417.
2. J. W. Krause, C. M. Duarte, I. A. Marquez, P. Assmy, M. Fernández-Méndez, I. Wiedmann, P. Wassmann, S. Kristiansen and S. Agustí, Biogenic silica production and diatom dynamics in the Svalbard region during spring, *Biogeosciences*, 2018, **15**(21), 6503–6517.
3. H. J. Brasser, G. C. Krijger, T. G. van Meerten and H. T. Wolterbeek, Influence of Silicon on Cobalt, Zinc, and Magnesium in Baker's Yeast, Saccharomyces cerevisiae, *Biol. Trace Elem. Res.*, 2006, **112**(2), 175–190.
4. H. Zamani, A. Jafari, S. M. Mousavi and E. Darezereshki, Biosynthesis of silica nanoparticle using Saccharomyces cervisiae and its application on enhanced oil recovery, *J. Pet. Sci. Eng.*, 2020, **190**, 107002.
5. S. Angural, A. Kumar, D. Kumar, R. Warmoota, S. Sondhi and N. Gupta, Lignolytic and hemicellulolytic enzyme cocktail production from Bacillus tequilensis LXM 55 and its application in pulp biobleaching, *Bioprocess Biosyst. Eng.*, 2020, **43**(12), 2219–2229.
6. S. Angural, I. Bala, A. Kumar, D. Kumar, S. Jassal and N. Gupta, Bleach enhancement of mixed wood pulp by mixture of thermo-alkali-stable xylanase and mannanase derived through co-culturing of Alkalophilic Bacillus sp. NG-27 and Bacillus nealsonii PN-11, *Heliyon*, 2021, 7(1), e05673.
7. D. Kumar, R. Bhardwaj, S. Jassal, T. Goyal, A. Khullar and N. Gupta, Application of enzymes for an eco-friendly approach to textile processing, *Environ. Sci. Pollut. Res.*, 2021, **30**(28), 71838–71848.
8. D. Kumar, A. Khullar, N. Sharma and N. Gupta, Immobilization of laccase from *Bacillus subtilis* DS on chitosan beads and standardization of process for biodegradation of textile dyes, *J. Chem. Technol. Biotechnol.*, 2022, **97**(2), 466–473.
9. J. W. Hunt, A. P. Dean, R. E. Webster, G. N. Johnson and A. R. Ennos, A Novel Mechanism by which Silica Defends Grasses Against Herbivory, *Ann. Bot.*, 2008, **102**(4), 653–656.

10. M. J. Defant and M. S. Drummond, Derivation of some modern arc magmas by melting of young subducted lithosphere, *Nature*, 1990, **347**(6294), 662–665.
11. D. J. Conley, Terrestrial ecosystems and the global biogeochemical silica cycle, *Global Biogeochem. Cycles*, 2002, **16**(4), 68-1–68-8.
12. A. Alexandre, J.-D. Meunier, F. Colin and J.-M. Koud, Plant impact on the biogeochemical cycle of silicon and related weathering processes, *Geochim. Cosmochim. Acta*, 1997, **61**(3), 677–682.
13. B. Jones and R. W. Renaut, Hot spring and geyser sinters: the integrated product of precipitation, replacement, and deposition, *Can. J. Earth Sci.*, 2003, **40**(11), 1549–1569.
14. H. S. Chafetz and S. A. Guidry, Bacterial shrubs, crystal shrubs, and ray-crystal shrubs: bacterial *vs.* abiotic precipitation, *Sediment. Geol.*, 1999, **126**(1–4), 57–74.
15. E. J. McKenzie, K. L. Brown, S. L. Cady and K. A. Campbell, Trace metal chemistry and silicification of microorganisms in geothermal sinter, Taupo Volcanic Zone, New Zealand, *Geothermics*, 2001, **30**(4), 483–502.
16. M. S. Jones, R. D. Wakefield and G. Forsyth, The occurrence of rare minerals on decayed medieval Scottish building stone colonized by biological growths, *Mater. Constr.*, 1999, **49**(256), 3–14.
17. V. Kolb, A. I. Philip and R. S. Perry, Testing the Role of Silicic Acid and Bioorganic Materials in the Formation of Rock Coatings, in *Instruments, Methods, and Missions for Astrobiology, VIII*, 2004, vol. 5555, pp. 116–125.
18. H. Alfredsson, W. Clymans, J. Stadmark, D. Conley and J. Rousk, Bacterial and fungal colonization and decomposition of submerged plant litter: consequences for biogenic silica dissolution, *FEMS Microbiol. Ecol.*, 2016, **92**(3), fiw011.
19. A. Pieła, E. Żymańczyk-Duda, M. Brzezińska-Rodak, M. Duda, J. Grzesiak, A. Saeid, M. Mironiuk and M. Klimek-Ochab, Biogenic synthesis of silica nanoparticles from corn cobs husks. Dependence of the productivity on the method of raw material processing, *Bioorg. Chem.*, 2020, **99**, 103773.
20. J. W. Krause, M. R. Stukel, A. G. Taylor, D. A. A. Taniguchi, A. De Verneil and M. R. Landry, Net biogenic silica production and the contribution of diatoms to new production and organic matter export in the Costa Rica Dome ecosystem, *J. Plankton Res.*, 2016, **38**(2), 216–229.
21. S. Zhang, H. Liu, Y. Ke and B. Li, Effect of the Silica Content of Diatoms on Protozoan Grazing, *Front. Mar. Sci.*, 2017, **4**, 202.
22. N. Llopis Monferrer, D. Boltovskoy, P. Tréguer, M. M. Sandin, F. Not and A. Leynaert, Estimating Biogenic Silica Production of Rhizaria in the Global Ocean, *Global Biogeochem. Cycles*, 2020, **34**(3), e2019GB006286.
23. C. O. Prempeh, S. Formann, T. Schliermann, H. B. Dizaji and M. Nelles, Extraction and Characterization of Biogenic Silica Obtained from Selected Agro-Waste in Africa, *Appl. Sci.*, 2021, **11**(21), 10363.
24. H. A. Alyosef, A. Eilert, J. Welscher, S. S. Ibrahim, R. Denecke, W. Schwieger and D. Enke, Characterization of Biogenic Silica Generated

by Thermo Chemical Treatment of Rice Husk, *Part. Sci. Technol.*, 2013, **31**(5), 524–532.
25. Y. Basaran Elalmis, S. Yucel and I. Aydin, Amorphous biogenic silica production and utilization in experimental dental composites: Effect of silica gel formation pH on silica and composite properties, *Polym. Compos.*, 2021, **42**(10), 5111–5124.
26. U. Kalapathy, A simple method for production of pure silica from rice hull ash, *Bioresour. Technol.*, 2000, **73**(3), 257–262.
27. A. Sola-Rabada, P. Sahare, G. J. Hickman, M. Vasquez, L. T. Canham, C. C. Perry and V. Agarwal, Biogenic porous silica and silicon sourced from Mexican Giant Horsetail (Equisetum myriochaetum) and their application as supports for enzyme immobilization, *Colloids Surf., B*, 2018, **166**, 195–202.
28. A. Gendron-Badou, T. Coradin, J. Maquet, F. Fröhlich and J. Livage, Spectroscopic characterization of biogenic silica, *J. Non-Cryst. Solids*, 2003, **316**(2–3), 331–337.
29. N. Ben Yahia, D. Lerari, R. Bensouilah, S. Boussen, A. Sebei and F. Chaabani, Physicochemical characterization of biogenic silica of the upper Numidian babouchite siliceous rocks, northwestern Tunisia, *J. Afr. Earth Sci.*, 2022, **194**, 104608.
30. A. C. Fischer, O. M. Steinebach, K. R. Timmermans and H. T. Wolterbeek, A method for the destruction and analysis of biogenic silicon in two Antarctic diatom species: Thalassiosira sp. and Chaetoceros brevis, *J. Appl. Phycol.*, 2007, **19**(1), 71–77.
31. F. E. Round, R. M. Crawford and D. G. Mann, *The Diatoms. Biology and Morphology of the Genera*, Cambridge University Press, Cambridge, 1990, pp. 1–130.
32. D. Schneider, A. G. Attallah, S. Wassersleben, M. Wenzel, J. Matysik, R. Krause-Rehberg and D. Enke, Advanced textural characterization of biogenic silica by nitrogen physisorption, positron annihilation lifetime spectroscopy and hyperpolarized 129Xe NMR spectroscopy, *Microporous Mesoporous Mater.*, 2020, **307**, 110515.

CHAPTER 4

Biosilica Nanoparticles from Agricultural Waste and Their Applications

MOHINI YADAV, SHREYA GUPTA, SWATI SHARMA* AND NANCY GEORGE*

University Institute of Biotechnology, Chandigarh University, Mohali, Punjab 140413, India
*Emails: swati.e8750@cumail.in; nancy.george@cumail.in

4.1 Introduction

Agricultural waste is a significant portion of the waste present on earth having possible negative effects on humans and the environment, if not managed properly. Researchers have been working on developing novel, robust and convenient methods to treat agro-waste and turn it into useful and innovative products. One such pathway is the fabrication of silica nanoparticles from agricultural waste such as wheat husk,[1] rice husk/straw,[2] barley,[3] coconut husk, fruits,[4] sedge,[5] *etc.* A significant amount of work has been conducted in order to realize the potential of agro-waste to synthesize biosilica nanoparticles. Rice husk is the most studied substrate and has been used to produce biosilica nanoparticles by using different synthesis approaches such as the thermal combustion process or alkaline-grade solvent method.[2,6]

Other silica sources from agriculture have been discussed in Table 4.1. The silica obtained from agro-waste can be used in various fields such as biomedical,[7] environmental remediation,[8] construction material, different industries, and the engineering area.[9] The bio-nanosilica procured from

Biogenic Silica: Fundamentals and Applications
Edited by Ashok Kumar Nadda and Vijai Kumar Gupta

Published by the Royal Society of Chemistry, www.rsc.org

agro-waste possesses efficient absorptive and catalytic properties which increases the applicability of the silica nanoparticles. The production of highly pure biosilica from sugarcane with high adsorption ability and high surface area could be used to remove contaminants from the environment.[10] Thereby the biosilica nanoparticles derived from agro-waste hold certain properties which can be of vital importance.

The silica nanoparticles can be synthesized *via* various processes such as sol–gel,[11] acid, and thermal treatment,[12] alkaline method,[13] sequential approach,[14] novel magneisothermic reaction,[15] and green biotransformation method.[16] Of all the discussed methods, the biotransformation approach is the most desirable approach as it offers sustainable methodology without any use of toxic chemicals to develop the nanostructures while the other discussed methods in one way or the other cause noxious effects on nature. However, more in depth study is still needed to develop more applicable biotransforming procedures for the easy and considerable development of nanoparticles.

The waste generated from agricultural fields is the by-products of the production and processing of agricultural products that may include various useful substances of relatively lower economic value.[17] Agricultural-dependent industries generate ample quantity of waste residuals which lacks proper disposal methods as they are largely disposed of by landfilling, dumping, or burning. The inadvertent agro-waste disposal eventually leads to the production of greenhouse gases, and ultimately climate change, and also imposes detrimental effects on the environment and humans.[18] Environmentalists from around the globe have been concentrating to develop a comprehensive procedure for the synchronized and safe disposal of agricultural waste. As such, employment of agricultural waste to develop value-added entities could be a relevant remedial pathway to precisely deal with the problems of discarding agro-waste. Various studies have been based on the development of SiNP through a biological extraction approach. But still, the extraction by microbes is an unexplored field which requires a significant amount of work in order to employ this as a commercial extraction approach.

This review specifies several agro-wastes with promising applicability as inoculum or precursor material to generate biosilica nanoparticles. Several Si extraction mechanisms have been precisely discussed. Furthermore, different applications of biosilica nanoparticles have been reviewed, specifying their essential part in protecting the environment.

4.2 Different Agro-waste as a Source for Synthesis

Agro-waste is the form of waste that is produced during various agricultural field-based activities. The agricultural waste is generated in millions of tonnes per annum, which is responsible for causing detrimental effects on the environment and unwanted health issues, specifically in developing countries where most of this waste is either openly discarded or burned. The

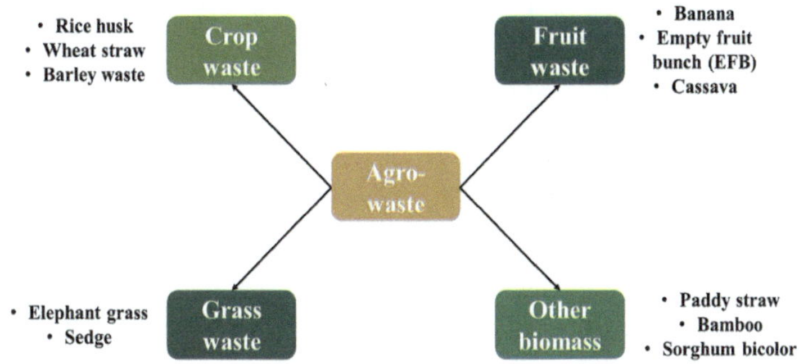

Figure 4.1 Different agro-waste as a source of synthesis.

physico-chemical characteristics of the produced agricultural waste suggest the strategy of disposal for the agricultural waste. Pollution and waste management are linked intricately, researchers are conducting research to find the best route to discard waste in such a possible way that the waste will not cause many hazards to the environment. One such possible way is combustion which requires less area and is also effective but is costly and causes air pollution. In this regard, nanotechnology has come up as the leading area of science, exhibiting excellent applications to manage agro-based waste and consequently converting the waste into value-added products. The silica, naturally available in agro-based waste is a cost-effective and safe pathway to fabricate SiNP. The SiNP attained by using agricultural waste is of a high grade and also easily available. The potential agro-based sources with applicability to generate SiNP are discussed below and shown in Figure 4.1 and Table 4.1.

4.2.1 Crop Waste

Different crops such as rice husk, wheat straw, barley husk, sugarcane bagasse, sorghum, and corn-cob husk can be utilized for the commercial production of SiNP. *Oryzasativa i.e.* rice belonging to the Gramineae family generates husk accounting for about 20–22% of its in-total production, which is majorly disposed of as waste by the rice-mill industry.[17] Rice straw and husk are rice agriculture's by-products and are widely studied for relatable industrial applications. Rice husk is the highly available rice by-product comprising 80–90% of silica. Patil *et al.* 2018 synthesized SiNP from rice husks *via* different analytical techniques. At 700 °C for 2 hours resulted in the highest yield of Si (89.28% ± 0.03%). Characterization analysis revealed the hydrophilic nature of the particles with an average diameter of 93.24 nm. The availability of hydrogen-bonded siloxane and silanol groups (Si–O–Si) along with spherical-shaped, evenly distributed SiNP in an agglomerated phase was also demonstrated through characterization.[19] Barley is also the ignored agro-waste with the potential capability to generate SiNP. Barley accounts for various inorganic (Na, Mg, SiO_2, K, Al, Fe, and Mn) and

Table 4.1 Characteristics of biosilica nanoparticles with the mode of synthesis and various extraction agro-waste resources.

S. no.	Agro-based waste	Synthesis	Optimized conditions	Characterization	Si yield	Applications	Ref.
1.	Rice husk	Acid and thermal treatment	700 °C, 2 h (calcination)	Spherical shape, diameter (93.24 nm)	(89.28% ± 0.03%)	Industrial and agricultural	19
2.	Barley husk	Acid treatment and combustion	700 °C (calcination)	150 nm, excellent mechanical strength	85%	Adsorbents	12
3.	Wheat husk	Facile precipitation treatment	Extract/water/HCl and ethanol was 1 : 1.5 : 0.5	Surface area (8.23 $m^2 g^{-1}$) and diameter (227 nm)	80–90%	Industrial	20
4.	Sugarcane bagasse	Magneisothermic reaction	800 °C for 5 h (calcination)	Amorphous, size (10 nm)	97%	Optical, biomedical, and electrical	15
5.	Ashes of sugarcane bagasse	Sol–gel route	A mixture of ethanol/ ammonia (30 : 90 mL)	Size (90 ± 10 nm), surface area (156 $m^2 g^{-1}$)	98.8%	Filler within the natural rubber	21
6.	Banana leaf	Sol–gel method	Precipitation at 80 °C for 24 h at pH 7	FTIR (590–3990 cm^{-1}) at 4 cm^{-1} resolution	0.3% silica content produced 0.14% SiNP	Adsorption of methyl orange and phenol	22
7.	Palm kernel	Sol–gel extraction	750 °C, 3 h (calcination)	Surface area (438 $m^2 g^{-1}$), size (50–98 nm)	54.35%	Highly advanced material	23

8.	Groundnut hull ash	Sol–gel method	338 K, 24 h (calcination)	Surface area (191 $m^2 g^{-1}$), size (14 nm)	80%	Engineering fields	24
9.	*Cassavaperiderm*	Modified sol–gel	700 °C (for calcination)	Particle size (3.12–50.75 nm)	~39%	Industrial usage	25
10.	Sedge	Combustion and acid treatment	600 °C	Size (4.1 nm), surface area (1174 $m^2 g^{-1}$)	98%	Industrial purposes	5
11.	Grass *Stenotaphrum secundatum*	Biotransformation	500 °C for 19 h (calcination)	Size (530 nm)	—	Novel biomaterial	16
12.	Wheat grass	Acid treatment and combustion	700 °C (calcination)	Size (102 nm)	85%	Nickel adsorption	26
13.	Elephant grass	Modified sol–gel	600 °C	Size (2–50 nm)	99.2%	Dyes adsorption	27
14.	*Sorghum bicolor*	Sequential method	600 °C, 2 h (calcination)	Size (30–90 nm), amorphous	9.34%	Food industry	14
15.	Bamboo waste	Alkaline extraction and combustion method	723 K, 2 h (calcination)	Diameter (25 nm), high specific surface area (428 $m^2 g^{-1}$)	50.2%	Construction material	13
16.	Maize stalk	Sol–gel method	270 min at 700 °C (calcination)	XRD confirmed the presence of siliceous and silica groups	7.01% for 4.80% of ash	Industrial	28
17.	Paddy straw	Acid–base hydrolytic process	pH 6.5 with constant stirring at 37 °C	Size (17 nm), spherical	9.26%	Industrial	29

organic components (hemicellulose, lignin, and cellulose). For the extraction of SiNP from barley husk, Akhayere in 2020 carried out a study under various temperature conditions. The temperature value of 600 and 700 °C resulted in higher SiO_2 yield with excellent mechanical strength. The produced nanoparticle composite consisted of carbon, oxygen and silicon thereby forming a suitable nanoparticle structure.[12] Wheat husk is the most common wheat milling by-product and also the most essential agro-waste as it accounts for 80–90% of silica composites. Cui *et al.* 2015 prepared SiO_2 microspheres from wheat husk through a facile precipitation method, involving sodium silicate extraction, dilution process of extracted sodium silicate with ethanol and distilled water, silica precipitation with HCl and finally the calcination process. The produced SiO_2 composites had amorphous structure with surface area of about 8.23 $m^2 g^{-1}$ and diameter of about 227 nm. The ideal volume ratio for extract/water/HCl and ethanol was 1 : 1.5 : 0.5 and led to the development of better silicon composites.[20] Further, sugarcane bagasse a waste material, is a cellulose rich fibre and a leftover following extraction of juice from sugarcane. It is one of the prominent sources of Si containing 50–97% w/w of Si composites hence can lead to more SiNP yield. Falk *et al.* 2019 produced SiO_2 nanoparticles from sugarcane bagasse *via* the sol–gel method. SiNP obtained through this method showed amorphous character, high purity level (97%), and even average size (10 nm). Magneisothermic reaction (conducted 800 °C for 5 hours) developed efficient silicon nanocomposites.[15] Similarly, Boonmee *et al.* 2020 reported SiNP synthesis from the ashes of sugarcane bagasse *via* the sol–gel method. The freeze dried SiNP showed higher porosity and surface area in comparison to standard heat dried SiNP. The SiNP were successfully prepared with an average size ranging between 90 ± 10 nm.[21] Thus, the crop waste is the major source of silica and their subsequent conversion into SiNP can decrease crop waste and its better utilization for various biosilica-based applications.

4.2.2 Fruit Waste

The consumption of fruits is quite high and so are the leftovers of fruits. Mostly the fruit waste is disposed by landfilling and burning. The waste produced from oranges, pomegranate, empty fruit branches, and coconut husk and banana exhibit a large probability to produce SiNPs. Banana accounts for 16% production globally.[17] The peel of banana has shown the presence of silica which can be extracted to develop SiNP. So, researchers have worked to extract silica from banana peel. For instance, Mohamad *et al.* 2019 produced mesoporous SiNP from banana peel *via* the sol–gel method to adsorb phenol and methyl orange. The study revealed that 3200 g of banana peel when converted into ash weigh around 428.3 g with an ash content of 13.38% which produced 1.5 g of silica (0.35% of silica content). The SiNP considerably adsorbed the contaminants and showed better usage.[22] Palm kernel shell (PKS), a part of palm fruit is also a good source of silica.

Patrick *et al.* 2020 produced SiNP from PKS through an enhanced sol–gel extraction technique. The microstructural examination revealed a high surface area of 438 $m^2 g^{-1}$ and the size of the extracted SiNP was about 50–98 nm. Amorphous SiNP were developed with negligible mineral pollutants and a good yield of silica (54.35%) was achieved by using PKS waste.[23] Groundnut shells are one of the most commonly used nut products which is also a good source of silica. Guzel *et al.* 2020 synthesized silica xerogel from ground-hull ash through the sol–gel process. The obtained amorphous silica demonstrated promising characteristics like, lower thermal conductivity, bulk density and higher thermal stability. The modification of silica xerogel by alkyl-ammonium salt revealed better performance of the silica nanocomposites.[24] Other fruit sources like *Cassava periderm* is also a good silica source. Adebisi *et al.* 2018 extracted silica from *Cassava periderm* through a modified sol–gel technique. The waste was initially treated with HCl solution to eliminate soluble metallic pollutants. The calcination of the waste was carried out at 700 °C to procure ashes which were leached by HCl and then sol-gelled to develop ethylene glycol modified silica gel nanocomposite. A highly pure silica with the size ranging between 3.12–50.75 nm was developed by using *Cassava periderm.*[25] As a result, development of SiNP is the best possible way to make vital use of fruit waste and reduce the adverse effects of this waste on the environment.

4.2.3 Grass Waste

Grass waste consists of edible-grown grasses and livestock farming, which are also a part of agricultural waste. Various grasses have been worked upon to extract SiNP. Sedge is the most common grass waste having a relatively higher silica concentration. Ghorbani *et al.* 2013 prepared mesoporous amorphous silica composites from sedge through acid treatment and combustion. The ash of the sedge majorly contained SiO_2 while a few traces of Al_2O_3, MgO, K_2O, and CaO were also available. A highly pure silica composite (98%) was procured by applying combustion pre-treatment with leaching followed by reflux of the sedge ashes as post-treatment.[5] On the same line, Gonzalez *et al.* 2014 extracted silica particles from grass *Stenotaphrum secundatum via* an annelid-dependent biotransformation method. This biosynthesis was possible when the grass extracts were turned into humus *via* vermicompost, followed by the calcination process and acid treatment of the obtained silica particles. The silica entities were crystalline, spherical with size of ∼530 nm. This suggested a cost-effective and convenient method to synthesize biosilica.[16] Biosilica nanocomposites can also be obtained from wheat grass. This is one of the agro-wastes with a high silica concentration. Akhayere *et al.* 2020 developed SiNP from barley husk. The magnetic nanosilica obtained were 85% efficient in removing the petroleum pollutants from wastewater and have a size of 162 nm with good stability. In another study, the size of the SiNP obtained from wheat and barley grass was about 102 nm and showed significant results in adsorbing nickel from wastewater.[26] Elephant grass waste is also a good source of silica. Akpotu *et al.* 2018 used this grass waste to produce mesoporous silica with the ability to form SiNP to silica nanotubes. The silica composites with the best all-around textural properties and higher surface area were produced at 600 °C and were 99.2% pure silica with a diameter ranging between 2–50 nm and had an efficient absorption capability to eliminate contaminants such as dyes from the aqueous solutions.[27] Therefore, grass waste is a convenient and cost-effective source to produce SiNP with a green approach.

4.2.4 Other Biomass

Other biomass such as sorghum, maize stalk, bamboo leaves and paddy straw can also be used as a source of biosilica nanocomposites. *Sorghum bicolor* is a member of the poaceous family and is a gramineous plant. It has large proportions of silica which are available in the epidermis layer of the root and leaves endodermis. Athinarayanan *et al.* 2020 extracted SiNP from sorghum residues by employing a sequential process. The developed nanocomposites were amorphous, spherical shaped and had a diameter size of 30–90 nm. Also sorghum-derived Si exhibited good thermal stability with the presence of functional groups which enhanced the efficiency of the nanocomposite.[14] Bamboo is a high yielding and fast-growing green resource with ample availability as a construction material. Biosilica nanoparticles were synthesized from bamboo waste *via* an alkaline extraction and combustion method. The particles had a diameter of about 25 nm with a spherical structure and a high specific surface area of 428 $m^2 g^{-1}$. The purity of yielded silica was about 50.2%, comparatively less than rice husk (62.1%). The nanocomposites showed non-toxic effects and upheld the potential for the large-scale production of biosilica nanoparticles.[13] Maize stalk is another agro-waste which is a source of SiNP. Adebist *et al.* 2019 fabricated SiNP from maize stalk *via* the sol–gel method. The purity of the silica obtained from maize stalk ash was about 7.01% for 4.80% of ash content. Characterization results showed the presence of silicates and Si forms with availability of SiO_2 (38%). The best calcination time and temperature noted for best Si yield was 270 min at 700 °C, respectively.[28] Paddy straw is a tough and bulky biomass and has high deposits of silica along with cellulosic and lignin components. Kauldhar *et al.* 2018 recovered silica from paddy straw through an acid–base hydrolytic process. A highly pure silica (9.26%) was obtained with negligible presence of mineral pollutants. The particles size was about 17 nm and were an even spherical structure.[29]

Thus, the agro-waste can help in the progressive enhancement of sustainable development by reducing the waste produced by agriculture and to effectively convert this waste into a more useful product *i.e.*, biosilica nanoparticles. The use of agro-waste can also positively impact the economy with decreased environmental pollution. However, specific study needs to be conducted in order to commercialize such effective technology throughout the world.

4.3 Biogenic Synthesis of SiNP from Agro-waste

4.3.1 Mechanism of Biobased Silica Solubilization

In the natural environment, microbes are essential in synchronizing the geochemical silicon availability. The ecosystem of silicon has abundant microbes available with the ability to perform the silicate solubilization process thereby making it available for plants. On the basis of diverse microbial species and a preferable concentration of dissolved Si, plants up-take various silicon concentrations from the soil.[30,31] Moreover, bioweathering is the conventional geochemical cycle, responsible for Si extraction from silicates (insoluble). Depolymerisation/dissolution of silicates are carried out by an expanded range of microbes and some microbes can be utilized for bonded Si extraction from agricultural waste. Considerable amount of silica solubilizing microbes, especially bacteria are present in various Si minerals including, muscovite, calcium aluminosilicate and phyrosil.[32] The solubilization process of silica occurs by various processes which consists of:

4.3.1.1 Exopolysaccharides Production, Sulphide, Hydrogen and Siderophores

Microbes exhibit a wetting property, in which produced exopolysaccharides are implemented for silicate disintegration and rock weathering. Exopolysaccharides absorb and consequently bind with inorganic silicates because of the availability of reactive-species phosphates and carboxylates in Gram positive bacteria (teichoic acid) and Gram-negative bacteria (teichuronic acid) which can solubilize Si. Also, some bacteria hold the ability to generate hydrogen sulphide during a sulphate reduction step, and the reduced form of sulphur can interact with cations such as, iron, and calcium (within silicate minerals) to release the silica. Hydrogen peroxide is the prominent part of the rice ecosystem so they can interact with cations to solubilize Si.[33,34] Although this phenomenon is still to be well drafted. On the other hand, siderophores are the chelators known to combine with di- or trivalent metals to solubilize Si within acidic soils.

4.3.1.2 Carbon Dioxide (CO_2) Hydration and Acid Generation

The hydration process of respiratory (CO_2) generated from microorganisms present in soil, leads to carbonic acid formation, which interacts with silicate composites to form soluble Si. Microbial metabolites that have metal complexing abilities like, inorganic acids, organic acids, amino acids and phenolic compounds bind with iron and aluminium present in silicates, leaving behind soluble Si. The organic acids/inorganic acids carry out two processes simultaneously, which are protonation and chelation which encourages the dissolution process.[35] Among the known acids, 2-ketogluconic acid (produced by bacteria), has shown significant silicate solubilizing efficiency. Organic matter breakdown and biological nitrogen-fixation by microflora generates amines and ammonia which also plays a considerable role in the silicate solubilization process. Fungi has also shown remarkable silicate solubilizing effects, relatively more in comparison to bacteria prominently by forming various complexing agents and organic acids which helps in silicate solubilization.[17,36]

4.3.1.3 Enzyme-dependent Mechanism

In *Suberites domuncula* (marine sponge), a silica-anabolic silicase enzyme has been reported which can mediate the interconversion reaction between silica and silicic acid. Silicase enzyme has the ability to hydrolyse crystalline and/or amorphous silicon dioxide composites to generate silicic acid (free), and as the reaction is a reversible process, silicases can generate silicium, silicones and silicone dioxide compounds (Figure 4.2). Silicase enzyme has excellent structural similarity with carbonic-anhydrases. Also, sponge silicase consists of amino acids which can be found in eukaryotic-kind carbonic anhydrases.[37,38] Other than silicases, bacterial species (*Actinobacter*) also uses an enzyme-dependent mechanism to form silicon/silica nanocomposites when interacted with precursor solution involving potassium fluorosilicate.

Of all the discussed mechanisms, organic acid generation is the most common employed method to produce a sufficient amount of free silica by following a comparatively convenient method.

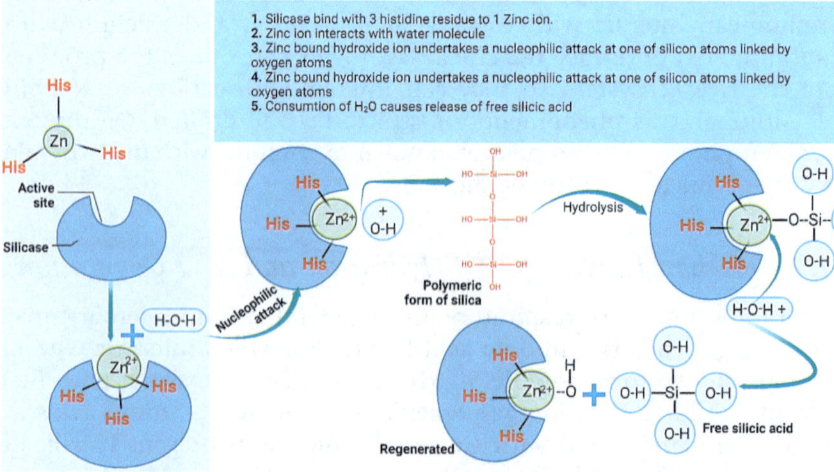

Figure 4.2 Enzyme-dependent mechanism for Si solubilization.[17] Reproduced from ref. 17 with permission from Elsevier, Copyright 2022.

4.3.2 Microbes Involved in Extraction of Si from Agro-waste

Biological synthesis of nanosilica particles is the most desirable extraction process for it involves the least contamination and is a renewable source of Si synthesis. Various microbes have been worked upon to study their effects in the Si solubilization process. Fungi and bacteria are the commonly employed microbial species that are used for Si solubilization as shown in Table 4.2. Pineda *et al.* 2012 extracted silica particles from rice husks by using *Fusarium oxysporum* species. The ashes of the rice husk were incubated at pH 6.8 and 28 °C with *F. oxysporum,* which lead to silica solubilization. The micrographs showed changes in the morphology of the silica as the size decreased from 600 nm to 5 μm and also an increase in surface area was seen from 115 to 125 $m^2 g^{-1}$ and a reduction in pore volume from 0.026 to 0.013 $cm^3 g^{-1}$ was observed.[39] Likewise, Zielonka *et al.* 2018 synthesized nanosilica particles from *Aspergillus parasiticus* leading to the development of more value added substances. This strain was capable of biotransforming accumulates of amorphous Si into structured nanocomposites. The nanoparticles were developed extracellularly and the size of these particles ranged from 3 to 400 nm on the basis of biotransforming conditions and also the supernatant post biotransformation processing. There forms of silica were obtained, spherical (24 ± 8 nm), pyramidal (400 nm) and cubical (85 nm).[40] On the same line, Singh *et al.* 2008 synthesized Si nanocomposites from

Table 4.2 Microbes for silicon synthesis from agro-waste.

Microbial species	Raw material	Incubation conditions	Ref.
Fusarium oxysporum mycelia	Rice husk	pH 6.8 and 28 °C	39
Aspergillus parasiticus	Rice husk	pH 7.2 and 27 °C	40
Actinobacter sp.	Soil (K_2SiF_6 precursor)	24 h and 37 °C	41
BKH1	Soil (magnesium tri-silicate, tetraethyl orthosilicate)	6–7 days at 60 °C	42
Actinomycetes	Rice husk	48 h	45
Actinomycetes	Wheat bran	48 h	45
Trichoderma sp.	Rice husk	24 h	45
Trichoderma sp.	Wheat bran	24 h	45
Enterobacter ludwigii GAK2	Soil (paddy or forest)	30 °C for approx. 3 days	43
Rhizobium, Pseudomonas, Streptomycetes, Microbacterium, Ensifer, Bacillus, Micribacteium, Aeromonas, Flavobacterium, Paracoccus and *Cellvibrrio*	*Pheretimaguillelmi*	28 °C for 48 h	44

Actinobacter sp. The bacteria secrets oxidising and reductase enzymes in the presence of silicates (K_2SiF_6) in the presence of optimized conditions to form the desired nanoparticles. The nanocomposites did not demonstrate cytotoxic impact towards human cell lines and suggested the contact of 1 mM composite concentration for the perfect growth of cell lines.[41] Another green approach to synthesise nanosilica from BKH1 was proposed by Show *et al.* in 2014. The relevant value of zeta potential revealed the stability of bacteria derived nanoparticles. These particles exhibited two prominent luminescence peaks (UV-visible regions) thereby depicting their applicability as an optical probe system for biomedical applications. The bacterial synthesis of the nanoparticles was a convenient, cost-effective and environmentally friendly protocol conducted under precise temperature conditions.[42] Similarly, Kaur *et al.* 2016 synthesized Si nanocomposites using *Actinomycetes* and *Trichoderma* sp. from wheat bran and rice husks. It was studied that *Trichoderma* sp. and *Actinimycetes* took 48 hours to convert the amorphous phase of Si into a crystalline phase for the rice husk, while the same organisms took 24 hours to bio transform the amorphous silica into crystalline silica. The particles were spherical and the spherical structure increased with the contact time. Also, blue emission with a band edge ranging between 230–250 nm was observed.[43] It has also been reported that bacteria present in the gut of earthworms can also cause silicon solubilization hence increasing the availability of the silica to the plants. Comparatively more bacterial species were present in the gut of earthworms than the surrounding soil. Twenty different strains of bacteria were isolated from the earthworm gut including, *Rhizobium*, *Pseudomonas*, *Streptomycetes*, *Microbacterium, Ensifer, Bacillus, Micribacteium, Aeromonas, Flavobacterium, Paracoccus* and *Cellvibrrio*.[33,44]

Therefore, the presence of earthworms in the soil acts as a bio fertilizer which can relatively increase the silicon bioavailability. The presence of silica solubilizing microbes inside the soil is an important factor for plant growth and development as Si is responsible for carrying out various important functions such as regulation of plant regulators, growth hormones and all-in-all healthy, disease-free growth of plants.

4.3.3 Screening of Silica Solubilizing Microbes

Recently there has been a lot of interest in silica solubilizing microbes, due to their ability to dissolve different silicate minerals thus making silica and potassium available for absorption by crops and lowering the requirement of potash fertilizers. Research has revealed that in addition to releasing calcium, iron, phosphate and potassium from silicate minerals present in soil, these microbes are also able to solubilize silica. Therefore, researchers became interested in these microorganisms and began to promote them as potassium-mobilizing biofertilizers.[46] Numerous advantageous microorganisms have been documented for their helpful effects on plants under various stress environments by improving their absorption of minerals. It is well known that microbial-generated organic acids induce the solubilization of both insoluble silicon and phosphate, increasing their availability to plants.[47]

First solubilization of Si was presented by Aleksandrov *et al.* in 1967 when they were working on phosphoric-acid liberation *via* silicate bacteria.[48] Several microbes have been reported which can solubilize/depolymerize Si *viz.*: *Pseudomonas*, *Burkholderia* and *Rhizobium*. A few studies also proved fungi to be an active Si solubilizing microbe like, *Fusarium oxysporum*. The rhizosphere in plants is the prominent silicate solubilizing source. The screening process of the silicate solubilizers employs a conventional method involving the cultivation of test microbes isolated from the rhizospheres in different nutrient mediums supplemented with generally an inorganic source of Si *i.e.*, dimagnesium trisilicate and then selecting the microbes which are capable of forming a clearance zone. A pH indicator-based dye is used to sense the alkaline- or acid-based microorganism actions to solubilize silicate.[49,50] The presence of solubilized Si can be confirmed by plasma-mass spectrometry, emission/absorption spectrometry, and colorimetry.

Isolation of Si solubilizing microbe:[47]

- The rhizosphere soil of plants, sea sand, aquatic silt and mineral ores, all of these act as a source for the isolation of silica solubilizing microbes.
- Following the collection of samples an aliquot (~0.1 mL) of the appropriate serial dilution is inoculated onto a nutrient medium. Nutrient agar supplemented with 0.25% dimagnesium trisilicate can be used to separate silica solubilizing microbes from the total microbes.
- Another nutrient medium has been utilized by researchers known as NBRISSM, having the capability of distinguishing silica solubilizers from phosphate solubilizers. This medium has dimagnesium trisilicate in the concentration of 26.63–37.50 $\mu g\, mL^{-1}$. Early color change from blue to yellow (~3rd day after inoculation) indicates the existence of Si solubilzers and the change lasts for up to 7 days. Thus, a clear zone of solubilization indicates the presence of Si solubilizing microbes.

Screening of Si solubilizing microbes

- Following isolation, the microbes were spot inoculated into silicate, glucose agar medium supplemented with 0.25% dimagnesium trisilicate. After spot inoculation, the plates were incubated at 28 °C for approximately 7 days.
- Similarly, another medium Bunt and Rovira agar were used and incubated at 30 °C for 4 days. The ability of the microbe to solubilize silica was indicated by formation of a clear zone on the screening plate.

- Also, the silica solubilizing activity can be estimated in broth culture by carrying out the modified Elliot and Synder method (1991), by quantifying the amount of Si released through the formation of silico-molybdate complex and taking the absorbance spectrophotometrically at 650 nm.

4.4 Application of Silica Nanoparticles

Silica is a significant inorganic substance with a wide range of uses in the textile and automotive industries, as well as in biology, medicine, adsorbents, drug delivery systems, *etc.* (Figure 4.3). It also has an advantage over traditional precursors due to its abundance and low cost as a substrate. SiNPs high surface area and reactivity across a wide range make them potentially useful in a variety of industries. With applications in the detection of platelet-derived microparticles and the identification of leukaemia cells, these adaptable features increase their potential for constructing biosensors and biomarkers.[51] SiNPs and super absorbent polymers work together to reduce plastic shrinkage. Additionally, a shorter setting time and reduced calcium leaching losses for cement-based materials are made possible by SiNPs exceptional capacity to display the nucleation effect and pozzolanic activity. Salimian *et al.* 2018 also looked at the usefulness of SiNPs as fillers in nanosilica composites. SiNPs special property tends to increase its catalytic and photocatalytic utility for removing an organophosphate insecticide, removing heavy metals from wastewater, treating textile effluents, and decolorizing dyes.[52]

4.4.1 Environmental Remediation

The structural versatility of silica nanoparticles lures investigators to use them in a wide range of environmental remediation projects also. Some of the explored environmental applications of SiNPs are:

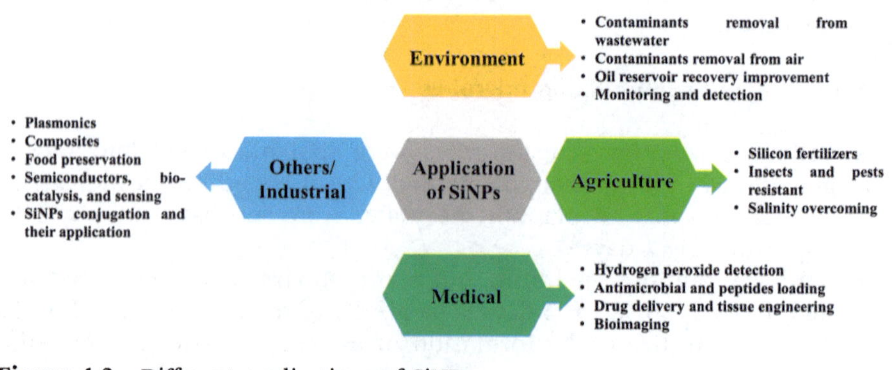

Figure 4.3 Different applications of SiNPs.

4.4.1.1 *Removal of Contaminants from Wastewater*

Biogenic SiNPs have great potential to be used in cleaning contaminated sites and environmental remediation. The specific interactions between the pollutant and the functional groups that are present on the surface of nanomaterials leads to adsorption of contaminants on the surface and subsequent removal. Silica nanoparticles have surface hydroxyl groups known as silanols on their surface that can selectively adsorb compounds and metal ions.[36] Owing to their chemical stability and ease of surface modification, silica and hybrid silica-based nanomaterials are the most extensively utilised as nano-adsorbents for environmental remediation.

Rovani *et al.* 2018 used bare silica particles made from sugarcane bagasse to successfully eliminate acid orange from water.[10] Also, cadmium (Cd) and Congo red (CR) dyes were removed from aqueous solutions using bamboo leaf-derived silica nanoparticles. At varying initial concentrations, the produced silica nanoparticles efficiently removed Cd and CR with optimal conditions and 30 min of contact time. These silica nanoparticles showed maximum adsorption capacity for 133 $mg\,g^{-1}$ for Cd and 172 $mg\,g^{-1}$ CR.[53] In another study banana peel ash was used to make mesoporous silica nanoparticles, which were tested for their ability to adsorb methyl orange and phenol under a variety of circumstances, including pH, adsorbent concentration, starting concentration, and temperature. Each of the parameters was tested in a batch experiment. The optimal pH for the removal of phenol and methyl orange was determined to be pH 5 and pH 3, respectively, with an adsorbent dosage of 0.4 $g\,L^{-1}$ utilising 10 ppm of phenol and methyl orange, yielding 92.6% and 91.1%, respectively. The results of their findings indicated that nanoparticles produced from banana peel ash can be utilized to effectively remove phenol and methyl orange from aqueous media.[22] Similarly, in a study conducted by Yu *et al.* 2018, natural diatomite was successfully used to create low-cost mesoporous silica nanoparticles and used for the adsorption of methylene blue from aqueous solutions. The highest amount of dye adsorption onto mesoporous silica was 347.2 $mg\,g^{-1}$. Furthermore, mesoporous silica may be effectively regenerated and can be reused five times without substantial loss of adsorption capacity using a simple heating process.[54] In a study by Singh *et al.* 2022, agro-waste rice straw was used to synthesize nanosilica and were used for photocatalytic degradation of methylene blue dye. Results showed that in the first 10 min 66.26% of dye was degraded and it followed first order kinetics with a half-life of 6.79 min.[51] Pt nanoparticles were assembled over biogenic porous silica particles from three sources to form Pt/biogenic SiO_2 hybrid structures. The ability of the nanocomposites to degrade 4-nitrophenol (4-NP), a typical pollutant found in wastewater from the textile and pharmaceutical industries, was examined. When compared to Ag or Au nanoparticles, the Pt/SiO_2 catalyst showed better catalytic activity and kinetic rate constants. Pt particles arranged densely provided active sites that allowed for a high turnover capacity of the catalyst. The hybrid catalyst's excellent reusability for purifying 4-NP contaminated water in a "filtering and catalyzing" device has been proven.[55]

4.4.1.2 Removal of Toxic Contaminants from Air

SiNPs have also been investigated for use to remove heavy metal contamination from environmental samples. Lead (Pb) is a basic metal element that is found in nature and is one of the most prevalent heavy metal pollutants due to its widespread use in numerous sectors. Cars, lead smelters, metal processing companies, and incinerators are the main sources of Pb pollution, which pollutes drinking water and the atmosphere. The potential use of SiNPs as an adsorbent to capture excess Pb in the air from two polluting electric battery facilities in Shenzhen, China, was investigated by Yang and his team. The results reveal that SiNPs acquire more Pb after weeks of exposure to workshop air than either the blank control or micro-sized SiO_2 particles. This discovery suggests that biogenic SiNPs could be used as an adsorbent to extract heavy metals from polluted air, implying a potentially innovative application for SiNPs in air pollution treatment.[56] This discovery suggests that SiNPs could be used as an adsorbent to extract heavy metals from polluted air, implying a possibly unique application for SiNPs in environmental pollution treatment.[56]

The release of volatile organic compounds (VOC) like carboxylic acid, methanol, benzene, toluene, formaldehyde, acetone, *etc.* into the environment leads to continuous deterioration of water and air quality. Therefore, it is mandatory to remove VOCs to improve the quality of the environment. Formaldehyde, a type of VOC, can be removed by using amine functionalized silica nanoparticles. These polar VOCs interact with the hydroxyl groups on silica and carbon, thus improving the VOC adsorption capacity of silica and carbon.[57]

4.4.1.3 Improving Oil Reservoir Recovery

Oil recovery denotes the process of extraction of crude oil from an oil field which cannot be achieved otherwise. Improving the oil reservoir recovery depends upon the changes in the wettability properties of the oil. Altering wettability is highly effective in enhanced oil recovery. Wettability alterations changes the wetting state of rock to make it more water-wet. The oil contains compounds with one polar end that absorbs on the surface of rock and exposes the hydrocarbon end, causing the rock surface to become more oil-wet. Silica nanoparticles can help in improving the oil recovery from reservoirs. In the study done by Maghzi and co-workers 2012, dissolved biogenic SiNPs in water (DSNW), were used to see the effect on the wettability property of oil. It was demonstrated that covering the micromodel glass made it oil-wet, and coating it with DSNW made it strongly water-wet. They also studied the capacity of SiNPs to transform wettability from oil-wet to water-wet on a tiny level thus improving the recovery.[58]

In another study, *Saccharomyces cerevisiae* yeast was employed to produce silica nanoparticles using sodium silicate as a precursor. The nanoparticles were then introduced to injectable solutions in an oil-wet micromodel with varying concentrations to study the influence of this nanoparticle on oil production efficacy. The difference between the influence of biochemically generated nanoparticles and commercial nanoparticles on the analysis of oil production efficacy was found to be roughly 5–7% higher for biologically synthesized nanoparticles than for commercial nanoparticles.[59] Thus, using nanotechnology to improve oil reservoir recovery could be a smart method to meet energy demand while also lowering production costs and increasing production of oil.

4.4.1.4 Monitoring and Detection

Silica nanoparticles can be tagged with antibodies, conjugated with enzymes, coated with fluorophores or doped with metal ions for detection of pesticides and other contaminants. These can be conjugated with enzymes like acetylcholinesterase and organophosphate hydrolase and can then be used for the detection of pesticides like organophosphates and carbamates. Detection of these pesticides in the form of a sensor is based on the changes in the current, potential difference or absorbance.[60] The carboxyesterase enzyme (derived from the bug *Spodoptera litura*) mounted on SiNPs was applied to monitor organophosphate pesticide Malathion and a carbamate Diethofencarb.[61] Similarly, silica nanoparticles can also be used for the detection of paraoxon[62] and methyl parathion.[63] Using silica sol–gel technology, an electrochemical immunosensor for the detection of pesticide carbofuran has been developed. Carbofuran was detected directly from cabbage and lettuce samples, with recoveries ranging from 89.14 to 106.8%.[64]

Thus, SiNPs are important in the development of extraction, detection, and degradation systems for environmental contaminants. SiNPs are a potential tool for constructing devices for 'on-site' applications because of their chemically inert nature and adaptability to surface changes. Further research and translating laboratory-based results to actual *in situ* conditions will lead to the effective application of SiNPs in environmental remediation.

4.4.2 Agricultural Applications

The soil contains significant levels of silicon, which is the second most common element in the Earth's crust.[65] Silicon is a semi-essential element for plants; it is not involved in their growth or metabolism, but a lack of it can lead to a number of abnormalities in their growth and development. Due to the existence of inherent anti-stress mechanisms that permit higher yields and higher-quality end products, research has demonstrated that the presence of silicon enhances plants' resistance to difficult climatic conditions. This strengthens the relationship between plants and their environment.[66]

The toxicity of metals, salinity, dryness, and stress brought on by high temperatures have all been demonstrated to be reduced by silicon.[49] For crops that require large nitrogen inputs, like rapeseed, alternative strategies must be developed in order to reduce the environmental harm caused by excessive nitrogen fertiliser consumption.[67] Nanoparticles made from silica are complex functional materials with a wide range of applications due to their high specific surface area, dispersion, and other characteristics.[68]

4.4.2.1 Silicon Fertilizers

One of the applications of silica nanoparticles is the use of these in the form of fertilizers to reduce the negative environmental effects caused by conventional nitrogen fertilizers. Considering SSB12 has a high potential for dissolving insoluble silica under *in vitro* conditions, it was used in a study to solubilize silica. *Serratia surfactantfaciens* was identified as the isolate SSB12's molecular species by the full-length 16S rRNA gene sequencing (Accession number MW308551). The application of SiO_2 nanoparticles significantly increased shoot length, root length, root number, leaves number, chlorophyll, proline, and silica content in 14-day-old maize seedlings after a thorough evaluation of the biologically produced SiO_2 nanoparticles.[30] Additionally, in all SiO_2 nanoparticle treatments, there is a noticeably greater expression of stress-related enzymes. Furthermore, when SiO_2 nanoparticle treatments were compared to control plantlets, it was found that the expression of stress-related enzymes was significantly higher in all cases, with the maximum expression shown at higher concentrations of 200 $mg\,L^{-1}$ SiO_2 nanoparticles. In addition to reducing stress, maize's greater production of these antioxidant/defense enzyme activities will promote plant development. Due to their favorable effects on maize seedlings, these SiO_2 nanoparticles have promising use in the form of nano-fertilizers.[17]

4.4.2.2 Resistance to Insects and Pests

Increased agricultural output assures the survival of the world's expanding population. Unfortunately, this yield can be greatly decreased by insect and microbial attacks. Using pesticides to remedy this problem has worsened the situation due to health and environmental issues. Pesticide abuse is bad for the environment and people's health in addition to the soil. Pesticides applied directly to crops run the risk of being absorbed and remaining intact within the plants, whereupon they could enter people's bodies and cause a variety of illnesses like cancer, reproductive abnormalities, neurological issues, and respiratory ailments.[69] This has prompted scientists to create an environmentally friendly insecticide. Furthermore, boosting seedlings resistance to microbial assaults is another strategy for eliminating pesticides. Using mesoporous silica nanoparticles (MSNPs) has made this possible.[70,71]

Similar to this, Siddiqui *et al.* 2014 were able to increase germination rate by 22.16% by employing 8 $g\,L^{-1}$ of SiNPs. This demonstrates how the concentration of silica nanoparticles affects the rate of germination.[72] Additionally, Maryam *et al.* 2021 had a recent noteworthy study in which the researchers created poly-ethyleneimine (PEI) coated MSNPs. The particles included genetic material (pDNA) that could be successfully transfected into plant cells when subjected to ultrasonic treatment.[73]

4.4.2.3 Overcoming Salinity

The ability of plants to thrive under stressful circumstances like salinity is exceedingly tough. This issue can be rectified by encoding the protein with the gene which will increase the plant's tolerance to the crucial salinity conditions. However, this procedure is time- and money-consuming. As a result, researchers have demonstrated the viability of incorporating the gene into silica nanoparticles. Kalteh *et al.* 2014, for instance, employed SiNPs on basil plants under more saline circumstances. It was discovered that the SiNPs improved the plants' ability to withstand stress and reduced sodium toxicity.[74] On the other hand, another study used SiNPs to cure algae (*Scenedesmus obliquus*). In contrast to other studies, it was demonstrated that the increased SiNPs concentration decreased the algae's chlorophyl content. However, at a moderate SiNP concentration, good stress tolerance was seen. Overall, silica nanoparticle use strengthens plants, improves their ability to withstand stress, and is non-hazardous.[75]

4.4.3 Medical

Silica nanoparticles are also employed in a variety of other fields, including electrical, textile, and rubber. Its role in biomedicine and bioimaging is highly significant (Figure 4.4). SiNPs can be used to diagnose, and control genetic abnormalities like they can be used for targeted imaging of *in vivo*

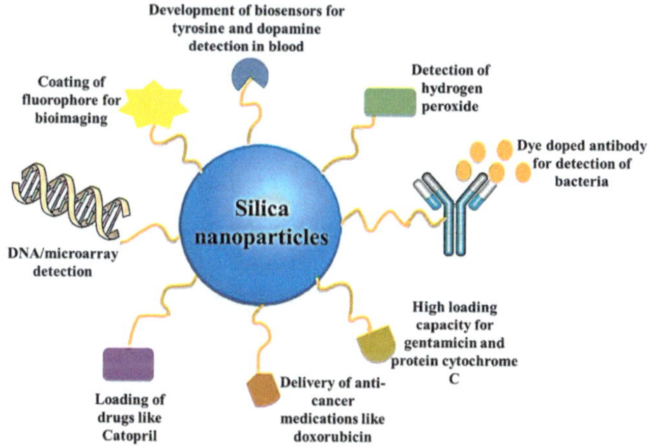

Figure 4.4 Medical application of SiNPs.

early gastric cells,[76] controlled release of drugs like glutathione gentamicin, protein cytochrome *C*[77,78] and anti-tuberculosis therapy.[79] Based on distinct multi component nano-dimensional structures with diversified capabilities (such as intrinsic directed energy and electron transfer processes) silicon dioxide nanoparticles are also used as light absorption and signal amplification devices,[59] aiding in bioimaging processes.

4.4.3.1 Detection of Hydrogen Peroxide

In order to detect hydrogen peroxide (H_2O_2) without the need of enzymes, graphene oxide and silica nanoparticles were combined. With a low detection limit of 2.6 μM and great reproducibility, amperometric tests demonstrate that composites of graphene oxide and silica nanoparticles improve the electrochemical activity of hydrogen peroxide detection in alkaline medium. According to the findings, graphene oxide can improve the catalytic capability of silica nanoparticles toward H_2O_2, making these composites attractive new materials for the creation of nonenzymatic sensors.[80]

4.4.3.2 Loading of Antimicrobials and Peptides

The production of nanoparticles by plants is a cutting-edge method with several uses in farming, food processing, and medicine. This work shows that *Punica granatum* leaf extract can produce silica nanoparticles in an environmentally responsible manner using a straightforward, affordable, and effective process. Following investigation and characterization, it was discovered that the particles were of an amorphous, spherical shape, and had an average size of 12 nm. By using the well diffusion method, silica nanoparticles aided by *P. granatum* were evaluated for their antibacterial efficacy against two-Gram negative bacterial pathogens (*E. coli* and *Salmonella typhi*) having zones as 10 mm and 20 mm, respectively. According to antibacterial investigations, *P. granatum* helped silica nanoparticles develop strong antibacterial capabilities. These investigations will aid in the search for a novel nano-drug or medication to treat bacteria that are resistant to numerous drugs.[81]

4.4.3.3 Role in Drug Delivery and Tissue Engineering

Drug delivery's primary goal is to deliver the therapeutic substance to the desired location without causing any damage to its structure or function. Another requirement for creating a systematic drug delivery carrier is the controlled release of the load molecules from their nanocarrier.[82] Nanomedicine is rapidly developing drug delivery systems based on silica nanoparticles; however, this requires investigation of the toxicity, biodistribution, clearance, and immune response to silica and modified silica.[83] MSNPs are frequently employed in applications for medication delivery. They can be functionalized to adjust the load and release of pharmaceuticals since they have surface silanol groups. MSNPs that have been amine-functionalized are frequently employed.[80] There is evidence of high drug loading and sluggish drug release. Such a method has been used for a number of medications, including heparin,[84] hydrocortisone,[85] metoprolol,[86] salidroside,[87] venlafaxine,[88] vancomycin,[89] and methotrexate.[80]

Drugs like allantoin are delivered *via* composites made of polycaprolactone nanofibers and silica nanoparticles generated from rice husk.[90] 5-Fluorouracil was attached to chitosan-coated biomass silica nanoparticles to lower the price and improve biocompatibility. The drug conjugation mediated by chitosan exhibits a prolonged release at an acidic pH, making it appropriate for delivery to cancer cells. Chitosan-mediated drug conjugation on the surface of silica nanoparticles is less toxic to fibroblast cell lines and more toxic to cancer cell lines than direct 5-fluorouracil conjugation, according to *in vitro* cell culture studies conducted on MC3T3 fibroblast cell lines, MCF-7, and A549 cancer cell lines.[91]

Regenerative medicine holds great promise for curing diseases and replacing missing or damaged tissue. MSNPs are highly sought-after in the field of tissue construction and have a remarkable range. MSNPs include a flexible drug delivery system and can be used as development platforms for young microorganisms or even as a source of signals for developing cells to control the arrangement of undifferentiated organisms. MSNPs could play a significant role in future biomaterial innovation.[92,93]

Mesoporous silica and magnetic carbon nanotubes have a high loading capacity for medicinal compounds like the medicine gentamicin and the protein cytochrome *C*.[78]

SiO_2 xerogels were created as a carrier for the controlled release of cargo molecules in order to accomplish this goal. MSNPs have received substantial research as an alternative to conventional drug nanocarriers for the delivery of hydrophobic medicines.[94] The effectiveness of hydrophobic medicines was constrained during oral administration due to their poor solubility and low absorption. SiNPs wide surface area and pore capacity make it easier to trap the drug. Particularly improved performance as a hybrid drug delivery system capable of controlling the release of various drugs in a variety of media that resemble complex biological environments has been shown for porous silica particles grafted with thermo-responsive polymers with thermally triggered on/off gating mechanisms.[82]

In order to test its cytotoxicity, curcumin was administered coupled with SiNPs to HeLa cell lines and fibroblast cells.[95] A poorly soluble anticancer medication called camptothecin (CPT) was administered using a fluorescent mesoporous silica nanoparticle (FMSN) that was created. The addition of CPT to FMSNs improves the drug's solubility while simultaneously causing anticancerous effects. In order to create MSNPs in both discrete and continuous modes, silica extracted from rice and wheat husks was used.[1] The MSNPs that were created were then used as a vehicle for anticancer medications (Doxorubicin). The results showed that MSNPs generated in discrete mode at pH 5.4 exhibited a larger cumulative released drug and a high-efficiency anticancer effect on the MCF-7 cell line when compared to MSNPs manufactured in continuous mode at pH 5.4.[82] Alshatwi *et al.* in 2014 developed highly pure biogenic silica nanoparticles (bSiNPs) from rice husk. The biocompatibility of the rice husk-derived bSiNPs with human liver cancer cells (hLFCs) was tested using an *in vitro* model. Based on these *in vitro* cell-based methods, it is showed that there was no significant incompatibility between the bSiNPs. Thus, bSiNPs could be safely applied as an anti-caking agent in the food industry and for bone tissue engineering in biomedical applications.[96] In another study, bSiNPs were synthesized from rice husk. The rice husk was pretreated with acid to eliminate inorganic contaminants and facilitate the hydrolysis of organic materials. Using human mesenchymal stem cells (hMSC) and the MTT test, the biological characteristics of bSiNPs were investigated. The results of the cell viability assays demonstrated the exceptional biocompatibility between bSiNPs and hMSC, proving the potential use of these bSiNPs in bone tissue engineering.[97] In a study, synthesized AgNPs were coated with silica for dental antimicrobial applications. The fabricated Ag@SiO_2 NPs showed potent antimicrobial action against *S. mutans* biofilm. No cytotoxicity was observed with Ag@SiO_2 NPs at MIC concentrations. It was demonstrated that the fabricated Ag@SiO_2 NPs are promising agents for dental materials to prevent *S. mutans* biofilm formation.[98] In a study, bSiNPs were synthesized using sugarcane bagasse, a low-cost, fibrous material. The biocompatibility of the bSiNPs was evaluated; at high doses, there was a minor affect in mitochondrial membrane potential but no effect on cell viability or morphology. The study suggests bSiNPs may be suitable for biomedical applications.[99]

4.4.4 Bioimaging

Optical imaging techniques like fluorescence, magnetic resonance imaging (MRI) and bioluminescence, radionuclide imaging, *e.g.*, positron emission tomography (PET) and single-photon emission computed tomography (SPECT), computed tomography (CT), ultrasound, photoacoustic imaging and Raman imaging can easily utilize silica nanoparticles.[100] For encapsulation of contrast agent particles like iron oxide, organic dyes, silver and quantum dots, silica nanoparticles can be used as contrasting agents for carrying out medical imaging.[80]

Biomass-derived silica nanoparticles were combined with a fluorescent substance to boost biocompatibility and employed for biological imaging applications. Here, green fluorescent carbon dots were hot injected into mesoporous silica nanoparticles made from rice husk. In comparison to carbon dots alone, the complex exhibits great drug loading efficiency and brighter fluorescence. Thus, theragnostic applications for this complicated material are possible. Microwave-assisted combustion was used to create a silica composite made from rice husk that contains rare earth elements and europium/gadolinium ions.[101] The silica matrix contained europium and gadolinium ions, which each had the potential for T_1-weighted magnetic resonance imaging and fluorescence imaging, respectively. As a result, rice husk silica nanoparticles containing rare earth ions are an excellent substitute for creating bioimaging contrast agents that are also affordable.[76]

4.4.5 Others/Industrial Applications

Without a doubt, nanosized materials have drawn a lot of attention due to their exciting potential applications, where their exceptional properties open the door for novel technologies and uses. As a result, current advancements in nanomaterial technologies hold the promise of resolving numerous difficulties in the application of advanced sciences. Applications for silica nanoparticles include enhanced catalysis, biological and medication delivery, environmental remediation, and wastewater treatment. As a result, silica nanomaterials are among the most widely utilised nanomaterials. In 2016, the demand for these nanoparticles reached 2.8 million metric tonnes, growing at a rate of 5.6% annually. SiNPs have a high surface area and a mesoporous structure that make them excellent for a variety of industrial applications.[102] These applications are expanding as a result of the exceptional qualities that these particles possess. Some of the industrial applications of SiNPs are given below:

4.4.5.1 SiNPs in Plasmonics

A wide range of applications, including telecommunications, imaging, chemical and biological sensing, biomedicine, and quantum computing, are made possible by photonics and plasmonics, which enable light manipulation at the nanoscale.[103] The control of light propagation in particular spectral ranges, the creation of waveguide bends, splitters, and nanocavities, as well as the amplification of optical fields with inherently low losses, are all made possible by the use of properly designed dielectric, periodic nanostructures, such as photonic crystals (PhCs). These capabilities can then be used to improve the interaction of radiation with the analyte in an optical sensing system.[104]

Diatomite nanoparticles (DNPs) are 3D porous silica nanostructures that are easily produced from diatomite natural deposits or recently cultured diatom frustules through efficient methods. Diatomite, often known as diatomaceous earth (DE), is a porous siliceous sedimentary rock that is soft and friable. It is mostly created by the long-term accumulation of diatom remnants that have been crushed into a fine powder. Since it is primarily made up of readily accessible diatom fossils, its primary chemical composition is amorphous silica. Other minor inorganic components, such as alumina, calcite, iron oxide, and magnesia, which depend on the sedimentation process, environmental conditions and organic matter content, make up the remaining 0.1%.[105]

Plasmonic NPs-decorated diatomite biosilica allows for real-time monitoring of drug release and direct observation of nano-vectors in living cells. In a recent study one-pot liquid-phase synthesis was proposed to create diatomite nanoparticles (PEG-DNPs) modified with polyethylene glycol (PEG) and embellished with gold nanoparticles (AuNPs) for use in medicine. As determined by nitrogen adsorption/desorption isotherm analysis, the inclusion of AuNPs had no effect on the DNPs' loading capacity or pore availability. The substances demonstrated excellent stability, effective absorption in human cervical epithelioid carcinoma (HeLa) cells for up to 48 h, and extremely minimal *in vitro* cytotoxicity at concentrations of around 400 $g\,mL^{-1}$ for 72 h.[106]

Diatomite's periodic nanopores and metallic nanoparticles (NPs) nearby or inside them can combine to generate hybrid photonic–plasmonic modes. Additionally, diatomite is simple to spin-coat to create a consistent layer on a substrate. According to a study, both characteristics can be utilised to create a low-cost lab-on-chip system that enables ultra-sensitive SERS detection of the target molecules. With ultra-high detection sensitivity down to 1 ppm, DNPs decorated with 60 nm AuNPs have been utilised to separate and detect tiny compounds from combination samples. The created hybrid plasmonic-diatomite biosilica has been used as the SERS substrate for separating and detecting phenethylamine (PEA) from plasma as well as the matrix for on-chip chromatography.[107]

Ma *et al.* 2021 have suggested another use for diatomite-based sensors in the field of food safety. The core of the colorimetric sensor created for quantitative monitoring of pesticides in tea is diatomite and plasmonic nanoparticles. A hydrothermal process was used to create a porous chitosan/partially reduced graphene oxide/diatomite composite (CS/prGO/DM), and the relative concentrations of the three materials were then experimentally assessed for effective tea interferent adsorption and to enhance the signal of the analyte molecules. The sulfuric groups in the insecticides have a great affinity for gold surfaces through Au–S bonding. These robust interactions cause the AuNPs to aggregate and, as a result, the distance between them decreases, which is accompanied by a change in the tea solution's hue to blue. The suggested sensing platform was employed to detect phosalone and thiram in a complex tea matrix with detection limits of 90 nM and 13.8 nM, respectively, using UV-vis spectroscopy and smartphone assisted RGB value analysis. Materials based on diatomite are appealing for on-site colorimetric quantitative detection in a number of fields, including bioanalysis and food sensing.[107]

4.4.5.2 SiNPs in Composites

Since ancient times, people have utilised metals with inherent antibacterial action to ward off illnesses, such as copper (Cu) and silver (Ag). However, they have weak pharmacokinetic properties and a high cytotoxicity against mammalian cells. Therefore, using these metals as co-delivery payloads with

SiNPs for antibacterial applications has been successful. Using *Bacillus anthracis* and *Escherichia coli* as examples, an MCM-41 type MSNP with encapsulated silver nanocrystals (Ag@MESs) in a yolk/shell pattern was produced and evaluated. Ag@MESs were found to be able to stop bacterial growth in both species. While the nanocomposite was able to totally suppress growth in *B. anthracis* at 100 g mL^{-1} and only significantly delay it at 20 g mL^{-1}, it had no discernible impact on *E. coli*.[79]

4.4.5.3 SiNPs Conjugation and Their Application

The demand for understanding biology and medicine at the molecular level has been driven by advanced bioanalysis, including precise quantification. Bio-conjugated silica nanoparticles may be able to handle this new problem. There have also been discovered uses for silica nanoparticles in the transport of genes and drugs, as well as some very exciting diagnostic and therapeutic uses in cancer and infectious diseases. Cell-membrane proteins can be labelled by attaching receptor ligands and recognition peptides to NPs. To target SCC-9 cancer cells, which overexpress folate receptors, folic acid, a tiny vitamin molecule identified by cancer cells, was coupled to dye-doped silica NPs in a study.[108] The effectiveness and robustness of the detection system for human squamous cancer detection were confirmed by laser scanning confocal images, which showed that folate-fluorescent NPs were internalised by the SCC-9 cells *via* a receptor-mediated endocytosis process in significantly greater amounts than the normal human dermal fibroblast cells.[108]

Also, for the safety of food and water, clinical diagnosis, and the avoidance of unintentional outbreaks, it is crucial that trace levels of organisms, such as harmful bacteria, can be quickly and accurately detected. Early detection of tiny levels of harmful germs is crucial in the case of deliberate terrorist acts. The ability of microbes to develop into visible colonies over time on specialised growth media, which may take 1–5 days, is what the existing technologies for detecting bacteria rely on. Under these circumstances, it would take much longer to detect traces of bacteria, which normally call for amplification or enrichment of the target bacterium in the sample. Time restrictions and the complexity of on-site analysis are thus significant barriers to the best bacterium detection. A new kind of NP-based bacterium detection technique has now been created to get over these restrictions. Particularly, dye-doped silica NPs (diameter: 60 nm $\pm$ 4 nm) were coupled to antibodies against *E. coli* O157:H7.[109]

For improving their intracellular absorption in cancer cells, DNPs were created and modified using a dual-biofunctionalization technique *via* polyethylene glycol (PEG) covering and cell-penetrating peptide (CPP) bioconjugation. The surface modification enhanced *in vitro* biocompatibility, as the modified DNPs demonstrated high hemocompatibility after 48 h of incubation with erythrocytes and negligible cytotoxicity after 24 h of incubation with MCF-7 and MDA-MB-231 breast cancer cells, at concentrations up to 200 g mL^{-1}. Additionally, the dual surface modification of DNPs improved

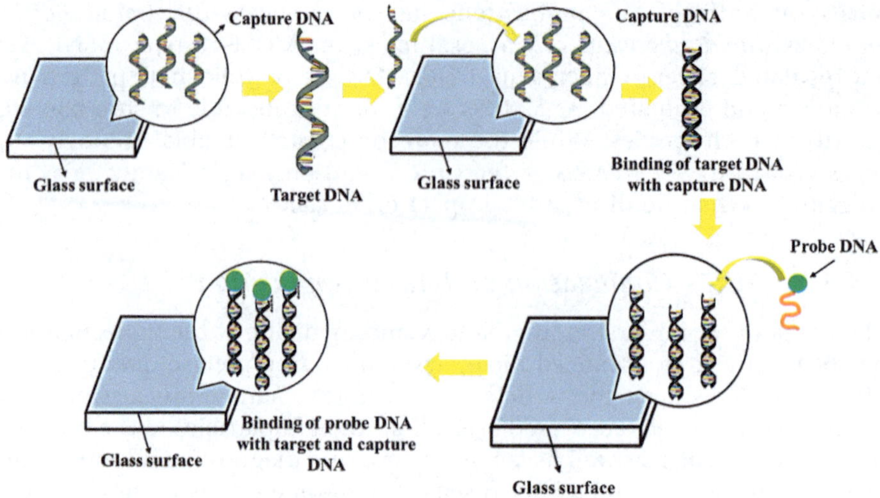

Figure 4.5 DNA/microarray detection using SiNPs.

the drug release profiles in aqueous solutions as well as the loading of Sorafenib, a poorly water-soluble anticancer medication, with a loading capacity up to 22 wt%.[104]

Similarly, using the conjugation principle, fluorescent silica NPs have been used as probes for DNA/microarray detection in order to improve the fluorescence signal of the gene target (Figure 4.5). Sandwich assays served as the foundation for the first lab-based trial. In order to hybridise with unlabelled target DNA that was complementary to both the capture and probe DNA sequences, capture DNA was immobilised onto a glass surface. Then, NPs with probe DNA attached were included. One dye-doped NP (60 nm in diameter) is immobilised on the surface when one probe DNA hybridises with one target DNA, bringing thousands of dye molecules for signalling. The DNA detection limit can be lowered to sub-picomolar levels by employing this method. The detection of single nucleotide polymorphisms is also possible. Additionally, cyanine Cy3- and Cy5-doped Au/silica core–shell particles have been created and employed in a sandwich assay format for two-color microarray detection.[109]

4.4.5.4 SiNPs in Food Preservation

Silica and its derivatives have been used in several food and nutrition-related fields as nanosized particles. These particles have not, however, been included in food products as an ingredient. They have been used in the analysis of food and nutrient components as a support material for column chromatography or to boost the sensitivity or effectiveness of glucose biosensors; as fillers or protectors in food packaging to extend the shelf life of food products; as edible (flavour) carriers; as an adsorbent of moisture; or as a clarifying agent to get rid of impurities and unwanted particles still present in beverages.[110]

In food items, amorphous silica nanoparticles are frequently utilised as thickeners, anti-caking agents, and transporters for tastes and perfumes. Food products containing nanosilica (E551) were examined and it was discovered that they contained silica nanoparticles at concentrations ranging from 0.1 to 6.9 $mg\,g^{-1}$ product, with particle sizes ranging from 30 to 200 nm. Nanosilica (E551) is a food additive that is registered in the European Union. A 60 kg person's estimated daily consumption of silicon from silica used as an addition is 20–50 mg silicon, according to the Scientific Committee on Food of the European Food Safety Authority (EFSA).[111]

Evaluation of the toxicity, metabolism, excretion, absorption, and distribution of 110 nm silica nanoparticles in mice after various exposure routes was conducted. These researchers discovered that intramuscular or hypodermic injection of nanoparticles allowed them to cross biological barriers like the gut. Even though there was little absorption, the particles were nevertheless delivered to the liver. Orally administered silica nanoparticles were shown to enter the digestive tract. They also showed that intravenously delivered nanoparticles were primarily found in the liver and spleen and were eliminated in urine and faeces.[111] Athinarayanan *et al.* in 2020 developed biogenic silica nanostructures (BSNs) by using sorghum residues as a precursor. According to the cytotoxic analysis, it is proven that BSNs are not causing death of the colon cells. The findings collectively indicated that BSNs show good compatibility with colon cells and could be use in the food industry as an anti-caking agent.[14] In a study researchers focused on extracting cellulose fiber and biogenic silica from rice husks and preparing silica-reinforced composite bioplastic films for packaging applications. The results show increased barrier properties with increased silica content, with the highest tensile strength at 1347 $kN\,m^{-2}$. The biodegradability of these films is significantly higher than conventional plastics, making them a promising green packaging material.[112] Ojha and Das, synthesized bSiNPs using groundnut shells for the fabrication of novel PHBV/SiO_2 biodegradable nanocomposites. Improved physical, chemical, mechanical, thermal, barrier, antibacterial, and biocompatible properties were observed in these nanocomposites. There was significant improvement in mechanical, thermal, heat resistance, and barrier properties with the addition of bSiNPs. These nanocomposites are eco-friendly and cost-effective alternatives to synthetic plastic food packaging materials.[113]

4.4.5.5 Others Such as Semiconductors, Bio-catalysis, Sensing

Due to their long cycle life performance, high charge/discharge speeds, and high-power density, supercapacitors have generated interest. Supercapacitors are seen to be a very promising energy storage technology that could ultimately replace or supplement the batteries used in electric and hybrid cars, wearable electronics, and other portable devices. According to the energy storage mechanisms, supercapacitors can be divided into two main categories: electric double-layer capacitors (EDLC) and pseudo-capacitors. It was discovered that graphene oxide-wrapped conducting polyaniline wires wrapped around mesoporous silica nanoparticles increased the surface area, allowed ions to travel freely, and allowed charge transfer.[80]

In the field of producing sustainable and renewable energy, the beneficiation of agricultural waste is a pressing concern. Also, silicon nanoparticles used in the manufacture of solar cells not only conserves resources but also increases the effectiveness of the solar cell device. Performance is anticipated to improve compared to randomised particles thanks to the precise tuning of periodic nanoparticle arrays to the absorption properties of the amorphous silicon material. To permit long carrier diffusion length in solar cells, it is generally accepted that silicon purity (Si) must be greater than 99.99% (weight%). The following benefits of silicon nanowires and nanoparticles are possible: improved absorption, effective carrier extraction, and lowered requirements for material quality in solar photovoltaics.[114]

4.5 Conclusion

In nanotechnology, silica nanoparticles have already made a substantial contribution. Mesoporous structure, high surface area, variable particle size, pore size and shape, and biocompatibility are just a few of the exceptional qualities that give this material considerable advantages in a variety of applications. In the areas of agriculture, food preservation, biomedicine, and catalytic reactions, these nanoparticles have made notable contributions. For a wide range of bioactive compounds, SiNPs have been demonstrated to be good encapsulating agents. They have previously been safely demonstrated for targeted drug administration. SiNPs ability to combine with various polymeric and non-polymeric materials to create hybrid composites has also expanded the functionalities that they can be used in. Mesoporous silica nanoparticles have already demonstrated outstanding carrier qualities, which are used in chemotherapy to target particular cancer cells. In conclusion, it can be said that SiNPs have been demonstrated to be a secure, dependable, and functional replacement that is finding increasing application in every industry.

Acknowledgements

Authors are thankful to Chandigarh University, Mohali, India for providing necessary infrastructure and facilities.

References

1. S. Porrang, N. Rahemi, S. Davaran, M. Mahdavi and B. Hassanzadeh, Preparation and *in-vitro* evaluation of mesoporous biogenic silica nanoparticles obtained from rice and wheat husk as a biocompatible

carrier for anti-cancer drug delivery, *Eur. J. Pharm. Sci.*, 2021, **163**, 105866.

2. M. Tutak and J. Brodny, Determination of particular endogenous fires hazard zones in goaf with caving of longwall, *IOP Conf. Ser.: Earth Environ. Sci.*, 2017, **95**, 42026.
3. H. Gomaa, M. A. T. Hussein and M. M. Motawea, *et al.*, A hybrid mesoporous CuO@barley straw-derived SiO_2 nanocomposite for adsorption and photocatalytic degradation of methylene blue from real wastewater, *Colloids Surf., A*, 2022, **644**, 128811.
4. D. M. Abo-El-Matty, M. Z. Rizk, H. F. Aly, H. I. Abd-Alla, S. M. Saleh and E. A. Younis, Role of Fruit Waste and Flavanones-Loaded Silica Nanoparticles in Ameliorating Oxidative Stress and Histological Changes in Rat Brain Induced by Acrylamide, *J. Mater. Environ. Sci.*, 2018, **9**(6), 1817–1828.
5. F. Ghorbani, H. Younesi, Z. Mehraban, M. S. Çelik, A. A. Ghoreyshi and M. Anbia, Preparation and characterization of highly pure silica from sedge as agricultural waste and its utilization in the synthesis of mesoporous silica MCM-41, *J. Taiwan Inst. Chem. Eng.*, 2013, **44**(5), 821–828.
6. A. Mahmud, P. S. M. Bt Megat Yusoff and F. Ahmad, Processing of rice husk tobio-silica nanoparticlesthrough thermal combustion, *ARPN J. Eng. Appl. Sci.*, 2016, **11**(12), 7650–7654.
7. N. Shafiei, M. Nasrollahzadeh and S. Iravani, Green Synthesis of Silica and Silicon Nanoparticles and Their Biomedical and Catalytic Applications, *Comments Inorg. Chem.*, 2021, **41**(6), 317–372.
8. E. Akhayere, A. Vaseashta and D. Kavaz, Novel magnetic nano silica synthesis using barley husk waste for removing petroleum from polluted water for environmental sustainability, *Sustainability*, 2020, **12**(24), 1–16.
9. H. Nguyen, M. Jamali Moghadam and H. Moayedi, Agricultural wastes preparation, management, and applications in civil engineering: a review, *J. Mater. Cycles Waste Manage.*, 2019, **21**(5), 1039–1051.
10. S. Rovani, J. J. Santos, P. Corio and D. A. Fungaro, Highly Pure Silica Nanoparticles with High Adsorption Capacity Obtained from Sugarcane Waste Ash, *ACS Omega*, 2018, **3**(3), 2618–2627.
11. S. S. Owoeye, S. M. Abegunde and B. Oji, Effects of process variable on synthesis and characterization of amorphous silica nanoparticles using sodium silicate solutions as precursor by sol–gel method, *Nano-Struct. Nano-Objects*, 2021, **25**, 100625.
12. E. Akhayere, D. Kavaz and A. Vaseashta, Synthesizing nano silica nanoparticles from barley grain waste: Effect of temperature on mechanical properties, *Pol. J. Environ. Stud.*, 2019, **28**(4), 2513–2521.
13. L. Bai, C. Li and C. Korte, *et al.*, Effects of silica-based nanostructures with raspberry-like morphology and surfactant on the interfacial behavior of light, medium, and heavy crude oils at oil-aqueous interfaces, *Appl. Nanosci.*, 2017, 7, 947–972.

14. J. Athinarayanan, S. A. A. H. Jaafari, V. S. Periasamy, T. N. A. Almanaa and A. A. Alshatwi, Fabrication of Biogenic Silica Nanostructures from Sorghum bicolor Leaves for Food Industry Applications, *Silicon*, 2020, **12**(12), 2829–2836.
15. G. Falk, G. P. Shinhe, L. B. Teixeira, E. G. Moraes and A. P. N. de Oliveira, Synthesis of silica nanoparticles from sugarcane bagasse ash and nano-silicon *via* magnesiothermic reactions, *Ceram. Int.*, 2019, **45**(17), 21618–21624.
16. A. Espíndola-Gonzalez, R. Fuentes-Ramirez, A. L. Martínez-Hernández, V. M. Castaño and C. Velasco-Santos, Structural characterization of silica particles extracted from grass stenotaphrum secundatum: Biotransformation *via* annelids, *Adv. Mater. Sci. Eng.*, 2014, **2014**, 956945.
17. M. Yadav, V. Dwibedi, S. Sharma and N. George, Biogenic silica nanoparticles from agro-waste: Properties, mechanism of extraction and applications in environmental sustainability, *J. Environ. Chem. Eng.*, 2022, 108550.
18. N. Sarkar, S. K. Ghosh, S. Bannerjee and K. Aikat, Bioethanol production from agricultural wastes: An overview, *Renewable Energy*, 2012, **37**(1), 19–27.
19. N. B. Patil, H. Sharanagouda, S. R. Doddagoudar, C. T. Ramachandra and K. T. Ramappa, Biosynthesis and Characterization of Silica Nanoparticles from Rice (Oryza sativa L.) Husk, *Int. J. Curr. Microbiol. Appl. Sci.*, 2018, 7(12), 2298–2306.
20. J. Cui, H. Sun, Z. Luo, J. Sun and Z. Wen, Preparation of low surface area SiO_2 microsphere from wheat husk ash with a facile precipitation process, *Mater. Lett.*, 2015, **156**, 42–45.
21. A. Boonmee and K. Jarukumjorn, Preparation and characterization of silica nanoparticles from sugarcane bagasse ash for using as a filler in natural rubber composites, *Polym. Bull.*, 2020, **77**(7), 3457–3472.
22. D. F. Mohamad, N. S. Osman and M. K. H. M. Nazri, *et al.*, Synthesis of Mesoporous Silica Nanoparticle from Banana Peel Ash for Removal of Phenol and Methyl Orange in Aqueous Solution, *Mater. Today: Proc.*, 2019, **19**, 1119–1125.
23. P. E. Imoisili, K. O. Ukoba and T. C. Jen, Green technology extraction and characterisation of silica nanoparticles from palm kernel shell ash *via* sol-gel, *J. Mater. Res. Technol.*, 2020, **9**(1), 307–313.
24. G. Guzel Kaya and H. Deveci, Modified silica xerogel derived from groundnut hull ash by alkyl-ammonium salt for epoxy nanocomposites: Synergistic effects on thermal stability and flame retardancy, *Thermochim. Acta*, 2020, **689**, 178637.
25. J. A. Adebisi, J. O. Agunsoye and S. A. Bello, *et al.*, Extraction of silica from cassava periderm using modified sol-gel method, *Niger. J. Technol. Dev.*, 2018, **15**(2), 57.
26. E. Akhayere, E. A. Essien and D. Kavaz, Effective and reusable nano-silica synthesized from barley and wheat grass for the removal of nickel

from agricultural wastewater, *Environ. Sci. Pollut. Res.*, 2019, **26**(25), 25802–25813.

27. S. O. Akpotu and B. Moodley, Effect of synthesis conditions on the morphology of mesoporous silica from elephant grass and its application in the adsorption of cationic and anionic dyes, *J. Environ. Chem. Eng.*, 2018, **6**(4), 5341–5350.
28. J. A. Adebisi, J. O. Agunsoye and S. A. Bello, *et al.*, Extraction of Silica from Sugarcane Bagasse, Cassava Periderm and Maize Stalk: Proximate Analysis and Physico-Chemical Properties of Wastes, *Waste Biomass Valorization*, 2019, **10**(3), 617–629.
29. B. S. Kauldhar and S. K. Yadav, Turning waste to wealth: A direct process for recovery of nano-silica and lignin from paddy straw agro-waste, *J. Cleaner Prod.*, 2018, **194**, 158–166.
30. K. K. Garg, D. Jain and D. Rajpurohit, *et al.*, Agricultural Significance of Silica Nanoparticles Synthesized from a Silica Solubilizing Bacteria, *Comments Inorg. Chem.*, 2022, **42**(4), 209–225.
31. N. Vasanthi, L. M. Saleena and S. A. Raj, Silica Solubilization Potential of Certain Bacterial Species in the Presence of Different Silicate Minerals, *Silicon*, 2018, **10**(2), 267–275.
32. X. F. Sheng, F. Zhao, L. Y. He, G. Qiu and L. Chen, Isolation and characterization of silicate mineral-solubilizing Bacillus globisporus Q12 from the surfaces of weathered feldspar, *Can. J. Microbiol.*, 2008, **54**(12), 1064–1068.
33. G. Raturi, Y. Sharma and V. Rana, *et al.*, Exploration of silicate solubilizing bacteria for sustainable agriculture and silicon biogeochemical cycle, *Plant Physiol. Biochem.*, 2021, **166**, 827–838.
34. K. Rohatgi, S. V. Prasad and P. K. Rohatgi, Release of silica-rich particles from rice husk by microbial fermentation, *J. Mater. Sci. Lett.*, 1987, **6**(7), 829–831.
35. H. Ehrlich, K. D. Demadis, O. S. Pokrovsky and P. G. Koutsoukos, Modern Views on Desilicification: Biosilica and Abiotic Silica Dissolution in Natural and Artificial Environments, *Chem. Rev.*, 2010, **110**(8), 4656–4689.
36. S. A. Jadhav, H. B. Garud and A. H. Patil, *et al.*, Recent advancements in silica nanoparticles based technologies for removal of dyes from water, *Colloid Interface Sci. Commun.*, 2019, **30**, 100181.
37. H. C. Schröder, A. Krasko and G. Le Pennec, *et al.*, Silicase, an enzyme which degrades biogenous amorphous silica: contribution to the metabolism of silica deposition in the demosponge Suberites domuncula, *Prog. Mol. Subcell. Biol.*, 2003, **33**, 249–268.
38. W. E. G. Müller, A. Boreiko and X. Wang, *et al.*, Silicateins, the major biosilica forming enzymes present in demosponges: protein analysis and phylogenetic relationship, *Gene*, 2007, **395**(1–2), 62–71.
39. T. Pineda, C. H. L. Soares, D. Hotza, A. E. Casas-Botero, M. Ramírez-Carmona and M. Torres-Taborda, Extracellular synthesis of silica oxide particles by fusarium oxysporum from rice husk ash, *Mater. Sci. Forum*, 2012, **727–728**, 1153–1157.

40. A. Zielonka, E. Żymańczyk-Duda, M. Brzezińska-Rodak, M. Duda, J. Grzesiak and M. Klimek-Ochab, Nanosilica synthesis mediated by Aspergillus parasiticus strain, *Fungal Biol.*, 2018, **122**(5), 333–344.
41. S. Singh, U. M. Bhatta, P. V. Satyam, A. Dhawan, M. Sastry and B. L. V. Prasad, Bacterial synthesis of silicon/silica nanocomposites, *J. Mater. Chem.*, 2008, **18**(22), 2601–2606.
42. S. Show, A. Tamang, T. Chowdhury, D. Mandal and B. Chattopadhyay, Bacterial (BKH1) assisted silica nanoparticles from silica rich substrates: A facile and green approach for biotechnological applications, *Colloids Surf., B*, 2015, **126**, 245–250.
43. K. E. Lee, A. Adhikari and S. M. Kang, *et al.*, Isolation and characterization of the high silicate and phosphate solubilizing novel strain enterobacter ludwigii GAK2 that promotes growth in rice plants, *Agronomy*, 2019, **9**(3), 144.
44. L. Hu, M. Xia and X. Lin, *et al.*, Earthworm gut bacteria increase silicon bioavailability and acquisition by maize, *Soil Biol. Biochem.*, 2018, **125**, 215–221.
45. T. Kaur, G. P. Singh, G. Kaur, S. Kaur and P. K. Gill, Synthesis of biogenic silicon/silica (Si/SiO_2) nanocomposites from rice husks and wheat bran through various microorganisms, *Mater. Res. Express*, 2016, **3**(8), 1–11.
46. K. Maghsoudi, Y. Emam and M. Ashraf, Foliar application of silicon at different growth stages alters growth and yield of selected wheat cultivars, *J. Plant Nutr.*, 2016, **39**(8), 1194–1203.
47. N. Amaresan, P. Patel and D. Amin, *Practical Handbook on Agricultural Microbiology*, Springer, 2022.
48. V. G. Aleksandrov, R. N. Blagodyr' and I. P. Il'ev, Phosphorus acid isolation from apatite produced by silicate bacteria, *Mikrobiol. Zh.*, 1967, **29**(2), 111–114.
49. A. Artyszak, Effect of Silicon Fertilization on Crop Yield Quantity and Quality-A Literature Review in Europe, *Plants*, 2018, 7(3), 54.
50. C. Kaya, A. L. Tuna, O. Sonmez, F. Ince and D. Higgs, Mitigation Effects of Silicon on Maize Plants Grown at High Zinc, *J. Plant Nutr.*, 2009, **32**(10), 1788–1798.
51. G. Singh, H. B. Dizaji, H. Puttuswamy and S. Sharma, Biogenic nanosilica synthesis employing agro-waste rice straw and its application study in photocatalytic degradation of cationic dye, *Sustainability*, 2022, **14**(1), 539.
52. S. Salimian, A. Zadhoush, M. Naeimirad, R. Kotek and S. Ramakrishna, A review on aerogel: 3D nanoporous structured fillers in polymer-based nanocomposites, *Polym. Compos.*, 2018, **39**(10), 3383–3408.
53. K. Durairaj, P. Senthilkumar, P. Velmurugan, K. Dhamodaran, K. Kadirvelu and S. Kumaran, Sol-gel mediated synthesis of silica nanoparticle from Bambusa vulgaris leaves and its environmental applications: kinetics and isotherms studies, *J. Sol-Gel Sci. Technol.*, 2019, **90**, 653–664.

54. Z. H. Yu, S. R. Zhai and H. Guo, *et al.*, Removal of methylene blue over low-cost mesoporous silica nanoparticles prepared with naturally occurring diatomite, *J. Sol-Gel Sci. Technol.*, 2018, **88**, 541–550.
55. N. K. R. Bogireddy, P. Sahare, U. Pal, S. F. O. Méndez, L. M. Gomez and V. Agarwal, Platinum nanoparticle-assembled porous biogenic silica 3D hybrid structures with outstanding 4-nitrophenol degradation performance, *Chem. Eng. J.*, 2020, **388**, 124237.
56. X. Yang, Z. Shen and B. Zhang, *et al.*, Silica nanoparticles capture atmospheric lead: implications in the treatment of environmental heavy metal pollution, *Chemosphere*, 2013, **90**(2), 653–656.
57. S. P. Singh and N. Endley, Fabrication of nano-silica from agricultural residue and their application, in *Nanomaterials for Agriculture and Forestry Applications*, Elsevier, 2020, pp. 107–134.
58. A. Maghzi, S. Mohammadi, M. H. Ghazanfari, R. Kharrat and M. Masihi, Monitoring wettability alteration by silica nanoparticles during water flooding to heavy oils in five-spot systems: A pore-level investigation, *Exp. Therm. Fluid Sci.*, 2012, **40**, 168–176.
59. H. Zamani, A. Jafari, S. M. Mousavi and E. Darezereshki, Biosynthesis of silica nanoparticle using Saccharomyces cervisiae and its application on enhanced oil recovery, *J. Pet. Sci. Eng.*, 2020, **190**, 107002.
60. G. Bapat, C. Labade, A. Chaudhari and S. Zinjarde, Silica nanoparticle based techniques for extraction, detection, and degradation of pesticides, *Adv. Colloid Interface Sci.*, 2016, **237**, 1–14.
61. J. Diao, G. Zhao, Y. Li, J. Huang and Y. Sun, Carboxylesterase from Spodoptera Litura: Immobilization and use for the degradation of pesticides, *Procedia Environ. Sci.*, 2013, **18**, 610–619.
62. A. K. Singh, A. W. Flounders, J. V. Volponi, C. S. Ashley, K. Wally and J. S. Schoeniger, Development of sensors for direct detection of organophosphates. Part I: immobilization, characterization and stabilization of acetylcholinesterase and organophosphate hydrolase on silica supports, *Biosens. Bioelectron.*, 1999, **14**(8–9), 703–713.
63. S. Chen, J. Huang and D. Du, *et al.*, Methyl parathion hydrolase based nanocomposite biosensors for highly sensitive and selective determination of methyl parathion, *Biosens. Bioelectron.*, 2011, **26**(11), 4320–4325.
64. X. Sun, S. Du, X. Wang, W. Zhao and Q. Li, A label-free electrochemical immunosensor for carbofuran detection based on a sol-gel entrapped antibody, *Sensors*, 2011, **11**(10), 9520–9531.
65. H. F. Bakhat, N. Bibi and Z. Zia, *et al.*, Silicon mitigates biotic stresses in crop plants: A review, *Crop Prot.*, 2018, **104**, 21–34.
66. A. C. Marafon and L. Endres, Silicon: fertilization and nutrition in higher plants, *Rev. Cienc. Agrar.*, 2013, **56**(4), 380–388.
67. P. Laîné, C. Haddad, M. Arkoun, J. C. Yvin and P. Etienne, Silicon promotes agronomic performance in Brassica napus cultivated under field conditions with two nitrogen fertilizer inputs, *Plants*, 2019 **8**(5), 137.

68. F. Yan, J. Jiang, X. Chen, S. Tian and K. Li, Synthesis and characterization of silica nanoparticles preparing by low-temperature vapor-phase hydrolysis of $SiCl_4$, *Ind. Eng. Chem. Res.*, 2014, **53**(30), 11884–11890.
69. A. S. Jatoi, Z. Hashmi and R. Adriyani, *et al.*, Recent trends and future challenges of pesticide removal techniques–a comprehensive review, *J. Environ. Chem. Eng.*, 2021, **9**(4), 105571.
70. M. El-Shetehy, A. Moradi and M. Maceroni, *et al.*, Silica nanoparticles enhance disease resistance in Arabidopsis plants, *Nat. Nanotechnol.*, 2021, **16**(3), 344–353.
71. A. do Espirito Santo Pereira, H. Caixeta Oliveira, L. Fernandes Fraceto and C. Santaella, Nanotechnology potential in seed priming for sustainable agriculture, *Nanomaterials*, 2021, **11**(2), 267.
72. M. H. Siddiqui and M. H. Al-Whaibi, Role of nano-SiO_2 in germination of tomato (Lycopersicum esculentum seeds Mill.), *Saudi J. Biol. Sci.*, 2014, **21**(1), 13–17.
73. M. Zolghadrnasab, A. Mousavi, A. Farmany and A. Arpanaei, Ultrasound-mediated gene delivery into suspended plant cells using polyethyleneimine-coated mesoporous silica nanoparticles, *Ultrason. Sonochem.*, 2021, **73**, 105507.
74. M. Kalteh, Z. T. Alipour, S. Ashraf, M. M. Aliabadi and A. F. Nosratabadi, Effect of silica nanoparticles on basil (Ocimum basilicum) under salinity stress, *J. Chem. Health Risks*, 2014, **4**(3), 49–55.
75. C. Wei, Y. Zhang, J. Guo, B. Han, X. Yang and J. Yuan, Effects of silica nanoparticles on growth and photosynthetic pigment contents of Scenedesmus obliquus, *J. Environ. Sci.*, 2010, **22**(1), 155–160.
76. Z. Zhou, C. Zhang and Q. Qian, *et al.*, Folic acid-conjugated silica capped gold nanoclusters for targeted fluorescence/X-ray computed tomography imaging, *J. Nanobiotechnol.*, 2013, **11**, 1–12.
77. Y. Cui, H. Dong, X. Cai, D. Wang and Y. Li, Mesoporous silica nanoparticles capped with disulfide-linked PEG gatekeepers for glutathione-mediated controlled release, *ACS Appl. Mater. Interfaces*, 2012, **4**(6), 3177–3183.
78. R. K. Singh, K. D. Patel and J. J. Kim, *et al.*, Multifunctional hybrid nanocarrier: magnetic CNTs ensheathed with mesoporous silica for drug delivery and imaging system, *ACS Appl. Mater. Interfaces*, 2014, **6**(4), 2201–2208.
79. V. Selvarajan, S. Obuobi and P. L. R. Ee, Silica nanoparticles—a versatile tool for the treatment of bacterial infections, *Front. Chem.*, 2020, **8**, 602.
80. S. Prabha, D. Durgalakshmi, S. Rajendran and E. Lichtfouse, Plant-derived silica nanoparticles and composites for biosensors, bioimaging, drug delivery and supercapacitors: a review, *Environ. Chem. Lett.*, 2021, **19**(2), 1667–1691.
81. R. Periakaruppan, S. Manju Praveena, R. Priya, S. Gokul Raj and J. Danaraj, Biosynthesis of silica nanoparticles using the leaf extract of

Punica granatum and assessment of its antibacterial activities against human pathogens, *Appl. Biochem. Biotechnol.*, 2022, **194**(11), 5594–5605.

82. J. Sarkar, D. Mridha and J. Sarkar, *et al.*, Synthesis of nanosilica from agricultural wastes and its multifaceted applications: A review, *Biocatal. Agric. Biotechnol.*, 2021, **37**, 102175.
83. V. Biju, Chemical modifications and bioconjugate reactions of nanomaterials for sensing, imaging, drug delivery and therapy, *Chem. Soc. Rev.*, 2014, **43**(3), 744–764.
84. M. M. Wan, J. Y. Yang and Y. Qiu, *et al.*, Sustained release of heparin on enlarged-pore and functionalized MCM-41, *ACS Appl. Mater. Interfaces*, 2012, **4**(8), 4113–4122.
85. T. Lopez, E. Ortiz, R. Alexander-Katz, E. Basaldella and X. Bokhimi, Cortisol controlled release by mesoporous silica, *Nanomedicine*, 2009, **5**(2), 170–177.
86. Z. Guo, Y. Du, X. Liu, S. C. Ng, Y. Chen and Y. Yang, Enantioselectively controlled release of chiral drug (metoprolol) using chiral mesoporous silica materials, *Nanotechnology*, 2010, **21**(16), 165103.
87. H. Peng, R. Dong and S. Wang, *et al.*, A pH-responsive nano-carrier with mesoporous silica nanoparticles cores and poly (acrylic acid) shell-layers: fabrication, characterization and properties for controlled release of salidroside, *Int. J. Pharm.*, 2013, **446**(1–2), 153–159.
88. J. Tang, I. I. Slowing and Y. Huang, *et al.*, Poly(lactic acid)-coated mesoporous silica nanosphere for controlled release of venlafaxine, *J. Colloid Interface Sci.*, 2011, **360**(2), 488–496.
89. C. Y. Lai, B. G. Trewyn and D. M. Jeftinija, *et al.*, A mesoporous silica nanosphere-based carrier system with chemically removable CdS nanoparticle caps for stimuli-responsive controlled release of neurotransmitters and drug molecules, *J. Am. Chem. Soc.*, 2003, **125**(15), 4451–4459.
90. M. Ke, J. A. Wahab and B. Hyunsik, *et al.*, Allantoin-loaded porous silica nanoparticles/polycaprolactone nanofiber composites: fabrication, characterization, and drug release properties, *RSC Adv.*, 2016, **6**(6), 4593–4600.
91. D. Dhinasekaran, R. Raj and A. R. Rajendran, *et al.*, Chitosan mediated 5-Fluorouracil functionalized silica nanoparticle from rice husk for anticancer activity, *Int. J. Biol. Macromol.*, 2020, **156**, 969–980.
92. J. M. Rosenholm, T. Gulin-Sarfraz and V. Mamaeva, *et al.*, Prolonged Dye Release from Mesoporous Silica-Based Imaging Probes Facilitates Long-Term Optical Tracking of Cell Populations In Vivo, *Small*, 2016, **12**(12), 1578–1592.
93. J. J. Xu, W. W. Zhao, S. Song, C. Fan and H. Y. Chen, Functional nanoprobes for ultrasensitive detection of biomolecules: an update, *Chem. Soc. Rev.*, 2014, **43**(5), 1601–1611.
94. G. V. Deodhar, M. L. Adams and B. G. Trewyn, Controlled release and intracellular protein delivery from mesoporous silica nanoparticles, *Biotechnol. J.*, 2017, **12**(1), 1600408.

95. R. K. Gangwar, G. B. Tomar, V. A. Dhumale, S. Zinjarde, R. B. Sharma and S. Datar, Curcumin conjugated silica nanoparticles for improving bioavailability and its anticancer applications, *J. Agric. Food Chem.*, 2013, **61**(40), 9632–9637.
96. A. A. Alshatwi, J. Athinarayanan and V. S. Periasamy, Biocompatibility assessment of rice husk-derived biogenic silica nanoparticles for biomedical applications, *Mater. Sci. Eng., C*, 2015, **47**, 8–16.
97. J. Athinarayanan, V. S. Periasamy, M. Alhazmi, K. A. Alatiah and A. A. Alshatwi, Synthesis of biogenic silica nanoparticles from rice husks for biomedical applications, *Ceram. Int.*, 2015, **41**(1), 275–281.
98. M. C. Rodrigues, W. R. Rolim and M. M. Viana, *et al.*, Biogenic synthesis and antimicrobial activity of silica-coated silver nanoparticles for esthetic dental applications, *J. Dent.*, 2020, **96**, 103327.
99. J. Athinarayanan, V. S. Periasamy, M. Alhazmi and A. A. Alshatwi, Synthesis and biocompatibility assessment of sugarcane bagasse-derived biogenic silica nanoparticles for biomedical applications, *J. Biomed. Mater. Res., Part B*, 2017, **105**(2), 340–349.
100. L. Tang and J. Cheng, Nonporous silica nanoparticles for nanomedicine application, *Nano Today*, 2013, **8**(3), 290–312.
101. P. Araichimani, K. M. Prabu and G. S. Kumar, *et al.*, Rare-earth ions integrated silica nanoparticles derived from rice husk *via* microwave-assisted combustion method for bioimaging applications, *Ceram. Int.*, 2020, **46**(11), 18366–18372.
102. A. A. Nayl, A. I. Abd-Elhamid, A. A. Aly and S. Bräse, Recent progress in the applications of silica-based nanoparticles, *RSC Adv.*, 2022, **12**(22), 13706–13726.
103. A. F. Koenderink, A. Alù and A. Polman, Nanophotonics: Shrinking light-based technology, *Science*, 2015, **348**(6234), 516–521.
104. E. De Tommasi and A. C. De Luca, Diatom biosilica in plasmonics: applications in sensing, diagnostics and therapeutics, *Biomed. Opt. Express*, 2022, **13**(5), 3080–3101.
105. M. S. Aw, S. Simovic, J. Addai-Mensah and D. Losic, Silica microcapsules from diatoms as new carrier for delivery of therapeutics, *Nanomedicine*, 2011, **6**(7), 1159–1173.
106. X. Kong, E. Li and K. Squire, *et al.*, Plasmonic nanoparticles-decorated diatomite biosilica: extending the horizon of on-chip chromatography and label-free biosensing, *J. Biophotonics*, 2017, **10**(11), 1473–1484.
107. G. Ma, J. Cao and G. Hu, *et al.*, Porous chitosan/partially reduced graphene oxide/diatomite composite as an efficient adsorbent for quantitative colorimetric detection of pesticides in a complex matrix, *Analyst*, 2021, **146**(14), 4576–4584.
108. S. Santra, B. Liesenfeld and D. Dutta, *et al.*, Folate conjugated fluorescent silica nanoparticles for labeling neoplastic cells, *J. Nanosci. Nanotechnol.*, 2005, **5**(6), 899–904.
109. L. Wang, W. Zhao and W. Tan, Bioconjugated silica nanoparticles: development and applications, *Nano Res.*, 2008, **1**, 99–115.

110. M. R. Kasaai, Nanosized particles of silica and its derivatives for applications in various branches of food and nutrition sectors, *J. Nanotechnol.*, 2015, **2015**, 852394.
111. K. Higashisaka, Y. Yoshioka and Y. Tsutsumi, Applications and safety of nanomaterials used in the food industry, *Food Saf.*, 2015, **3**(2), 39–47.
112. A. E. Karaca, C. Özel, A. C. Özarslan and S. Yücel, The simultaneous extraction of cellulose fiber and crystal biogenic silica from the same rice husk and evaluation in cellulose-based composite bioplastic films, *Polym. Compos.*, 2022, **43**(10), 6838–6853.
113. N. Ojha and N. Das, Fabrication and characterization of biodegradable PHBV/SiO_2 nanocomposite for thermo-mechanical and antibacterial applications in food packaging, *IET Nanobiotechnol.*, 2020, **14**(9), 785–795.
114. F. Farirai, M. Ozonoh and T. C. Aniokete, *et al.*, Methods of extracting silica and silicon from agricultural waste ashes and application of the produced silicon in solar cells: a mini-review, *Int. J. Sustainable Eng.*, 2021, **14**(1), 57–78.

CHAPTER 5

Biogenic Silica Nanoparticles and Composites for Supercapacitors

ABHINAY THAKUR[a] AND ASHISH KUMAR*[b]

[a] Department of Chemistry, School of Chemical Engineering and Physical Sciences, Lovely Professional University, Phagwara, Punjab, 144411, India;
[b] Nalanda College of Engineering, Bihar Engineering University, Department of Science, Technology and Technical Education, Government of Bihar, 803108, India
*Email: drashishchemlpu@gmail.com

5.1 Introduction

Supercapacitors are energy storage devices that have various benefits above traditional batteries. They are often termed electrochemical capacitors or ultracapacitors. These devices store energy by electrostatically adsorbing ions at the interaction site across the electrolyte and the electrode, rather than through chemical reactions.[1–4] This unique energy storage mechanism results in extended cycle life, elevated power density and fast discharging/charging capabilities, making supercapacitors ideal for a broad array of applications. Supercapacitors can also store and release energy quickly, making them ideal for applications that require rapid energy storage and release, such as consumer electronics, power tools, and hybrid electric vehicles.[5–10] Supercapacitors can also operate over a broad array of temperatures, making them suitable for utilization in extreme environments.[11–14] They can also be utilized in integration with batteries to enhance the overall functionality of energy storage systems.[14–23] Supercapacitors offer a long cycle life and high

Biogenic Silica: Fundamentals and Applications
Edited by Ashok Kumar Nadda and Vijai Kumar Gupta

Published by the Royal Society of Chemistry, www.rsc.org

reliability, making them ideal for these critical applications.[24–28] In a supercapacitor, there are two electrodes separated by an electrolyte. Each electrode consists of a current collector (usually made of a conductive material like carbon or metal) coated with a material that can adsorb ions from the electrolyte. At the electrode–electrolyte interface, the adsorbed ions create an electrochemical double layer that serves as the foundation for C_{dl}. The ions in the electrolyte flow toward the electrode that is negatively charged when a voltage is supplied to the supercapacitor, accumulating at the electrode–electrolyte interface to form an electric field between the two electrodes.[29,30] This electric field stores energy electrostatically, similar to how static electricity is stored on a capacitor. In addition to electrostatic C_{dl}, some supercapacitors also use pseudocapacitance, which is the reversible faradaic redox reaction across the electrode material and the ions in the electrolyte. A reverse charge transfer across the electrolyte and the electrode occurs in pseudocapacitance as a consequence of a chemical reaction across the electrolyte and electrode material. Due to the electrode material itself contributing to the supercapacitor's ability to store energy, this reaction results in a higher capacitance than electrostatic double-layer capacitance.[10,31,32]

In past decades, researchers have been exploring various electrode materials to enhance the functionality and performance of supercapacitors.[33–35] Biogenic silica, derived from natural sources such as plants, algae, and diatoms, plays a significant role in advancing the field of supercapacitors.[36] As the demand for efficient energy storage solutions continues to grow, researchers are exploring novel materials with unique properties to enhance the performance and sustainability of supercapacitor devices. Biogenic silica nanoparticles (BSNPs) and their composites have emerged as promising candidates for supercapacitor applications due to their distinct characteristics, including high surface area, biocompatibility, low toxicity, and eco-friendliness. This chapter explores the role of biogenic silica in supercapacitors, focusing on its synthesis, fabrication, properties, and applications.

5.2 Role of BSNPs in the Supercapacitor

In past decades, researchers have been exploring various electrode materials to enhance the functionality and performance of supercapacitors. One such material is biogenic silica nanoparticles (BSNPs). BSNPs are synthesized from natural sources such as rice husk, diatoms, and other biogenic materials. BSNPs have unique physical and chemical characteristics that render them feasible for numerous applications, including supercapacitors. Due to their special features and possible uses in the sector of energy storage, specifically in the creation of supercapacitors, biogenic silica nanoparticles and their composites have attracted a lot of interest in past years. These materials are very appealing for usage in energy storage applications due to their many beneficial qualities, which include substantial surface region, porosity, stability, electrical conductance, and tunable properties. The large surface region to volume proportion of biogenic silica nanoparticles is one of their main benefits.[37–40] This means that they provide a large area for electrolyte ions to interact with, which is important for energy storage in supercapacitors. The enlarged surface region of the BSNPs allows for an increased capacitance, or the ability to store charge, which can lead to high energy densities in the supercapacitor. In addition, the porosity of the BSNPs allows for the rapid dispersion of ions, which can enhance the charging and discharging rates of the supercapacitor. This is important for applications that require rapid energy storage and release, such as electric vehicles or consumer electronics. Another advantage of biogenic silica nanoparticles and their composites is their high stability and resistance to chemical and thermal degradation. These materials are highly resistant to corrosion, oxidation, and other forms of degradation, which makes them suitable for usage in a broad array of environments and conditions. This stability also means that biogenic silica nanoparticle-based supercapacitors can have a long cycle life, which is important for applications that require a reliable and long-lasting power source.

Biogenic silica nanoparticles and their composites could also be doped using conductive materials, including graphene or carbon nanotubes, to improve their electrical conductivity. This is important for the efficient transfer of electrons during the discharge and charge phases of the supercapacitor. The inclusion of conductive materials could also facilitate the enhancement of the mechanical strength and stability of the composite. Furthermore, the properties of biogenic silica nanoparticles and their composites could be possibly tuned by modifying the shape, size and composition of the nanoparticles. This allows for the optimization of the material properties for specific applications. For example, the addition of a polymer coating to the nanoparticles can help to improve their stability in harsh environments or increase their affinity for specific electrolytes. Supercapacitors made of biogenic silica nanoparticles have numerous potential uses in the realm of energy storage. Supercapacitors could be utilized in electric vehicles, where they can deliver great power density and quick charging times, making this one of the most promising applications. Biogenic silica nanoparticle-based supercapacitors can also be used in renewable energy systems, such as wind or solar power, to store energy and provide a stable power supply.[41–44] Additionally, these materials could find use in consumer electronics, such as smartphones or laptops, to provide fast charging and long-lasting battery life. In medical devices, such as pacemakers, biogenic silica nanoparticle-based supercapacitors could provide a reliable and long-lasting power source. Furthermore, biogenic silica nanoparticle-based supercapacitors could find applications in aerospace and defense, such as satellite power systems or unmanned aerial vehicles, owing to their enhanced power density and stability in extreme environments. In order to create ultrafine Co_3O_4 grains over reduced oxide graphene nanosheets (Co_3O_4/rGO-C) for an improved supercapacitor electrode, Liu *et al.*[45] introduced a dual-template nano casting approach. The method uses silicate molecules and cetyltrimethylammonium bromide as the hard and soft templates, accordingly. The resultant Co_3O_4/rGO-C electrode displayed an impressive specific capacitance (709.1 F g^{-1} at 1 A g^{-1}) and long-lasting durability (91.2% after 6000 cycles). Additionally, a built Co_3O_4/rGO-C‖rGO solid-state asymmetric device produces a super-high energy-density of 48.2 W h kg^{-1} at 750.5 W kg^{-1}, allowing two devices to illuminate a red light-emitting diode for 340 s. The outcomes show that using nanocasting technology to develop enhanced electrode materials for energy storage devices is an effective technique.

Sun *et al.*[46] mentioned that *in situ* sulfurating the prefabricated Ni/CRs (carbon nanorods) resulted in a phase-controlled production of α-NiS nanoparticles imbedded in CRs. The production of α-NiS and inhibiting its conversion to β-phase depends on the confining of carbon substrate nanopores. Elevated capacitor achievement resulted owing to the ultra-small α-NiS NPs scattered in CRs' hybrid electrochemical charge storage, whereas elevated electrochemical stabilization resulted owing to the supercapacitor-battery electrode, that also created a synergistic effect between the elements of the battery (NiS NPs) and capacitor (CRs). This method not only provides a means of producing nickel sulfide but also opens the door to the development, installation and production of innovative nickel-based/carbon hybrid supercapacitor-battery electrode components. The simultaneous sulfuration of Ni/CRs in glycerol over an hour only allowed for the production of α-NiS/CRs composites. On either side, after sulfuration, the mixture of Ni nanoparticles and mesoporous CRs (MCRs) created a composite called β-NiS/CRs. After removing the Ni nanoparticles with Ni/CRs utilized 3 M HCl and letting it sit for the night, MCRs were produced. The electrode displayed a specific capacitance of 1092, 946, 835, and 740 F g^{-1} for current densities of 1, 2, 5, and 10 A g^{-1}, correspondingly. Following 2000 charge/discharge sessions, the electrode retains about 100% of its specific capacitance.

A honeycomb-like film made of reduced graphene oxide (RGO) and Si NPs composite was created by Yue *et al.*[47] utilizing a vacuum filter construction procedure without the use of a binder or conductive additives. The ultra-thick composite sheet created after the Si NPs were coiled around the RGO sheets was employed as an anode for lithium-ion batteries. The electrode demonstrated outstanding rate capability, a capacity greater than 1000 mA h g^{-1} at a current density of 4200 mA g^{-1} following 500 cycles, and a significant reversible capacity of 2370 mA h g^{-1} over 50 cycles at a current of 210 mA g^{-1}. These findings indicate that the RGO/Si NP composite film, which is pliable, has large areal content, binder-free, and unrestricted, has a lot of possibilities for usage in high-power lithium-ion batteries.

To synthesize biogenic silica nanoparticles, one common method is to extract them from diatomaceous earth, a sedimentary rock composed of fossilized diatoms. The diatoms are first cleaned to remove any impurities and then treated with acid to remove any organic material. The resulting silica skeleton is then ground into a fine powder, which can be further processed to create nanoparticles. Once the biogenic silica nanoparticles have been synthesized, they can be incorporated into composites for use in supercapacitors. One common approach is to mix the nanoparticles with a conductive material, including carbon nanotubes and graphene, to manufacture a nanocomposite.[47–52] The conductive material helps to enhance the electrical conductivity of the composite, that is important for its use in supercapacitors. Another strategy is to synthesize additional substances using the biogenic silica nanoparticles as a template. For instance, a layer of conducting polymer can be applied to the nanoparticles to create a core–shell structure. This could enhance the supercapacitor's composite's performance and stability.

Biogenic silica nanoparticles and their composites have several characteristics that render them significant for usage in supercapacitors. Some of these properties are:

- High surface area: biogenic silica nanoparticles have an extremely large surface region to volume proportions, which means that they provide a large area for electrolyte ions to interact with. This is important for energy storage in supercapacitors.
- High porosity: the porous structure of biogenic silica nanoparticles and their composites allows for the rapid dispersion of ions, which can enhance the charging and discharging rates of the supercapacitor.
- High stability: biogenic silica nanoparticles are highly stable and resistant to chemical and thermal degradation. This means that they can be used in a broad array of environments and circumstances.
- Good electrical conductivity: biogenic silica nanoparticles can be doped with conductive materials, including CNT or graphene, to improve their electrical conductivity. This is important for the efficient transfer of electrons during the discharge and charge phases of the supercapacitor.
- Tunable properties: by adjusting the size, shape, and content of the nanoparticles, the properties of biogenic silica nanoparticles and their composites can be easily modified. This allows for the optimization of the material properties for specific applications.

5.3 Synthesis Methods of Biogenic Silica Nanoparticles (BSNPs)

Biogenic silica nanoparticles (BSNPs) can be synthesized from various natural sources, including rice husk, diatoms, and other biogenic materials. There are several methods for synthesizing BSNPs, each with its advantages and disadvantages.

5.3.1 Sol–Gel Method

For the synthesis of BSNPs, the sol–gel method is frequently employed. In order to create a sol, silicon alkoxides are hydrolyzed and condensed in the influence of a catalyst, including ammonia or sodium hydroxide. The gel created after the sol has been aged and dried is calcined to create BSNPs.[46,53–55] The three phases of the sol–gel technique are gelation, condensation, and hydrolysis. The addition of silicon alkoxide to an aqueous solution containing the catalyst during the hydrolysis stage results in the production of silanol groups (Si–OH) on the particle surfaces. The hydrolysis reaction can be represented as follows:

$$Si(OR)_4 + 2H_2O \rightarrow Si(OH)_4 + 4ROH$$

where R is an alkyl group.

During the condensation stage, the silanol moieties over the substrate of the particles undergo a condensation reaction to form Si–O–Si bonds, which results in the production of a 3D network of particles. The condensation reaction might be depicted:

$$Si(OH)_4 \rightarrow Si\text{–}O\text{–}Si + H_2O$$

During the gelation stage, the sol is transformed into a gel, which is a solid-like material that retains the shape of the container. The gelation process is regulated by the quantity of the sol, the pH of the solution, and the temperature. The solvent is then heated out from the resultant gel to create a porous substrate. The dried gel is then calcined at high temperatures to remove any residual organic material and form BSNPs. The calcination temperature and time could be adjusted to regulate the particle size and topography of the BSNPs. The sol–gel method offers several advantages over other methods for synthesizing BSNPs. It permits for the regulation of particle size and morphology, along with the ability to incorporate dopants or modifiers into the particles. It also produces BSNPs with high purity and crystallinity. However, the sol–gel method also has some limitations. It can be a time-consuming process, requiring several days or weeks for the sol to age and form a gel. It also requires the use of toxic solvents and high temperatures, which can limit the scalability and sustainability of the method. Despite these limitations, the sol–gel method remains a popular and versatile method for synthesizing BSNPs, with possibilities in numerous fields, such as energy storage, biomedicine, and catalysis.

5.3.2 Hydrothermal Method

The hydrothermal method is another frequently utilized technique for synthesizing boron silicate nanostructures (BSNs) or boron silicate nanoparticles (BSNPs). This method involves the reaction of silicon precursors with water under enhanced pressure and temperature conditions, resulting in the formation of BSNPs. The hydrothermal method could be dissociated into three phases: preparation of the precursor media, hydrothermal reaction, and post-treatment of the BSNPs. During the preparation of the precursor solution, a silicon source such as tetraethyl orthosilicate (TEOS) or sodium silicate is dissolved in water, along with a surfactant to regulate the particle morphology and size. The surfactant is typically a cationic or non-ionic surfactant such as cetyltrimethylammonium bromide (CTAB) or polyethylene glycol (PEG), respectively.[45,56,57] The precursor solution is heated to extreme temperatures and pressures in a sealed container during the hydrothermal reaction stage, typically at temperatures between 100 °C and 200 °C and pressures between 10 and 20 atm. In these circumstances, the silicon precursor passes through condensation and hydrolysis processes that lead to the synthesis of BSNPs.

During the post-treatment stage, the BSNPs are typically rinsed using water to eradicate any residual surfactants or contaminants and then dried at low temperatures. The resulting BSNPs are typically spherical or ellipsoidal and have a narrow size distribution. The hydrothermal method offers several advantages over other methods for synthesizing BSNPs. It allows for the formation of BSNPs at relatively low temperatures and pressures, which can improve the sustainability and scalability of the method. By altering the reaction conditions, the type of surfactants used, and their quantity, it also enables the regulation of particle form and morphology.[58–60]

However, the hydrothermal method also has some limitations. It can be a time-consuming process, requiring several hours or days for the reaction to take place. It also requires the use of specialized equipment and the generation of high-pressure conditions, which can limit the accessibility of the method. Despite these limitations, the hydrothermal method remains a popular and versatile method for synthesizing BSNPs, with applications in numerous sectors, such as catalysis, energy storage and biomedical imaging.

5.3.3 Microwave-assisted Method

It is a relatively new technique for synthesizing BSNPs. This method uses microwave radiation to promote the condensation and hydrolysis reactions of silicon precursors, resulting in the formation of BSNPs. The microwave-assisted method could be divided into two phases: preparation of the precursor media and microwave-assisted synthesis of BSNPs. During the preparation of the precursor solution, a silicon source such as tetraethyl orthosilicate (TEOS) or sodium silicate is dissolved in a solvent such as ethanol or water, along with a surfactant to regulate the particle morphology and size. The surfactant is typically a cationic or non-ionic surfactant such as cetyltrimethylammonium bromide (CTAB) or polyethylene glycol (PEG), respectively. In the microwave-assisted synthesis stage, the precursor solution is irradiated with microwave radiation at a frequency of 2.45 GHz.[61–63] Microwave radiation enables the condensation and hydrolysis reactions of the silicon precursor, resulting in the formation of BSNPs. The reaction time can be as short as a few minutes to several hours, relying on the reaction conditions and the desired size and morphology of the BSNPs.

After the microwave-assisted synthesis stage, the BSNPs are typically rinsed using water to eradicate any residual surfactants or pollutants and then dried at low temperatures. The resulting BSNPs are typically spherical or ellipsoidal and have a narrow size dispersion. The microwave-assisted techniques promote several advantages over other methods for synthesizing BSNPs. It allows for the formation of BSNPs at relatively low temperatures and pressures, which can improve the sustainability and scalability of the method. It also enables the regulation of the particle morphology and size by regulating the reaction conditions and the type and quantity of surfactants. Furthermore, the method is relatively fast and can be completed in a matter of minutes to hours. However, the microwave-assisted method also has some limitations. It requires the use of specialized equipment, including a microwave reactor, and can be difficult to scale up for industrial production.[36,64,65] Additionally, the method may require optimization of the reaction conditions to achieve the desired size and morphology of BSNPs. Despite these limitations, the microwave-assisted method shows promise for the synthesis of BSNPs with potential applications in several sectors such as catalysis, energy storage and biomedical imaging.

5.3.4 Ultrasonic-assisted Method

The ultrasonic-assisted technique is a novel technique for synthesizing BSNPs. This method utilizes ultrasonic waves to promote the condensation and hydrolysis reactions of silicon precursors, resulting in the formation of BSNPs. The ultrasonic-assisted method could be divided into two phases: preparation of the precursor media and ultrasonic-assisted synthesis of BSNPs. During the preparation of the precursor solution, a silicon source such as tetraethyl orthosilicate (TEOS) or sodium silicate is dissolved in a solvent such as ethanol or water, along with a surfactant to regulate the particle morphology and size. The surfactant is typically a cationic or non-ionic surfactant such as cetyltrimethylammonium bromide (CTAB) or polyethylene glycol (PEG), respectively.[66–68] In the ultrasonic-assisted synthesis stage, the precursor solution is irradiated with ultrasonic waves at a frequency of 20–100 kHz. The ultrasonic waves promote the hydrolysis and condensation reactions of the silicon precursor, resulting in the formation of BSNPs. The reaction time can be as short as a few minutes to several hours, relying on the reaction conditions and the desired size and morphology of the BSNPs. After the ultrasonic-assisted synthesis stage, the BSNPs are typically rinsed using water to eradicate any residual surfactants or contaminants and then dried at low temperatures. The resulting BSNPs are typically spherical or ellipsoidal and have a narrow size dispersion.

The ultrasonic-assisted technique provides several benefits over other methods for synthesizing BSNPs. It allows for the formation of BSNPs at relatively low temperatures and pressures, which can improve the sustainability and scalability of the method. It also enables the regulation of particle size and morphology by regulating the reaction conditions and the type and quantity of surfactants. Furthermore, the method is relatively fast and can be completed in a matter of minutes to hours. However, the ultrasonic-assisted method also has some limitations. It requires the use of specialized equipment, including an ultrasonic reactor, and can be difficult to scale up for industrial production. Additionally, the method may require optimization of the reaction conditions to achieve the desired size and morphology of BSNPs.[69–72] Despite these limitations, the ultrasonic-assisted method shows promise for the synthesis of BSNPs with potential applications in several sectors such as catalysis, energy storage and biomedical imaging. It offers an alternative method for synthesizing BSNPs with high yields and good reproducibility.

5.3.5 Emulsion Method

The emulsion method is a versatile technique for synthesizing BSNPs. This method involves the emulsification of a silicon precursor in a liquid phase followed by hydrolysis and condensation reactions to form BSNPs. The emulsion method could be divided into three phases: preparation of the precursor media, formation of the emulsion, and synthesis of BSNPs. During the preparation of the precursor solution, a silicon source such as tetraethyl orthosilicate (TEOS) or sodium silicate is dissolved in a solvent such as ethanol or water, along with a surfactant to regulate the particle morphology and size. The surfactant is typically a cationic or non-ionic surfactant such as cetyltrimethylammonium bromide (CTAB) or polyethylene glycol (PEG), respectively. In the formation of the emulsion stage, the precursor solution is mixed with an immiscible liquid phase such as oil or hexane. The mixture is then homogenized or sonicated to form an emulsion, where the precursor solution is dispersed in small droplets throughout the liquid phase.[1,73–75]

In the synthesis of BSNPs stage, the emulsion is added to a solution containing a catalyst such as ammonia or an acid, which initiates the hydrolysis and condensation reactions. BSNPs are then formed at the interface between the precursor droplets and the liquid phase, resulting in the production of BSNPs having a narrow size dispersion. After the production of BSNPs stage, the resulting BSNPs are typically rinsed utilizing water to eradicate any residual surfactants or contaminants and then dried at low temperatures. The resulting BSNPs are typically spherical or ellipsoidal and have a narrow size distribution. The emulsion method offers several advantages over other methods for synthesizing BSNPs. It allows for the production of BSNPs having a narrow size dispersion and high yields. Modifying the reaction conditions and the kind and amount of surfactants, also enables the control of particle morphology and size.

Furthermore, the emulsion method is relatively simple and can be completed using standard laboratory equipment. It is also scalable and can be adapted for industrial production. However, the emulsion method also has some limitations. It can be time-consuming to form a stable emulsion, and the emulsification process can be sensitive to changes in temperature, pressure, and surfactant concentration. Additionally, the emulsion method may require optimization of the reaction conditions to achieve the desired size and morphology of BSNPs. Despite these limitations, the emulsion method shows promise for the synthesis of BSNPs with potential applications in numerous sectors such as catalysis, energy storage and biomedical imaging. It offers an alternative method for synthesizing BSNPs with high yields and good reproducibility.

5.3.6 Freeze-drying Method

The freeze-drying method is a commonly used technique for synthesizing BSNPs. This technique involves the synthesis of a silica precursor media, followed by the formation of BSNPs through a freeze-drying process. The freeze-drying method could be deliberated into three phases: preparation of the precursor media, freezing of the solution, and drying of the frozen sample. During the preparation of the precursor solution, a silicon source such as tetraethyl orthosilicate (TEOS) or sodium silicate is dissolved in a solvent such as ethanol or water, along with a catalyst and a surfactant to regulate the particle morphology and size. The surfactant is typically a cationic or non-ionic surfactant such as cetyltrimethylammonium bromide (CTAB) or polyethylene glycol (PEG), respectively.[76–79] After the precursor solution is prepared, it is frozen by rapidly cooling the solution in a bath of liquid nitrogen or dry ice. The frozen sample is then transferred to a vacuum chamber where the pressure is reduced and the temperature is increased, causing the frozen solvent to sublime and leave behind a dry, porous matrix of silica nanoparticles. During the drying stage, the remaining water in the matrix is removed by sublimation, leaving behind BSNPs that are extremely permeable and have a wide surface region. The resulting BSNPs are typically spherical or ellipsoidal and have a narrow size distribution.

The freeze-drying method offers several advantages over other methods for synthesizing BSNPs. It allows for the formation of highly porous BSNPs having a wide surface region, which is desirable for many applications.[80–82] The method is also relatively simple and can be completed using standard laboratory equipment. Additionally, the freeze-drying method allows for the control of particle size and morphology by regulating the reaction conditions and the type and quantity of surfactants. However, the freeze-drying method also has some limitations. It can be time-consuming to prepare the precursor solution and to freeze the sample. Additionally, the freeze-drying process can be sensitive to changes in temperature and pressure, and it could be hard to achieve a consistent morphology and size of BSNPs. Despite these drawbacks, the freeze-drying technique is promising for producing BSNPs with the potential for use in a variety of industries, such as catalysis, catalysis, and biomedical imaging. It provides a different way to make BSNPs with a lot of surface area and permeability, and it may be used in industrial manufacturing.

5.4 Utilization of BSNs and Their Composites in Supercapacitors

Biogenic silica nanoparticles (BSNPs) have recently gained attention in various fields owing to their unique characteristics wide surface region, biocompatibility and thermal stability. In supercapacitors, BSNPs and their composites have shown promising results in improving energy storage capacity and cycling stability. BSNPs are obtained from renewable sources including sugarcane bagasse, rice husk and diatoms. The synthesis of BSNPs involves a straightforward and minimal price process that could be easily scaled up. To improve their compatibility with various electrolytes, the BSNPs can be functionalized with a variety of functional groups. It has been discovered that BSNP composites with carbon-based substances like graphene, carbon nanotubes, and activated carbon enhance the electrochemical characteristics of supercapacitors. The BSNPs increase the composite's surface area, offer electrolyte ions active sites, and enhance the composite's conductivity. Several studies have reported the successful applicability of BSNP-based composites in supercapacitors. For instance, a recent study reported the creation of a composite of BSNPs and graphene oxide (GO), which displayed decent cycling stability up to 5000 cycles and a large specific capacitance of 262 $F g^{-1}$ at a current density of 1 $A g^{-1}$. Carbon nanotube, graphene and activated carbon are a few examples of carbon-based materials that are frequently used to create composites with BSNPs.

5.4.1 BSNPs/Graphene Composites

BSNPs/graphene composites are a type of composite material that combines biogenic silica nanoparticles (BSNPs) with graphene. Graphene is a 2D carbon material having enhanced electrical conductivity and a large surface area, while BSNPs are silica nanoparticles derived from biological sources such as diatoms and radiolarians. A composite material with improved characteristics for usage in supercapacitors can be created by mixing BSNPs and graphene. Particularly, the enhanced surface region and the effective charge transmission between the BSNPs and graphene account for the composite's elevated specific capacitance. The utilization of BSNPs/graphene composites in high-performance supercapacitors is therefore intriguing. BSNPs/graphene composites are typically synthesized in two steps. A refined Hummers technique is used in the initial stage to create the graphene oxide (GO) from graphite. After that, a reducing agent like hydrazine is used to convert the GO into graphene. The second process involves adding the BSNPs to the graphene solution and sonicating the combination to create the composite material. By changing the BSNPs' surface, it is possible to further improve the properties of composites made of graphene and BSNPs. For instance, to increase the compatibility of the BSNPs with the graphene matrix and the charge transfer between the BSNPs and graphene, the surface of the BSNPs can be functionalized with organic molecules. The functionalization can also enhance the BSNPs' distribution in the graphene matrices, resulting in a composite material that is more homogeneous.

The working principle of BSNPs/graphene composites in supercapacitors involves the use of better cycling stability and elevated specific capacitance of the composite material to store and release electrical energy efficiently. In a supercapacitor, electrical energy is retained using the segregation of charge at the interface across an electrolyte and two electrodes. The electrodes are typically made of a high surface area material with good conductivity, such as graphene. Whenever a voltage is injected into the electrodes, the charge is transferred between the electrolyte and the electrodes, leading to the increase of charge at the electrolyte/electrode interaction. BSNPs/graphene composites are designed to enhance the function of supercapacitors by elevating the specific capacitance of the electrode material. The composite's BSNPs may enhance the electrode material's surface region, while graphene can make the composite more conductive. As a result, the composite material's overall capacitance rises, and the efficiency of charge transmission across the electrode/electrolyte interface is enhanced. The strong linkage across the BSNPs and graphene also contributes to the good cycling stability of the BSNPs/graphene composite. This means that the composite can maintain its electrochemical properties over multiple charge/discharge cycles, making it a reliable energy storage material for long-term use. Javed *et al.*[60] presented a new approach to synthesizing nanostructures for high-performance supercapacitors by preparing mesoporous silica nanoparticles (MSNs) wrapped with conducting polyaniline (PANI) wires and GO. X-ray diffraction (XRD) and transmission electron microscopy (TEM) were employed to analyze the shape and chemical composition of the resultant nanocomposites. Numerous tests, including electrochemical impedance spectroscopy (EIS) and cyclic voltammetry (CV), were used to examine the supercapacitive behavior of the nanocomposites. The 20%-GO@MSNs/PANI nanocomposite showed the greatest specific capacitance of 412 $F g^{-1}$, and also quick charge–discharge cycles and strong cycle durability, as contrasted to pure MSNs and PANI. Whereas the MSNs nanoparticles considerably improve the specific surface region for effective ion transit and charge transmission, the GO coating inhibits morphological disintegration and serves as an extremely conductive route. An effective method for creating innovative functional substances for electrochemical purposes is presented in this paper.

Rabia *et al.*[83] presents the synthesis and characterization of a flower-shaped CoS-Co_2O_3/G-C_3N_4 nanocomposite for the application of a two-symmetrical electrodes supercapacitor. Urea was directly burned in the air to create G-C_3N_4, and cobalt salts were used in a hydrothermal process to create the composite CoS-Co_2O_3/G-C_3N_4. SEM, XRD, XPS and TEM were used to characterize the nanocomposite, confirming its chemical makeup, functional units, and elemental composition. The supercapacitor was evaluated utilizing CV, charge/discharge, stability, and impedance experiments in both basic (6.0 M NaOH) and acidic (0.5 M HCl) mediums. In acidic and basic environments, the supercapacitor revealed an energy density of 30.2 and 28.7 $W h kg^{-1}$ and a specific capacitance of 92 and 361 $F g^{-1}$, correspondingly. Following 600 cycles, the supercapacitor maintained its stability at 100%, with a little drop to 98.5%. Figure 5.1 shows SEM pictures of the composites made of (a) G-C_3N_4 and (b) CoS-Co_2O_3/G-C_3N_4 both exhibit thin film topologies. The films of G-C_3N_4 possess a mean diameter and length of 80 nm and 170 nm, respectively, and exhibit 2D nanosheet characteristics, including folds and crumples in some places. The flattened, wrinkled, unevenly formed sheets are visible in the TEM picture (Figure 5.1c), and the black region encompassing the films verifies the integration of CoS-Co_2O_3 in the G-C_3N_4 sheet. The cross-section of the produced CoS-Co_2O_3/G-C_3N_4 composite, which has a relatively uniform ruggedness and is coated in minute particles, is depicted by theoretical modeling in Figure 5.1d. The results imply that supercapacitors might replace batteries as a power source owing to their simple and inexpensive production process.

A thermostable supercapacitor was developed by combining SnO_2-doped graphene aerogel electrodes with ionic liquid (IL)-based composite electrolytes, according to Kim *et al.*[58] Despite subsequent annealing, *in situ* synthesis chemically bonds the small SnO_2 nanoparticles to the graphene sheets and starts the electrochemical synergy impact at elevated temperatures. Fumed SiNPs were included to the composite electrolytes to enhance the thermostability and ionic conductivity of the polymer matrices. The attained

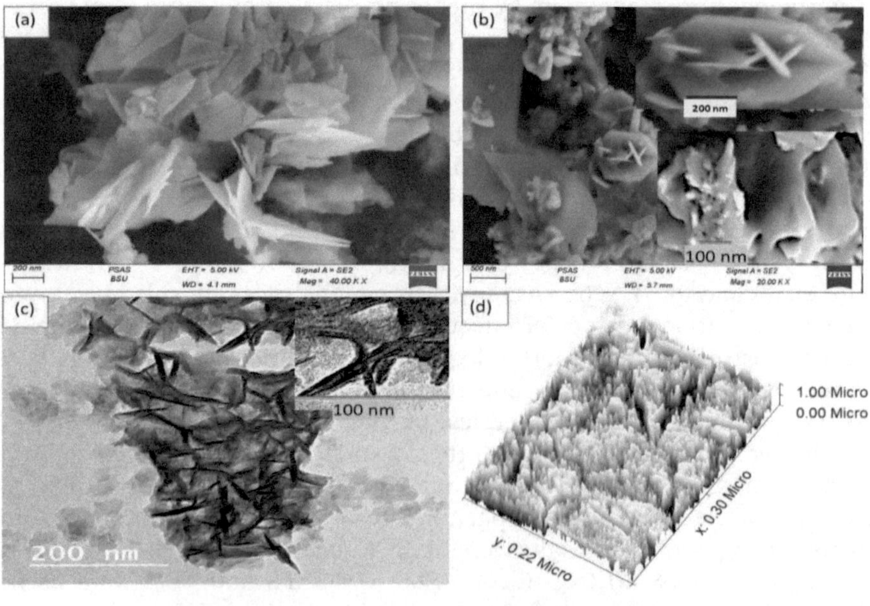

Figure 5.1 SEM of (a) G-C_3N_4 and (b) CoS-Co_2O_3/G-C_3N_4 composite. (c) TEM and (d) image of CoS-Co_2O_3/G-C_3N_4 composite. Reproduced from ref. 83, https://doi.org/10.3390/mi13122234, under the terms of the CC BY 4.0 license, https://creativecommons.org/licenses/by/4.0/.

supercapacitor has an energy density of 160 W h kg^{-1} at 125 °C, a large specific capacitance of 541 F g^{-1}, and a persistence of capacitance of 97.3% following 10 000 cycles. In order to create energy storage with superior efficiency, extended cyclic stability, and higher temperature which could generate electricity in challenging settings, this research offers a logical approach. Similarly, Sun *et al.*[41] published a method for creating a negative electrode for lithium-ion batteries made of SiNPs with NH_2-terminations adsorbed on GO (Si@NH_2/GO). Using hydrogen bonding and cross-linked engagements among NH_2-modified Si NPs and GO, as well as GO and sodium alginate, which serves as an adhesive, the Si@NH_2/GO composite was created. In the negative electrode arrangement, a strong network was created as a result, helping to sustain the electrode throughout the discharge and charge phases. The Si@NH_2/GO electrode offers excellent rate performance and capacity preservation, providing a reversible capacity of 1000 mA h g^{-1} following 400 cycles at a current of 420 mA g^{-1} and nearly 100% capacity preservation. According to the research, creating robust electrodes featuring better electrochemical performance requires a system-level approach.

With sucrose serving as the carbon precursor and silica nanoparticles-graphene oxide as the template, Bo *et al.*[82] reported on the production of macroporous carbon–graphene composites. The composites combine graphene's superior mechanical and electrical capabilities alongside macroporous carbon's permeable nature. Pt NPs having a mean size of 3.1 ± 0.7 nm were spread across the composite while utilized as a scaffold for Pt electrocatalysts. The combination exhibits enhanced molecular transport and reduced graphene sheet agglomeration. The Pt/MC-GR composite offers the ability to be used as a nonenzymatic glucose sensor and demonstrates considerable catalytic performance for the oxidation of methanol. It could produce steady-state signals in 3 seconds despite evident contamination from other substances.

5.4.2 BSNPs/Carbon Nanotube Composites

These composites are a type of composite material that combines biogenic silica nanoparticles (BSNPs) with CNTs. CNTs are a type of carbon-based material that has elevated electrical conductivity with a broad surface region, while BSNPs are silica nanoparticles derived from biological sources such as diatoms and radiolarians. The combination of BSNPs and CNTs can lead to the production of a composite material with enhanced properties for use in supercapacitors. Specifically, the increased specific capacitance of the composite owned to the increased substrate region and the efficient charge transit across the BSNPs and CNTs. This makes BSNPs/CNTs composites promising materials for utilization in effective supercapacitors. The production of BSNPs/CNTs composites typically involves the dispersion of CNTs in a BSNPs suspension followed by sonication. The BSNPs/CNTs composite is then obtained by evaporating the solvent and drying the resulting mixture. The characteristics of the BSNPs/CNTs composite could be further enhanced by regulating the amount of CNTs in the composite and optimizing the distribution of the CNTs in the BSNPs matrices. Owing to the significant association across the BSNPs and CNTs, the BSNPs/CNTs composite not only has a high specific capacitance but also good cycling stability. The two materials' strong interaction can increase the mechanical stability of the composite material and increase its resistance to deformation and fracture.

The working principle of BSNPs/carbon nanotubes (CNTs) composites in supercapacitors are similar to that of BSNPs/graphene composites. The composite material is designed to improve the efficacy of the supercapacitor by increasing its capacitance and improving its cycling stability. In a supercapacitor, the charge is retained at the electrolyte/electrode interface by separating positive and negative charges. Typically, CNTs or another extremely conductive material having a large surface area serves as the electrode material. The charge is exchanged between the electrolyte and the electrode material when a voltage is supplied to the electrodes, which causes an accumulation of charge at the electrode/electrolyte interface. By adding BSNPs to the CNT matrix, BSNPs/CNT composites are intended to raise the specific capacitance of the electrode material. While the CNTs offer high electrical conductivity, the BSNPs can enhance the surface area of the electrode material and produce extra charge storage sites. The combination of these two materials can lead to an efficient improvement in the overall capacitance of the composite substance, improving the energy storage capacity of the supercapacitor. Moreover, the BSNPs/CNTs composite also has good cycling stability owing to the robust interaction across the BSNPs and CNTs. The interaction between the two materials can lead to an enhancement in the mechanical stability of the composite, making it more resistant to deformation and fracture during charge/discharge cycling. For instance, in an experiment, an interlinked system of matched single-walled CNTs containing decreased GO films was used by Yu *et al.*[84] to propose a technique for creating hierarchically structured carbon microfibers. Mesoporous structures having a substantial specific surface region and good electrical conductivity characterized the produced fibers. Utilizing a silica capillary column that serves as a hydrothermal microreactor, the scientists created a sustainable way to constantly generate the fibers. The fibers have a significant specific volumetric capacity in sulphuric acid of 305 F cm,[23] or polyvinyl alcohol (PVA)/H_3PO_4 electrolyte of 300 F cm.[23] The researchers showed how fiber-based, all-solid-state micro-supercapacitors may charge an ultraviolet photodetector and an LED by being simply incorporated into miniature flexible devices. The micro-supercapacitors retain a power density more than two magnitudes greater than that of batteries and exhibited a volumetric energy density of ~6.3 mW h cm,[23] (similar to that of 4 V–500 mA h thin-film lithium batteries). They also possess a prolonged cycle life. Similarly, amorphous manganese silicate/multi-walled CNTs (AMSi/MWCNTs) composites were created and characterized by Wang *et al.*[85] who also discussed how to use them as electrode components for supercapacitors. The presence of MWCNTs was shown to affect the electrochemical efficiency of the composite, with the maximum specific capacitance (236 F g^{-1} at 0.5 A g^{-1}) being noticed at 40 wt% MWCNTs. Having a capacitance retention rate of 81% following 1000 cycles, an asymmetric supercapacitor made of AMSi/MWCNTs and activated carbon demonstrated specific capacitances of 146–99 F g^{-1} at current densities of 0.5–10 A g^{-1}. The findings point to the possibility of using AMSi/MWCNTs composites in high-performance energy storage systems.

Javed *et al.*[48] presented on making and characterizing a composite electrode substance for supercapacitors that consists of silica particles with amino and acid functionalities attached to MWCNTs. SEM, TEM, XRD, TGA, and FTIR were among the methods used to describe the composite structure. EIS, CV and galvanostatic charge–discharge were used to examine the electrochemical characteristics of the composite electrode. The findings demonstrated that the MWCNTs electrode outperformed the SiO_2/MWCNTs composite electrode in terms of specific capacitance, power density, and energy density, having a maximum specific capacitance of 128.4 F g^{-1} at a current rate of 2 A g^{-1}. At a current density of 2 A g^{-1}, the SiO_2/MWCNTs composite electrode's power density, which was highly reliant on current density, could approach up to 1000 W kg^{-1}. Therefore, the research points to the MWCNTs electrode as a promising contender for high-performance supercapacitor applications since it performs electrochemically superior to the SiO_2/MWCNTs composite electrode.

In order to develop Ni_3S_2/carbon nanotube composites for usage as cathode components in asymmetric supercapacitors, Dai *et al.*[86] investigated their synthesis and characterization. Utilizing a hydrothermal technique supported by glucose, Ni_3S_2 NPs having diameters varied from 10 to 80 nm were grown on the substrate of conductor MWCNTs to create the composites. The scientists discovered that following the composite electrode was activated by repeated cycles of CV in a 2 M KOH media, the Ni_3S_2 NPs accumulated on MWCNTs break down into narrower elements, growing their active surface region and improving the capacitive effectiveness of the composite electrode. The composite electrode's capacitive efficiency is greatly enhanced by the presence of MWCNTs and Ni_3S_2 NPs as additives, resulting in an impressive specific capacitance of 800 F g^{-1} and outstanding cycle stability even when subjected to a current density of 3.2 A g^{-1}. The scientists also created an asymmetric supercapacitor system with an anode of AC and a cathode of a composite of Ni_3S_2 and MWCNTs. It delivers a significant specific capacitance of 55.8 F g^{-1} at 1 A g^{-1} and may be controlled reversibly across 0 and 1.6 V. This results in a maximal energy density of 19.8 W h kg^{-1} at a power density of 798 W kg^{-1}. The asymmetric supercapacitor has outstanding stability as well; following performing 5000 rounds of galvanostatic discharge–charge at a current density of 4 A g^{-1}, the device still maintained 90% of its original capacitance.

5.4.3 BSNPs/Activated Carbon Composites

BSNPs/activated carbon composites are a type of composite material that combines biogenic silica nanoparticles (BSNPs) with activated carbon (AC). AC is an extremely permeable carbon material with a high surface area, while BSNPs are silica nanoparticles derived from biological sources such as diatoms and radiolarians. The combination of BSNPs and activated carbon can lead to the production of a composite material having enhanced characteristics for use in supercapacitors. Particularly, the enhanced surface region and the effective charge transmission across the BSNPs and activated carbon are to account for the composite's high specific capacitance. For usage in high-performance supercapacitors, BSNPs/activated carbon composites are hence intriguing materials. BSNPs are commonly mixed with an activated carbon solvent, accompanied with sonication and drying to create BSNPs/activated carbon composites. The properties of the BSNPs/activated carbon composite can be further enhanced by controlling the amount of BSNPs in the composite and optimizing the dispersion of the BSNPs in the activated carbon matrix. In addition to the high specific capacitance, the BSNPs/activated carbon composite also has good cycling stability owing to the robust interaction across the BSNPs and activated carbon. The strong interaction between the two materials can also enhance the mechanical stability of the composite material, making it more resistant to deformation and fracture. In addition, the utilization of activated carbon as the matrix component makes it simple to incorporate other functional components, like metal oxides, into the composite, which can improve its electrochemical capabilities.

The working principle of BSNPs/AC composites in supercapacitors are similar to that of BSNPs/graphene and BSNPs/CNTs composites. The composite material is designed to improve the capabilities of the supercapacitor by elevating its specific capacitance and improving its cycling stability. In a supercapacitor, the charge is retained at the electrolyte/electrode interface by separating positive and negative charges. The electrode component is typically consisting of a highly porous substance having a wide surface region, such as activated carbon. While a voltage is supplied to the electrodes, the charge is transferred between the electrolyte and the electrode material, leading to the deposition of charge at the electrolyte/electrode interface. BSNPs/AC composites are designed to enhance the specific capacitance by incorporating BSNPs into the AC matrix. The BSNPs can increase the surface region of the electrode substance and create additional charge retention sites, while the activated carbon provides an increased surface region and significant electrical conductance. The combination of these two materials can lead to a feasible improvement in the overall capacitance of the composite substance, improving the energy storage capacity of the supercapacitor. Moreover, the BSNPs/AC composite also has good cycling stability owing to the robust interaction across the BSNPs and AC. The interaction between the two materials can lead to an enhancement in the mechanical stability of the composite, making it more resistant to deformation and fracture during charge/discharge cycling. In an experiment, Xue *et al.*[87] addressed about utilizing rice husk as a naturally occurring silica template to efficiently synthesize AC electrode components. The research reveals that the silica template efficacy could be increased by selectively leaching the superfluous silica over the surface of the specimen utilizing NaOH solvent under mild circumstances. To expand the carbon structure, provide a more active surface area for hole creation, and increase the specific capacitance, the silica template that is present in the specimen could be maintained and used in combination using alkali activators. In-depth electrochemical characterization of the produced AC products revealed excellent specific capacitance and exceptional cycling stability. The research offers a practical method to speed up the commercial production of superior AC-based electrodes for cutting-edge supercapacitors. Guo *et al.*[59] discussed the hard template technique of producing nitrogen-enriched nanoporous carbon composites for supercapacitors. The hard template utilized by the researchers was made of SiNPs, and the precursors were CCL_4 and ethylenediamine. The nitrogen concentration was observed to grow with increased carbonization temperature, up to a threshold, after which it declined. The resultant mesoporous carbons displayed a great number of mesopores. Owing to its significant surface region (533 $m^2 g^{-1}$) and extremely high N concentration, the specimen carbonized at 700 °C (NC700) had the maximum capacitance of 306 $F g^{-1}$ (18.06%). With 96.6% capacitance retention following 10 000 cycles at a current density of 3 $A g^{-1}$, NC700 also displayed exceptional reliability. The scientists came to the conclusion that while the N element raised the pseudocapacitance and the wettability of the carbons, the elevation in specific surface regions enhanced the double-layer capacitance.

Liquid-phase plasma (LPP) was used by Lee *et al.*[88] to create silicon oxide-carbon composites (SiOCCs), which were then used as supercapacitor electrodes. Benzene and trimethoxyphenylsilane (TMPS) were carbonized in an aqueous solution utilizing LPP to create the SiOCCs. The SiOCCs created cluster chains that were primarily made of carbon and evenly dispersed oxygen and silicon. Following an elevation in silicon oxide concentration, the SiOCCs' specific capacitance and the specific surface region also improved. Furthermore, owing to silicon dioxide's density growth during the charge–discharge procedure, a larger silicon oxide concentration led to a greater capacitance deficit. Because of their strong cyclic stability, SiOCCs have the possibility to be used as supercapacitor electrodes. Furthermore, as an electrode component for supercapacitor purposes, Jaouadi *et al.*[89] demonstrated the improved electrocatalytic activity of AC generated from olive stones. To enhance the electrochemical and physicochemical features of the AC substrate, silica was used. According to the electrochemical studies, the constructed symmetric supercapacitor's specific capacitance rose dramatically following obtaining the silica coating, going from 11 $F g^{-1}$ to 141 $F g^{-1}$ at a current density of 0.5 $A g^{-1}$ whereas operating at a higher 1.25 V cell voltage. At a current density of 0.5 $A g^{-1}$, the AC-Si/AC-Si device displayed a specific power of 1661 $W kg^{-1}$ and a specific energy of 30.6 $W h kg^{-1}$.

According to the study's findings, silica-coated activated carbon made from olive stones is a viable electrode substance for supercapacitor devices.

5.4.4 BSNPs/Natural Composites

In addition to BSNPs, natural composites have also been explored for use in supercapacitors. Natural composites are materials that are composed of more than one component, where at least one component is derived from a natural source. Examples of natural composites that have been used in supercapacitors include cellulose-based composites, lignin-based composites, and chitin-based composites. The use of BSNPs and natural composites in supercapacitors has several advantages over traditional materials such as activated carbon. BSNPs and natural composites are renewable, biodegradable, and have a low environmental impact. They can also be synthesized using green methods that are cost-effective and environmentally friendly. Research has shown that BSNPs and natural composites could effectively enhance the efficacy of supercapacitors. For instance, BSNPs/cellulose composites have been shown to possess wide specific capacitance and excellent cycling stability. Similarly, chitin-based composites are shown to have high specific capacitance and good mechanical strength.

The working principle of biogenic silica nanoparticles (BSNPs) and their natural composites in supercapacitors is based on the electrostatic C_{dl} and pseudocapacitance mechanisms. The electrodes in a supercapacitor are typically constructed of a porous substance, like activated carbon or BSNPs, which offers a significant surface area for electrolyte ion adsorption. The electrolyte ions are drawn to the electrode surface and create an electrostatic double layer when a voltage is applied to the electrodes. As a result, energy is stored on the electrodes' surface in the form of an electrical charge. Some materials, including conducting polymers and transition metal oxides, can additionally experience pseudocapacitance in addition to the electrostatic double-layer capacitance. Pseudocapacitance is a phenomenon in which electrons are transferred across the electrode and the electrolyte as a result of redox processes occurring on the electrode surface. When compared to electrostatic double-layer capacitance, this technique can offer a better energy density. Depending on their characteristics and the nature of the electrolyte, BSNPs and their natural composites can exhibit both electrostatic double-layer capacitance and pseudocapacitance. Substantial surface areas and excellent electrical conductance of BSNPs and their natural composites can increase the electrostatic double-layer capacitance. Additionally, the existence of redox-active functional units in some natural composites, such as lignin-based composites, might cause them to display pseudocapacitance. When using acorn shells as a biomass precursor, Gopalakrishnan *et al.*[90] proposed an innovative way to create self-doped permeable carbon nanosheets having SiNPs imbedded in them. With a large surface region, a micropore–mesopore dispersion, and substantially graphitized sp^2 carbon molecules, the resultant activated carbon material exhibits these properties. The outstanding specific capacitance of the activated organic carbon (AOC) electrode—505 F g^{-1} at 1.6 A g^{-1} in three systems was greater than that of previously documented electrodes made from biomass. In comparison to conventional activated carbon, the symmetric supercapacitor constructed from the AOC electrode exhibits a noteworthy specific energy of 12.7 W h kg^{-1} at a specific power of 200 W kg^{-1}. It also has great rate capability and retains 96% of capacitance. Because of the electrode's huge surface region, ideal pores tortuosity for ion transit, excellent electrical conductance, and the existence of self-doped heteroatoms, it exhibits outstanding electrochemical efficiency. Ultimately, the scientists believe that this self-doped, adjustable permeable carbon substance with substantial specific energy could be used for commercial supercapacitors.

Wang *et al.*[61] researched the effectiveness of MnSi-C, a hybridized carbon substance made from naturally occurring bamboo leaves, as an electrode component for supercapacitors. Despite the usage of an extra regulating agent, manganese silicate was produced *in situ* using a straightforward hydrothermal process. When employed as a single electrode, the MnSi-C showed excellent capacitance retention of 85% following 10 000 cycles of charge–discharge and a capability of 162.2 F g^{-1} at a current density of 0.5 A g^{-1}. Asymmetric supercapacitors made of MnSi-C and $Ni(OH)_2$ had an energy density of 3 mW h cm^{-3} (24.6 W h kg^{-1}), a capacity of 438.5 mF cm^{-2} and a power density of 130.4 mW cm^{-3} (604.8 W kg^{-1}). Derived from empirical findings and accepted ideas, the article suggests a method for the *in situ* generation of manganese silicate upon the exterior of carbon. A potential electrode material for high-energy supercapacitors is MnSi-C.

5.5 Advantages and Disadvantages

5.5.1 Advantages

- High surface area: biogenic silica nanoparticles offer a wide surface area, which can increase the capacitance of supercapacitors.
- Biocompatibility: biogenic silica nanoparticles are biocompatible, making them safe for usage in biological systems and other beneficial sectors.
- Low toxicity: biogenic silica nanoparticles have low toxicity, making them environmentally friendly and safe for human health.
- Facile synthesis: biogenic silica nanoparticles can be synthesized easily and at a low cost, which makes them economically viable.
- Environmental friendliness: the synthesis of biogenic silica nanoparticles is environmentally friendly and does not utilize any hazardous chemicals.

5.5.2 Disadvantages

- Limited conductivity: biogenic silica nanoparticles have limited electrical conductivity, which can limit their use in certain supercapacitor applications.

- Limited mechanical strength: biogenic silica nanoparticles are fragile and can have limited mechanical strength, which can make them unsuitable for certain supercapacitor applications.
- Limited stability: biogenic silica nanoparticles may not be stable over long periods of time, which can limit their use in certain supercapacitor applications.
- Limited availability: biogenic silica nanoparticles are not yet widely available, which can limit their use in commercial supercapacitor applications.
- Limited compatibility: biogenic silica nanoparticles may not be compatible with certain supercapacitor electrolytes, which can limit their use in certain applications.

5.6 Challenges and Future Outlooks

The use of biogenic silica nanoparticles and their composites for supercapacitor applications has shown great promise. However, there are still challenges that need to be addressed before these materials can be widely used in commercial applications. One of the primary challenges is the limited conductivity of biogenic silica nanoparticles compared to other supercapacitor materials including carbon nanotubes and graphene. To overcome this challenge, researchers are exploring the use of conductive polymers and metals in combination with biogenic silica nanoparticles to improve conductivity. Another challenge is the limited mechanical strength of biogenic silica nanoparticles, which can make them unsuitable for certain supercapacitor applications that require high mechanical stability. To address this challenge, researchers are exploring the usage of composite materials, including metal oxide/biogenic silica nanoparticle composites, to enhance the mechanical stability of these substances.

In addition, biogenic silica nanoparticles may not be stable over long periods of time, which can limit their use in certain supercapacitor applications that require long-term stability. To overcome this challenge, researchers are exploring the use of surface modification techniques, such as functionalization with organic molecules or metal ions, to improve the stability of biogenic silica nanoparticles. Another challenge is the limited availability of biogenic silica nanoparticles, which can limit their use in commercial supercapacitor applications. To address this challenge, researchers are exploring the use of various biogenic sources, such as diatoms, sponges, and rice husks, to increase the availability of biogenic silica nanoparticles. Finally, biogenic silica nanoparticles may not be compatible with certain supercapacitor electrolytes, which can limit their use in certain applications. To address this challenge, researchers are exploring the use of different electrolytes and ionogels that are compatible with biogenic silica nanoparticles.

Despite these challenges, there are many potential future outlooks for biogenic silica nanoparticles and composites in supercapacitor applications.

One such outlook is the improved conductivity of these materials. Researchers are exploring various strategies to improve the conductivity of biogenic silica nanoparticles and composites, including the use of conductive polymers and metals, as well as doping with other elements such as nitrogen or carbon. These strategies could potentially enhance the performance of supercapacitors based on biogenic silica nanoparticles. Another potential outlook is the enhanced mechanical stability of biogenic silica nanoparticles. The use of composite materials, such as metal oxide/biogenic silica nanoparticle composites, can enhance the mechanical stability of these materials. Researchers are also exploring the use of other strategies such as crosslinking, to further improve the mechanical attributes of such substances. Furthermore, researchers are exploring ways to increase the stability of biogenic silica nanoparticles. Surface modification techniques such as functionalization with organic molecules or metal ions can improve the stability of biogenic silica nanoparticles. Furthermore, researchers are exploring other strategies such as the encapsulation of biogenic silica nanoparticles within polymeric matrices, which can provide additional protection against degradation and increase the long-term stability of these materials. Increasing the availability of biogenic silica nanoparticles is also a potential future outlook. The use of different biogenic sources, such as diatoms, sponges, and rice husks, can increase the availability of biogenic silica nanoparticles. Furthermore, researchers are exploring the use of alternative synthesis methods such as sol–gel, hydrothermal, and microwave-assisted synthesis, to increase the yield and scalability of these materials.

Finally, improving the compatibility of biogenic silica nanoparticles with different electrolytes and ionogels is another potential future outlook. Researchers are exploring the use of different electrolytes and ionogels that are compatible with biogenic silica nanoparticles, which could expand the range of applications for these materials.

5.7 Conclusion

In conclusion, biogenic silica nanoparticles and their composites have shown significant potential as an alternative substance for supercapacitor purposes. The novel characteristics of biogenic silica nanoparticles, including their wide surface region, low toxicity, and biodegradability, render them feasible for usage in various fields including energy storage. Although the challenges surrounding the utilization of BSNPs in supercapacitors such as limited conductivity and mechanical stability need to be addressed, researchers are exploring different strategies to overcome these challenges. Surface modification techniques, the use of conductive polymers and metals, and composite materials are some of the ways that researchers are working to enhance the functionality and stability of these substances. With ongoing research and development in this area, the potential future outlooks for biogenic silica nanoparticles and composites in supercapacitor applications are numerous. Improved conductivity, mechanical stability, and compatibility with different electrolytes and ionogels could lead to the widespread adoption of these materials in commercial applications in the future.

Additionally, biogenic silica nanoparticles and composites may find use in various other industries, including sensing, catalysis, and medication delivery. The versatility and unique properties of these materials make them an exciting area of research with many potential applications. Overall, biogenic silica nanoparticles and composites offer a significant area of research for the emergence of innovative energy storage solutions. Further research and development in this area could lead to the widespread adoption of these materials in various applications, contributing to the advancement of sustainable and renewable energy technologies.

References

1. W. Wang, S. Guo, I. Lee, K. Ahmed, J. Zhong, Z. Favors, F. Zaera, M. Ozkan and C. S. Ozkan, *Sci. Rep.*, 2014, **4**, 9–14.
2. X. Hou, T. Peng, J. Cheng, Q. Yu, R. Luo, Y. Lu, X. Liu, J. K. Kim, J. He and Y. Luo, *Nano Res.*, 2017, **10**, 2570–2583.
3. K. Li, Z. Hu, R. Zhao, J. Zhou, C. Jing, Q. Sun, J. Rao, K. Yao, B. Dong, X. Liu, H. Li, Y. Zhang and J. Ji, *J. Colloid Interface Sci.*, 2021, **603**, 799–809.
4. R. Oraon, A. De Adhikari, S. K. Tiwari and G. C. Nayak, *ACS Sustainable Chem. Eng.*, 2016, **4**, 1392–1403.
5. S. Meer, A. Kausar and T. Iqbal, *Fullerenes, Nanotubes Carbon Nanostruct.*, 2016, **24**, 507–519.
6. V. V. N. Obreja, *Phys. E*, 2008, **40**, 2596–2605.
7. I. Oh, M. Kim and J. Kim, *Appl. Surf. Sci.*, 2015, **328**, 222–228.
8. Y. J. Kang, H. Chung, C. H. Han and W. Kim, *Nanotechnology*, 2012, **23**, 065401.
9. Z. Lin, Z. Zeng, X. Gui, Z. Tang, M. Zou and A. Cao, *Adv. Energy Mater.*, 2016, **6**, 1600554.
10. X. Li, J. Zhu, L. Wang, W. Wu and Y. Fang, *Electrochim. Acta*, 2017, **258**, 291–301.
11. A. Thakur, A. Kumar, S. Sharma, R. Ganjoo and H. Assad, *Mater. Today: Proc.*, 2022, **66**, 609–621.
12. A. Thakur, S. Sharma, R. Ganjoo, H. Assad and A. Kumar, *J. Phys.: Conf. Ser.*, 2022, **2267**, 012079.
13. A. Thakur, S. Kaya, A. S. Abousalem and A. Kumar, *Sustainable Chem. Pharm.*, 2022, **29**, 100785.
14. A. Thakur and A. Kumar, *Sci. Total Environ.*, 2022, **834**, 155219.
15. A. Thakur, A. Kumar, S. Kaya, R. Marzouki, F. Zhang and L. Guo, *Coatings*, 2022, **12**, 1459.
16. A. Kumar and A. Thakur, *Encapsulated nanoparticles in organic polymers for corrosion inhibition*, Elsevier Inc., 2020.
17. A. Thakur, K. Savaş and A. Kumar, *Appl. Sci.*, 2023, **13**, 730.

18. A. Thakur, S. Kaya, A. S. Abousalem, S. Sharma, R. Ganjoo, H. Assad and A. Kumar, *Process Saf. Environ. Prot.*, 2022, **161**, 801–818.
19. S. Bashir, A. Thakur, H. Lgaz, I.-M. Chung and A. Kumar, *Arabian J. Sci. Eng.*, 2020, **45**, 4773–4783.
20. S. Bashir, A. Thakur, H. Lgaz, I. M. Chung and A. Kumar, *Surf. Interfaces*, 2020, **20**, 100542.
21. A. Thakur and A. Kumar, *Mater. Res. Express*, 2023, **10**, 034001.
22. S. Bashir, A. Thakur, H. Lgaz, I.-M. Chung and A. Kumar, *J. Mol. Liq.*, 2019, **293**, 111539.
23. A. Thakur, A. Kumar, S. Kaya, D. V. N. Vo and A. Sharma, *Fuel*, 2022, **312**, 122934.
24. Z. Sun, H. Qi, M. Chen, S. Guo, Z. Huang, S. Maganti, V. Murugadoss, M. Huang and Z. Guo, *Eng. Sci.*, 2022, **18**, 59–74.
25. M. Ates, A. A. Eker and B. Eker, *J. Adhes. Sci. Technol.*, 2017, **31**, 1977–1997.
26. U. Male, S. Uppugalla and P. Srinivasan, *J. Solid State Electrochem.*, 2015, **19**, 3381–3388.
27. J. Kim, M. Kim, S. Cho, C. M. Yoon, C. Lee, J. Ryu and J. Jang, *ChemNanoMat*, 2016, **2**, 236–241.
28. R. Amade, E. Jover, B. Caglar, T. Mutlu and E. Bertran, *J. Power Sources*, 2011, **196**, 5779–5783.
29. F. Su, X. Lv and M. Miao, *Small*, 2015, **11**, 854–861.
30. H. Gao, F. Xiao, C. B. Ching and H. Duan, *ACS Appl. Mater. Interfaces*, 2012, **4**, 7020–7026.
31. Y. Cheng, Y. Zhang and C. Meng, *ACS Appl. Energy Mater.*, 2019, **2**, 3830–3839.
32. X. Zhou, A. M. Cao, L. J. Wan and Y. G. Guo, *Nano Res.*, 2012, **5**, 845–853.
33. R. Mohammad-Rezaei and H. Razmi, *Int. J. Nanosci. Nanotechnol.*, 2016, **12**, 233–241.
34. M. Sarno, S. Galvagno, R. Piscitelli, S. Portofino and P. Ciambelli, *Ind. Eng. Chem. Res.*, 2016, **55**, 6025–6035.
35. Y. Cheng, Y. Zhang, Q. Wang and C. Meng, *Colloids Surf., A*, 2019, **562**, 93–100.
36. D. W. Kim, S. M. Jung and H. Y. Jung, *J. Mater. Chem. A*, 2020, **8**, 532–542.
37. Z. Yang, S. Chabi, Y. Xia and Y. Zhu, *Prog. Nat. Sci.: Mater. Int.*, 2015, **25**, 554–562.
38. S. Sun, S. Li, S. Wang, Y. Li, L. Han, H. Kong and P. Wang, *Mater. Lett.*, 2016, **182**, 23–26.
39. N. Kandhasamy, G. Ramalingam, G. Murugadoss, M. Rajesh Kumar, G. Manibalan, R. JothiRamalingam and H. M. Yadav, *J. Alloys Compd.*, 2021, **888**, 161489.
40. M. Fang, Z. Wang, X. Chen and S. Guan, *Appl. Surf. Sci.*, 2018, **436**, 345–353.
41. C. Sun, Y. Deng, L. Wan, X. Qin and G. Chen, *ACS Appl. Mater. Interfaces*, 2014, **6**, 11277–11285.
42. Q. Tang, X. Chen, D. Zhou and C. Liu, *Colloids Surf., A*, 2021, **620**, 126567.

43. D. Romero-Fierro, M. Bustamante-Torres, F. Bravo-Plascencia, H. Magaña and E. Bucio, *Polymers*, 2022, **14**, 1–23.
44. L. Wang, G. Duan, J. Zhu, S. M. Chen and X. Liu, *Electrochim. Acta*, 2016, **219**, 284–294.
45. C. Liu, A. Gao, F. Yi, D. Shu, H. Yi, X. Zhou, J. Hao, C. He and Z. Zhu, *Electrochim. Acta*, 2019, **326**, 134965.
46. C. Sun, M. Ma, J. Yang, Y. Zhang, P. Chen, W. Huang and X. Dong, *Sci. Rep.*, 2014, **4**, 1–6.
47. H. Yue, S. Wang, Z. Yang, Q. Li, S. Lin and D. He, *Electrochim. Acta*, 2015, **174**, 688–695.
48. M. Javed, S. M. Abbas, S. Hussain, M. Siddiq, D. Han and L. Niu, *Mater. Sci. Energy Technol.*, 2018, **1**, 70–76.
49. R. T. Woodward, F. Markoulidis, F. De Luca, D. B. Anthony, D. Malko, T. O. McDonald, M. S. P. Shaffer and A. Bismarck, *J. Mater. Chem. A*, 2018, **6**, 1840–1849.
50. M. Liu, M. Shi, W. Lu, D. Zhu, L. Li and L. Gan, *Chem. Eng. J.*, 2017, **313**, 518–526.
51. S. Y. Kim and B. H. Kim, *J. Power Sources*, 2016, **328**, 219–227.
52. Y. Huang, M. Zhong, Y. Huang, M. Zhu, Z. Pei, Z. Wang, Q. Xue, X. Xie and C. Zhi, *Nat. Commun.*, 2015, **6**, 10310.
53. H. Sai, J. Zhang, Z. Jin, R. Fu, M. Wang, Y. Wang, Y. Wang and L. Ma, *Polymers*, 2020, **12**, 807.
54. A. Alrahlah, R. Khan, A. B. Al-Odayni, W. S. Saeed, L. S. Bautista and F. Vohra, *Polymers*, 2020, **12**, 1–15.
55. R. Liang, Y. Du, P. Xiao, J. Cheng, S. Yuan, Y. Chen, J. Yuan and J. Chen, *Nanomaterials*, 2021, 11.
56. H. J. Liu, X. M. Wang, W. J. Cui, Y. Q. Dou, D. Y. Zhao and Y. Y. Xia, *J. Mater. Chem.*, 2010, **20**, 4223–4230.
57. D. Kumar and S. A. Hashmi, *J. Power Sources*, 2010, **195**, 5101–5108.
58. D. W. Kim, S. M. Jung and H. Y. Jung, *J. Power Sources*, 2020, **448**, 227422.
59. D. Guo, J. Qian, R. Xin, Z. Zhang, W. Jiang, G. Hu and M. Fan, *J. Colloid Interface Sci.*, 2019, **538**, 199–208.
60. M. Javed, S. M. Abbas, M. Siddiq, D. Han and L. Niu, *J. Phys. Chem. Solids*, 2018, **113**, 220–228.
61. Q. Wang, Y. Zhang, H. Jiang and C. Meng, *J. Colloid Interface Sci.*, 2019, **534**, 142–155.
62. Q. Liu, O. Nayfeh, M. H. Nayfeh and S. T. Yau, *Nano Energy*, 2013, **2**, 133–137.
63. X. L. Guo, M. Kuang, F. Li, X. Y. Liu, Y. X. Zhang, F. Dong and D. Losic, *Electrochim. Acta*, 2016, **190**, 159–167.
64. M. Y. Chong, A. Numan, C. W. Liew, H. M. Ng, K. Ramesh and S. Ramesh, *J. Phys. Chem. Solids*, 2018, **117**, 194–203.
65. J. Li and G. X. Xu, *Int. Eye Sci.*, 2017, **17**, 1655–1668.
66. B. Sriram, A. Sathiyan, S. F. Wang, E. Elanthamilan, X. B. Joseph, J. N. Baby, J. P. Merlin and J. C. Ezhilarasi, *J. Solid State Chem.*, 2021, **295**, 121931.

67. M. Hwang, J. S. Jeong, J. C. Lee, S. Yu, H. S. Jung, B. S. Cho and K. Y. Kim, *Korean J. Chem. Eng.*, 2021, **38**, 454–460.
68. P. Qin, Y. Yang, X. Zhang, J. Niu, H. Yang, S. Tian, J. Zhu and M. Lu, *Nanomaterials*, 2017, **8**, 4.
69. H. Lee, S. H. Park, S. J. Kim, Y. K. Park, B. J. Kim, K. H. An, S. J. Ki and S. C. Jung, *Int. J. Hydrogen Energy*, 2015, **40**, 754–759.
70. H. W. Jung, L. Hamenu, H. S. Lee, M. Latifatu, K. M. Kim and J. M. Ko, *Curr. Appl. Phys.*, 2015, **15**, 567–570.
71. Y. Li and X. Liu, *Mater. Chem. Phys.*, 2014, **148**, 380–386.
72. J. Xu, F. Liu, X. Peng, J. Li, Y. Yang, D. Jin, H. Jin, X. Wang and B. Hong, *ChemistrySelect*, 2017, **2**, 5189–5195.
73. C. Y. Bon, L. Mohammed, S. Kim, M. Manasi, P. Isheunesu, K. S. Lee and J. M. Ko, *J. Ind. Eng. Chem.*, 2018, **68**, 173–179.
74. A. Javaid, K. K. C. Ho, A. Bismarck, J. H. G. Steinke, M. S. P. Shaffer and E. S. Greenhalgh, *J. Compos. Mater.*, 2018, **52**, 3085–3097.
75. A. Borenstein, O. Hanna, R. Attias, S. Luski, T. Brousse and D. Aurbach, *J. Mater. Chem. A*, 2017, **5**, 12653–12672.
76. S. Chandra Sekhar, J. H. Lee, E. B. Cho and J. S. Yu, *J. Mater. Res. Technol.*, 2021, **13**, 1899–1907.
77. K. A. Ali and E. A. Ragab, *Appl. Chem. Eng.*, 2023, **6**, 12–19.
78. S. Sekar, T. A. Ahmed, A. I. Inamdar, Y. Lee, H. Im, D. Y. Kim and S. Lee, *Nanomaterials*, 2019, **9**, 1055.
79. H. Li, Y. Zhao, S. Liu, P. Li, D. Yuan and C. He, *Microporous Mesoporous Mater.*, 2020, **297**, 109960.
80. Q. Du, J. Sun, Y. Li, X. Yang, X. Wang, Z. Wang and L. Xia, *Chem. Eng. J.*, 2014, **245**, 99–106.
81. R. G. Shrestha, S. Maji, L. K. Shrestha and K. Ariga, *Nanomaterials*, 2020, **10**, 1–27.
82. X. Bo and L. Guo, *Electrochim. Acta*, 2013, **90**, 283–290.
83. M. Rabia, D. Essam, F. H. Alkallas, M. Shaban, S. Elaissi and A. Ben Gouider Trabelsi, *Micromachines*, 2022, **13**, 2234.
84. D. Yu, K. Goh, H. Wang, L. Wei, W. Jiang, Q. Zhang, L. Dai and Y. Chen, *Nat. Nanotechnol.*, 2014, **9**, 555–562.
85. Q. Wang, Y. Zhang, S. Jia, Y. Han, J. Xu, F. Li and C. Meng, *Colloids Surf., A*, 2018, **548**, 158–171.
86. C. S. Dai, P. Y. Chien, J. Y. Lin, S. W. Chou, W. K. Wu, P. H. Li, K. Y. Wu and T. W. Lin, *ACS Appl. Mater. Interfaces*, 2013, **5**, 12168–12174.
87. B. Xue, L. Jin, Z. Chen, Y. Zhu, Z. Wang, X. Liu and X. Wang, *J. Alloys Compd.*, 2019, **789**, 777–784.
88. H. Lee, I. S. Park, H. J. Bang, Y. K. Park, E. B. Cho, B. J. Kim and S. C. Jung, *Appl. Surf. Sci.*, 2019, **481**, 625–631.
89. M. Jaouadi, M. Marzouki, A. H. Hamzaoui and O. Ghodbane, *J. Appl. Electrochem.*, 2022, **52**, 125–137.
90. A. Gopalakrishnan, A. Yu and S. Badhulika, *Energy Fuels*, 2020, **34**, 11508–11518.

CHAPTER 6

Biogenic Silica for Biosensors and Bioimaging Applications

FULDEN ULUCAN-KARNAK,*[a] SACHIN MISHRA,[b] HIRESH MORADI[c] AND CANSU İLKE KURU[d,e]

[a] Ege University, Institute of Health Sciences, Medical Biochemistry Department, Izmir, Turkey; [b] Department of Electrical Electronic Engineering and Automation, Rovira I Virgili University, Spain; [c] Department of Environmental Engineering, Kwangwoon University, 01897, South Korea; [d] Graduate School of Natural and Applied Sciences, Department of Biotechnology, Ege University, Izmir, Turkey; [e] Buca Municipality Kızılçullu Science and Art Center, Izmir, Turkey
*Email: ulucanfulden@gmail.com

6.1 Introduction

The search for novel materials capable of revolutionising biosensors and bioimaging applications in modern biotechnology is never-ending.

Biogenic silica, also known as biogenic amorphous silica or biosilica, refers to silica (SiO_2) that is produced by living organisms, derived from the skeletons of diatoms and other silica-producing organisms, plays a crucial role in biosensors due to its unique properties. Biogenic silica emerges as a key material in biosensors,[1,2] which facilitates the development of extremely sensitive and selective detection systems. These structures possess a high surface area, excellent biocompatibility, and customizable surface chemistry, making them ideal candidates for immobilizing biomolecules like enzymes or antibodies. In biosensors, biogenic silica acts as a stable matrix for the attachment of biorecognition elements, facilitating specific molecular interactions with target analytes. This immobilization enhances the

Biogenic Silica: Fundamentals and Applications
Edited by Ashok Kumar Nadda and Vijai Kumar Gupta

Published by the Royal Society of Chemistry, www.rsc.org

sensor's sensitivity,[3,4] selectivity, and stability, thus improving its overall performance in detecting and quantifying various biological and chemical substances. Because biogenic silica structures are porous, there are many chances to immobilize biomolecules on their surface, such as enzymes, antibodies,[5–7] or nucleic acids, which allows for strong and precise analyte detection. Additionally, the porous nature of biogenic silica allows for the encapsulation of sensitive materials, protecting them from environmental interference and prolonging their functionality, making it a versatile and indispensable component in biosensor technology.

As research continues to uncover the limitless potential of biogenic silica, the range of biosensor and bioimaging[8,9] applications grows tremendously. From early illness diagnosis to customized medicine and environmental monitoring, biogenic silica has the potential to be a game changer,[9,10] bridging the gap between biology and technology with unparalleled efficacy and accuracy. Furthermore, biogenic silica nanoparticles outstanding surface area-to-volume ratio boosts signal transduction processes,[10,11] improving sensor performance[12–14] and requiring less sample, which is a significant benefit for environmental monitoring and point-of-care diagnostics. This introduction just touches the surface of the huge environment that awaits investigation, where innovation meets nature-inspired design to reshape the boundaries of biomedical engineering[15,16] and life sciences.

6.2 Synthesis and Modification of Biogenic Silica Nanoparticles

Biogenic silica is an excellent alternative to synthetic silica because of its structure, density, and composition. The naturally available agricultural bioresources materials like rice husk (RH), sugarcane bagasse, *etc.* is considered for non-metallic bio-precursors for biogenic silica nanoparticle synthesis.[17,18] Biogenic silica nanoparticles have been prepared using a number of methods, including sol–gel processes,[19,20] microwave hydrothermal processes,[21,22] flame synthesis,[23,24] and combustion synthesis.[25,26]

The sol–gel process is also known as *wet chemical method* to synthesize biogenic silica nanoparticles and is commonly based on the Stöber method that provides surfactant free nanoparticles.[27,28] Many parameters could affect the sol–gel process such as the ratio of alkoxysilane precursors to water, reaction mixture, pH, temperature, mixing speed and time. $pH = 7$ value is critical for the sol–gel formation. When $pH < 7$ chains were formed less tightly, $pH > 7$ allow gelation. Gel formation can be derived as xerogel, aerogel, nanospheres, fibers or coating form according to the pH changing.[29] In the sol–gel method, extraction from agricultural wastes and converted into the sodium silicate solution as mesoporous silica nanoparticles (MSNs) precursors.[30] In a study of Porrang *et al.* 2021, MSNs were prepared by adding sodium silicate from the rice husk and wheat husk with the sol–gel process to carry Doxorubicin (Dox), as a model anticancer drug.

The hydrophilicity, particle size and pore size could vary depending on the addition of the high concentration precursor in the reaction mixture.[31]

There is another approach to synthesize the biogenic silica nanoparticle using a microwave-assisted (MW) process, and has more satisfying advantages than conventional heating methods.[32] The microwave-assisted process could be chosen as an alternative to provide high value products from biological sources. Effective pretreatment may be carried out with a milling process in order to decrease silica particle sizes.[33] Using this technique in the leaching step could propose surface area, pore volume and distribution, purity improvement and adsorption capacity enhancing.[34]

The flame synthesis process can be regarded as a high temperature method that has two main strategies for injecting the used precursor in the flame: (i) in homogenous phase, by an evaporation system (FVS: Flame Vapor Synthesis) (ii) in a liquid phase (FSP: Flame Spray Pyrolysis) with a spray system. FSP provides flexibility for many types of precursors in order to use as flow control of the amount of produced material.[35] The chemical transition from precursor to oxide, the nucleation of oxide vapors, the collisions of oxide particles, the condensation of oxide vapor onto oxide particles, and the sintering of oxide particles are all critical stages in FVS.[36] The leftover remnant particles may be somewhat big depending on the parameters of the liquid precursor, such as volatility, concentration, and decomposition temperature.[37] The Flame Spray Pyrolysis process (FSP) has the benefit of allowing the use of a wide range of potentially low-cost precursors (mostly metal oxides such as TiO_2, Al_2O_3), resulting in a product with high purity and a reasonably narrow size distribution.[38]

The combustion process is chosen to provide intended properties of biogenic silica nanoparticles in terms of high quality with regards to pore size, high specific surface area and amorphous structure. In this procedure, the temperature of the raw material is changed in order to degrade organic components in either an oxygen or air environment (combustion, $\lambda \geq 1$), an inert atmosphere such as nitrogen or argon (pyrolysis, $\lambda = 0$), or stoichiometric proportions of air, oxygen, or CO_2 (gasification, $0 < \lambda < 1$). In terms of carbon content, increasing the combustion temperature (CT) and residence time (RT) enhances the purity of the biogenic silica. Quality parameters can be adjusted with the heating rate.[26] It is not possible to scale up such time-consuming consecutive combustion processes to industrial or bench-scales. A consecutive combustion process may also be built by pyrolyzing RH and then oxidizing it (combustion).[39]

The most effective synthesis method should be chosen according to yield, time, cost, experienced personnel requirements and intended use of synthesized nanoparticles.

6.3 Characteristics of Biogenic Silica Nanoparticles

Silica nanoparticles (SiO_2 NPs) have grown popular as nanostructuring, drug delivery, and optical imaging agents among other types of nanoparticles.

Silica nanoparticles are extremely stable and have a low toxicity. Because of their perfectly tuneable macroscopic form, chemical activity, and mesoporous structure, mesoporous silica materials with pore diameters ranging from 2 to 50 nm have gotten a lot of interest.[40,41] There are three main types of silica (silicon dioxide) as (1) crystalline silica, (2) amorphous silica and (3) synthetic amorphous silica (SAS) in many forms. Silica nanoparticles come in a variety of forms, including fumed silica, colloidal silica, silica gel, and silica aerogel, and can be made in a variety of ways. Different forms of silicas have different physical and chemical properties. Generally, SAS has a leaning to supersaturate and surface-treated hydrophobic SAS has lower solubility rather than its hydrophilic forms. In contrast to crystalline silica, SAS slowly dissolves in aqueous environments and body fluids.[17,42] The surface of these nanoparticles is rich with silanol groups that is very useful anchors for surface modifications. High surface area, high pore volume, stable mesoporous structure, adjustable pore diameter, adjustable particle size, and simple internal and exterior surface functionalization are some of the distinctive and useful structural properties of biogenic silica nanoparticles.[31,43] Pretreatment stragies such as heat treatment at high temperature, mechanical treatment and residual carbon, K^+ significantly affects the characteristics of the biogenic silica.[44,45]

Silica nanoparticles can be used as a probe in biomedical imaging applications. For these types of applications, they should have some characteristic such as:

(i) well-defined and tunable surface chemistry,
(ii) being a metal oxide material which could be chosen as a host material or encapsulation material for other organic or inorganic materials,
(iii) transparent to light and capable of doping conventional fluorescent dyes or QDs,
(iv) inert in the bioenvironment,
(v) biocompatible,
(vi) inexpensive, highly stable chemically, physically against microbial attack.[46]

Silica and biogenic silica have not only been used in biomedical applications, they have also been used for environmental pollution remediation, such as enhanced oil recovery to limit the release of brine, heavy metals, and radioactive chemicals into water, metals, non-metals, and radioactive materials removal, and water purification.[40,47] Again, because of its varied structure, density, and content, biogenic silica is a great alternative to synthetic silica.[17]

Plant husk (rice husk, rice raw, cornhusk, corncob, cassava peelings, coconut husks, yam peelings), is a cost-effective and non-metallic bio-precursor for biogenic silica nanoparticle production among the available agricultural bioresources.[47,48] Because of their biocompatibility and bioactivity, silica-based NPs are a diverse toolbox for biomedical applications. Cancer has been imaged, targeted, and treated using functionalized SiO_2

NPs.[17] The toxicity of biogenic silica that is derived from plant husk was investigated by So *et al.* 2008. According to the *in vivo* results, it is stated that "the nano-sized silica particle might have a toxic effect on the liver" even though there was no difference on health parameters after feeding a total amount of 140 g silica kg^{-1} mouse.[49] In another study of Sharma *et al.* 2019, biogenic silica nanopowder synthesized from brown rice husk ash (RHA) which had the smallest particle size, highest surface area and high porosity had the most superior bacterial compatibility.[50]

6.4 Applications of Biogenic Silica Nanoparticles

Silica nanoparticles have been widely used in theranostic, imaging, drug monitoring, drug delivery and biosensing applications in medicine.[51,52] Also they have been used in chemical, biotechnology, agriculture, environmental remediation and even wastewater purification.[53,54] In this chapter, we focused on applications of silica nanoparticles in biosensing and bioimaging fields.

6.4.1 Application of Biogenic Silica Nanoparticles in Biosensing

Physical properties of nanosilica particles alongside its biocompatibility have attracted the attention of researchers involved in biosensing to use this hydrophilic particulate in biosensors. The two key elements of a biosensor are biosensing element (*e.g.*, receptor, aptamer) and transducer. As a result, a biosensor can be classified based on sensing procedure or transducing procedure (*e.g.*, optical, electrochemical, mass sensitive, thermal).[55] In biosensors design the parameters like the interaction of the sensing part and transducer with the analyte, removing matrix effect, facilitated transfer of receptor response to the transducer and durability are of great importance. These parameters can affect the performance, detection limit and applicability of the sensors in biological measurements. The physical and chemical features of biogenic silica nanoparticles and their effect on the biosensing efficiency can be as follows:

6.4.1.1 High Specific Surface Area

Specific surface area (SSA) is defined as the surface area of a solid particle divided by the mass of the solid particle. The SSA (expressed in $m^2 g^{-1}$) as well as the surface to volume ratio (expressed in $nm^2 : nm^3$) of a particle is a function of porosity, pore size distribution, shape, size, and roughness. Different layer and pore distribution can be seen from Figure 6.1.[56]

SSA of a material, especially materials in the nanoscale, has a remarkable effect on the physicochemical properties of the material.[57] In fact, SSA determines the spatial and temporal transfer rate of matter and energy within an object and between an object and its surrounding environment. Since the

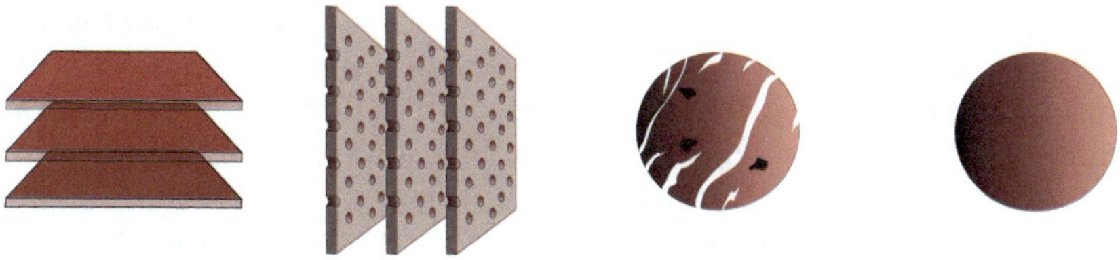

Figure 6.1 Different layer and pore distrubition of silica nanoparticles.

biosensing procedure is entangled with energy or material displacement, the selective and high transfer rate in the sensing part and transducer would result in more sensitivity and less measuring time. Among various nanostructured materials, the porous structure possessing a huge specific surface, is available to sense the probe-target interaction.[58–60] Although a great deal of recent studies in nanomaterials synthesis dedicated to achieving the high level of porosity and arrangement in nanostructured materials, preparation of desired structure is still demanding in case of instrumentation and cost. A feasible alternative which has received considerable attention in last decade, is to employ the nanostructure material which is achievable from natural resources.

Silica nanoparticles can be found in different sizes (10 to 500 nm), shape (*e.g.*, spherical, rod shape, layered structure), porosity (microporous with pore diameter less than 2 nm, mesoporous with pore diameter ranging from 2–50 nm, macroporous with pore diameter higher than 50 nm) and crystallinity (*e.g.*, monocrystalline, polycrystalline, and amorphous) in flora and fauna. Shape and crystallinity of the biogenic silica highly depends on the function of silica-based structure in the host plant or animal. For instance, the crystalline silica forms a solid structure which can serve as a frame to keep the shape of the plant parts (*e.g.*, horsetail and bamboo).[61–63] There are some organisms that have silicon dioxide-based structure in their body. The most well-known species are diatoms (the only algae with cell walls composed of transparent, opaline silica), radiolarians (a type of plankton), and hexacinellae (glass sponges). The nanostructured silica skeleton not only protects the living organism mechanically, but it also functions as an engineered structure to make a connection with the surrounding environment.[64] Having a high surface area, despite the source, silica nanoparticles have been employed in biosensing with two main approaches;

(i) Crystalline silica without optical, magnetic, or electrochemical activity has been employed for encapsulation or immobilization of sensing compounds on the porous surface.
(ii) Amorphous silica without optical, magnetic, or electrochemical activity allows the detection of an optical response of sensing molecules incorporated with silica nanoparticles.

High specific surface area has been used to incorporate sensitive and selective molecules on the surface of nanoparticles. Incorporation is usually achieved by the interaction between the specific molecules and the surface of the pore channels through covalent bonding and non-covalent bindings such as electrostatic interaction, hydrogen bonding, π–π stacking among others.[65] By immobilization or deposition of a higher amount of sensing compound due to the high available surface incorporation, the response signal which is produced by the sensing part or transducer will be amplified. As an outcome, the trace concentration of the analyte will be detected due to selectively amplified response in the sensing part or transducer. In this regard, for example, deposition of silver nanoparticles on the surface of diatoms silica skeleton resulted in a noticeable increase in detection efficiency. In this study, due to higher density of deposited silver nanoparticles on the surface, the detection limit of methanol *via* surface-enhanced Raman scattering (SERS) has been enhanced from 10^{-7} M (for deposition of Ag NPs on the glass slide) to 10^{-12} M by using diatoms skeleton as a substrate for silver deposition.[66] Furthermore, Kaminska *et al.* 2017 used diatoms derived silica as a substrate to use in the deposition of gold nanoparticles for detection of interleukins in blood plasma. Employing the SERS technique alongside a high density of gold nanoparticles on the surface, the detection limit of interleukin 8 enhanced from nanograms per mL to picograms per mL.[67]

6.4.1.2 *Sieving Effect*

Detecting and measuring of the target chemicals, especially dilute targets in biological samples, have been challenging tasks in biosensor design and application. In other words, the "matrix effect" in masking (*e.g.*, physically or chemically) the target analyte as well as high levels of noise that may occur due to the existence of non-target components, that are more abundant than the target analyte, require new techniques of detection. Nowadays, nanomaterials have been used in biosensors with two main objectives: first, facilitate the capturing process of target analyte; second, enhance the resulting signal that generated when the target analyte is detected by the sensing part. A remedy to decrease the noise from background components (differentiation between analyte and matrix components) is separation of the target analyte from the matrix. In this approach, depicted in Figure 6.2, the constituents of the matrix are excluded based on the size and physiochemical interaction of the components with the membrane like structure installed at the sensing part of the biosensors.[68,69]

Using a nanometric silica structure with nanopores (diameter = 3 nm, $h = 7$ nm), small nucleic acids (22-bp RNA and 25-bp DNA) discriminated by the biosensor benefited from the size exclusion of nanostructural membranes.[68] Synthetic highly permeable nanopore membranes are very thin with dimensions around 50 nm. Lacking mechanical properties, the application of these fragile membranes is limited.[70] However, biogenic silica materials possessing highly ordered porous hierarchical nanostructures,

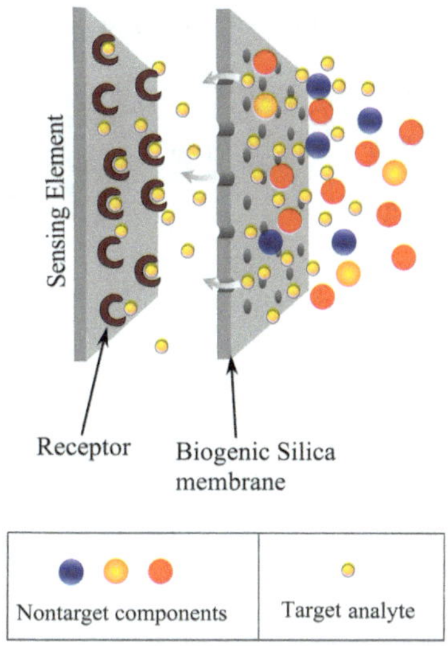

Figure 6.2 Sieving effect on silica nanoparticles.

show required mechanical stability. Thin nanopore membranes with short pores of a length of around 50 nm offer an advantage with respect to temporal and spatial resolution of particle and molecule translocation.[58,68] Enhanced diffusion of fluids in nano channels of biogenic silica has been employed to fabricate the highly sensitive biosensor for detection of two inflammatory proteins: C-reactive protein (CRP) and myeloperoxidase (MPO). In this study micro- and nanopores of *Coscinodiscus wailesii* (the largest diatom species found in the north) have been employed to form nanowells on the streptavidin-modified gold surface. Comparing biogenic silica with an alumina based sensor, a 100-fold increase in sensitivity is achieved by using a nanoporous bio-based silica nanostructure in electrochemical detection of cardiovascular biomarker proteins.[71]

6.4.1.3 *Optical Activity*

Dimension, geometry, crystallinity, and porosity are among the principal determining factors in the application of the silica nanoparticles in biosensing (Figure 6.3). The optical properties of silica particulate either natural or synthetic is highly entangled with its size, shape, and crystallinity. Silica nanoparticles, like other nanomaterials, can be categorized as 0D (all dimensions are in nanoscale, silica nanoparticles and core–shell), 1D (one dimension is not in nanoscale, like silica nanorods), 2D (only one dimension is in nanoscale, like silica nanosheets), and 3D (all dimensions are beyond 100 nm, like silica nanostructured skeleton of diatoms) based on their dimensionality.[72] Size dependence of optical properties of nanoparticles are due to two distinct phenomena: surface plasmons resonance (in metal nanoparticles) and increased energy level spacing due to the confinement of the

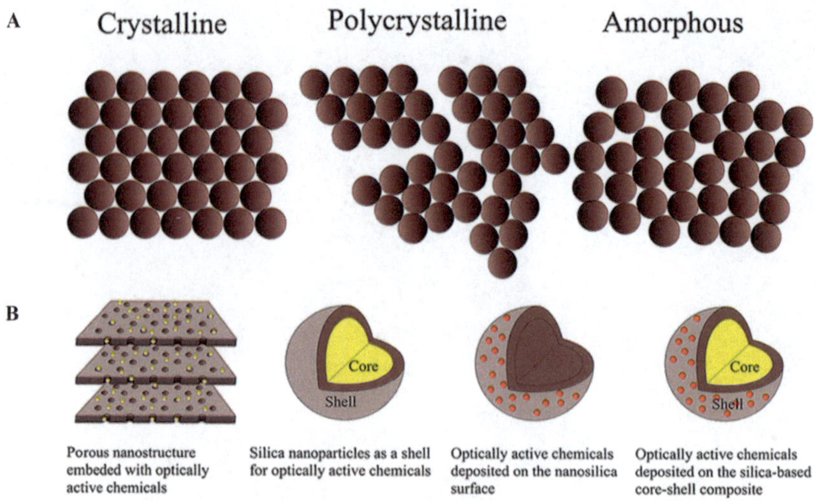

Figure 6.3 (A) Cristallinity of silica nanoparticles, (B) porosity effects on optically active silica nanoparticles.

delocalized energy states (most prevalent phenomenon in semiconductors).[72,73] Although crystalline silica nanostructure does not show light emittance,[58,74] it has been widely used as a substrate for immobilization of metals and organic compounds that can show optical response like photoluminescence.[75,76] Porous nanosilica and more especially amorphous silica beside its inherent optical properties has been adopted in preparing sensors as well as biosensors by incorporation of a required chemical and biochemical with amorphous silica nanoparticles and nanostructures.[77–79]

Recent studies on the optical properties of the nanostructured silica derived from marine species suggest that light propagation through this type of 3D structure not only depends on the interaction of silica with light but also it likely depends on the spatial distribution of the pores in the periodic layered structure. As a result, nanostructured biogenic silica can show different levels of light trapping and focusing properties regarding its nature, shape, and complexity. These features make biogenic silica an apt candidate to be employed in biosensing and bioimaging as a multifocal lens.[80–82]

6.4.1.4 *Surface Chemistry and Biocompatibility*

Composed of silicon and oxygen atoms solely, there are two main functional groups, silanol groups and siloxane groups at the surface of silica nanoparticles (Figure 6.4). The silanol groups are inherently hydrophilic and make silica particulates not only compatible with aqueous medium, but also more prone to be functionalized with numerous functional groups. The siloxane groups, on the other hand, are chemically inert and bring about biocompatibility, some degree of hydrophobicity, pH, heat, and mechanical resistance to the silicon dioxide nanoparticles. Biocompatibility and

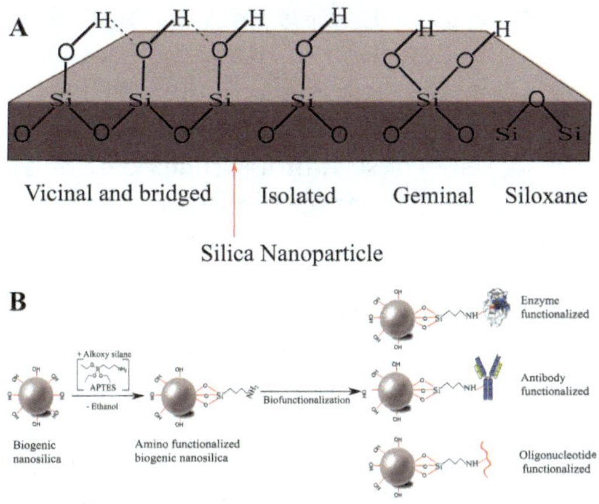

Figure 6.4 Biofunctionalized silica nanoparticles with an enzyme, antibody or oligonucleotide.

biodegradation of amorphous silica-based material used in bio applications has been investigated in recent years.[83,84] Surface silanol groups can be found in three different forms (a) isolated, (b) vicinal and bridged and (c) geminal form.[85]

The surface density of each silanol group will be different regarding the nature of the silica nanoparticles. Functionalization methods like layer-by-layer assembly, physical adsorption, and silanization have been employed to introduce various functional groups such as amino, carboxyl, thiol, *etc.* on the silica nanoparticles surface. Among these techniques, the silane functionalization method is widely used for silica surface modification. In this method the required function group is functionalized on the surface due to the covalent bond formation between the surface silanol groups and alkoxysilane group.[86–89] Molecular probes like enzymes, proteins, antibodies, DNA strands among others can be grafted to the surface of silica nanoparticles *via* further reaction of the surface functional groups with these probes. In a recent study, the surface of the silica nanoparticles has been functionalized with 3-mercaptopropyl trimethoxysilane and the concentration of L-tyrosine in a blood plasma sample was measured by differential pulse voltammetry.[90] The biosensor showed a linear response in the concentration range of 0.05–600 μM and a low detection limit about 0.02 μM because of the co-catalytic effect of enzyme and nanoparticles.

Here are some key applications of biogenic silica nanoparticles in biosensing aspects such as:

Biosensors for environmental monitoring: Biogenic silica nanoparticles have been utilized in the development of biosensors for detecting environmental pollutants such as heavy metals, pesticides, and toxins. Functionalized silica nanoparticles can enhance the sensitivity and selectivity of biosensors for rapid and efficient detection. For example, a study by Zheng *et al.* 2019 demonstrated the use of diatom-derived silica nanoparticles for the detection of heavy metals in water samples.[91]

Biomedical diagnostics: Biogenic silica nanoparticles have shown promise in biomedical diagnostics for the detection of biomarkers associated with various diseases. These nanoparticles can be functionalized with specific ligands to capture target biomolecules, enabling sensitive and selective detection. Li *et al.* 2017 reported the development of a biosensor based on diatomite-derived silica nanoparticles for the detection of prostate-specific antigen (PSA), a biomarker for prostate cancer.[92]

Food safety monitoring: Biogenic silica nanoparticles have been employed in biosensors for monitoring food safety by detecting pathogens, toxins, and contaminants in food samples. Surface-functionalized silica nanoparticles can provide high binding affinity towards target analytes, enabling rapid and specific detection. For instance, a study by Pang *et al.* 2018 utilized diatom-derived silica nanoparticles for the detection of Salmonella bacteria in food samples.[93]

Point-of-care diagnostics: Biogenic silica nanoparticles hold promise for point-of-care diagnostics due to their compatibility with portable and miniaturized biosensing platforms. These nanoparticles can be integrated into microfluidic devices or paper-based assays for on-site detection of target analytes. Guo *et al.* 2016 developed a point-of-care biosensor using diatom-derived silica nanoparticles for the detection of glucose levels in human saliva.[94]

Nucleic acid sensing: Biogenic silica nanoparticles have been explored for nucleic acid sensing applications, including the detection of DNA sequences and RNA molecules. Surface-functionalized silica nanoparticles can serve as efficient probes for hybridization-based detection of nucleic acids with high sensitivity and specificity. Zhang *et al.* 2018 demonstrated the use of diatomaceous earth-derived silica nanoparticles for DNA biosensing applications.[95]

6.4.2 Application of Nanosilica in Bioimaging

Bioimaging is an invasive process to visualize biological activity in real time. Moreover, it can be employed to achieve the information on the three-dimensional structure of the target specimen from the outside without physical interference.[96] Due to recent development in imaging, various methods like optical imaging (*e.g.*, fluorescence and bioluminescence), magnetic resonance imaging (MRI), computed tomography (CT), Raman imaging, *etc.* have been used in bioimaging. In bioimaging, presence and density of the traced analyte are recorded through the response of the analyte to the imaging technique. As a result, shape, activity, or concentration of the selected specimen (containing the analyte) can be visualized. Regarding to the recording method, biocompatibility, size, shape and surface properties of the amorphous biogenic silica nanoparticles and nanostructures can be utilized to enhance the sensitivity and reliability of the bioimaing results.[97–99] Generally, in these studies' magnetic properties of a metal (*e.g.*, europium, terbium) or photoluminescence of a dye (*e.g.*, rhodamine 101) or combination of these two types of properties have been employed by grafting, deposition, encapsulation or coprecipitation of these optically active elements with silica-based nanoparticles. Silicon nanocrystals show optical activity, and the photoluminescence (PL) quantum yields of these nanocrystals has been reported as 60%.[100,101] This unique size dependent feature of silica nanoparticles has been used in bioimaging.[102–104] Besides the inherent optical properties of the biogenic silica nanostructures, the transparency and biocompatibility of amorphous nanosilica have made it an exceptional cover or carrier for fluorophore compounds. Ow *et al.* 2005 used silica as a shell in preparing highly fluorescent and photostable core–shell nanoparticles in the size range of 20–30 nm. The prepared nanoparticles were 20 times brighter, and more photostable than their constituent fluorophore.[105] The silica-based shell provides this opportunity to functionalize its surface with a specific biochemical and target a particular cell or organism in the biological sample.[106] In a recent study by Prieto-Montero *et al.* 2020, rhodamine 101 was physically and chemically embedded in the silica nanoparticles structure with a size of 50–60 nm.[107] These functionalized fluorescent nanoparticles had enhanced brightness of 230–260-fold with respect to the free dye in solution. The nanoparticles were grafted by polyethylene glycol and folic acid and according to the obtained results folic acid functionalized nanoparticles showed a significant accumulation on HeLa cells and they can be used in targeting cancer cells. In another attempt Araichimani *et al.* 2020 used biogenic silica derived from rice husk to prepare contrast agents applicable in bioimaging. They used europium and gadolinium metals to embed these rare-earth metals with silica nanoparticles due to the electrostatic interaction with oxygen atoms. Strong red fluorescence of europium-integrated silica nanoparticles was attributed to 5D0→7F1 and 5D0→7F2 transitions of europium ions in the silica host matrix. Moreover, paramagnetism of gadolinium-integrated silica composite was ascribed to the magnetic moment originated from a spin–orbit coupling of partially filled 4f electrons of gadolinium in the silica host matrix.[108]

6.5 Conclusion

Nature is a great and diverse source of biological origined materials in micro or nanosizes. All nature derived materials have typical advantages. For the biogenic silica perspective, it is important to mention that biogenic silica's structure and morphology have been differed and their unique properties are based on the process of *in vivo* production of silica structures by a diatom without the involvement of complex chemical procedures. Biogenic silica obtained from the diatoms can be converted into fully biodegradable porous silicon replicas by a magnesiothermic reduction process, preserving the original dimension and morphology. They can also be scalable and tailored for a specific application with surface functionalization or a chemical modification. These processes help to obtain an extraordinarily cheap, green environmental and largely available material especially in order to use in healthcare nanodevices.[109,110]

6.6 Future Remarks

Climate is the key parameter affecting biological silica sources. It should be remembered that synthesis of biogenic silica is obviously connected and is dependent on the environment, biological producers and their predators by evoluationary ways. There are various inputs and/or outputs that remain uncertain and need to be optimized. In spite of the low chances of obtaining biogenic silica, it has opened new areas in material science and biomedical science.[111,112]

In the literature, it is stated that biosilica based materials are potential candidates especially for advanced drug carrier systems with the aid of their biocompatible and enhanced drug loading properties and significant improved cellular uptake profiles. Also, diatom template fabricated structures provide great opportunities in terms of high selectivity and sensitivity, fast response and low detection limits for fabrication of low-cost biosensing devices in biomedical applications.[113,114]

Diatom frustules, diatom biosilica, and diatomite NPs have great capabilities for both commercial/industrial applications in pharmaceutics and biomedicine, and would be preferred especially for high frequency applications; but there are still huge reqirements on detailed investigation of their bioavailability improvement, biodegradability analysis, their toxicity (*in vitro*/*in vivo* evaluations), pharmacokinetic and pharmacodynamic analysis. Also, biofuel production, gene delivery, separation, and remediation applications should be considered and investigated. As a consequence, the application of biogenic silicas on the scientific and industrial scales needs more study in terms of genetic modification/engineering, cultivation, metabolic/proteomic data, drying procedures, harvesting conditions, and growth rates/conditions in more detail.[58,115–117]

References

1. N. Bhalla, P. Jolly, N. Formisano and P. Estrela, Introduction to biosensors, *Essays Biochem.*, 2016, **60**, 1.
2. H. O. Fatoyinbo and M. P. Hughes, *Encyclopedia of Nanotechnology*, Springer, Netherlands, Dordrecht, 2016, p. 387.
3. S. Mishra, E.-S. Kim, P. K. Sharma, Z.-J. Wang, S.-H. Yang, A. K. Kaushik, C. Wang, Y. Li and N.-Y. Kim, *ACS Appl. Bio Mater.*, 2020, **3**(11), 7821–7830.
4. S. Mishra, Md. Ashaduzzaman, P. Mishra, H. C. Swart, A. P. F. Turner and A. Tiwari, *Biosens. Bioelectron.*, 2017, **89**(part 1), 305–311.

5. S. C. Davis, V. C. Sheppard, G. Begum, Y. Cai, Y. Fang, J. D. Berrigan, N. Kröger and K. H. Sandhage, *Adv. Funct. Mater.*, 2013, **23**, 4611.
6. J. Yang, L. Zhen, F. Ren, J. Campbell, G. L. Rorrer and A. X. Wang, *J. Biophotonics*, 2015, **8**, 659.
7. H. E. Townley, A. R. Parker and H. White-Cooper, *Adv. Funct. Mater.*, 2008, **18**, 369.
8. K. Kostarelos, A. Bianco and M. Prato, *Nat. Nanotechnol.*, 2009, **4**, 627.
9. X. Ji, H. Wang, B. Song, B. Chu and Y. He, Analysis, *Front. Chem.*, 2018, **6**, 1.
10. Y. He, C. Fan and S.-T. Lee, *Nano Today*, 2010, **5**, 282.
11. F. Peng, Y. Su, Y. Zhong, C. Fan, S.-T. Lee and Y. He, *Acc. Chem. Res.*, 2014, **47**, 612.
12. W. Wang, T. Gutu, D. K. Gale, J. Jiao, G. L. Rorrer and C. Chang, *J. Am. Chem. Soc.*, 2009, **131**, 4178.
13. D. Zhang, J. Pan, J. Cai, Y. Wang, Y. Jiang and X. Jiang, *J. Micromech. Microeng.*, 2012, **22**, 035021.
14. Y. Wang, J. Pan, J. Cai, A. Li, M. Chen and D. Zhang, *Chem. Lett.*, 2011, **40**, 1354.
15. Z. Bao, M. R. Weatherspoon, S. Shian, Y. Cai, P. D. Graham, S. M. Allan, G. Ahmad, M. B. Dickerson, B. C. Church, Z. Kang, H. W. Abernathy III, C. J. Summers, M. Liu and K. H. Sandhage, *Nature*, 2007, **446**, 172.
16. D. Losic, J. G. Mitchell and N. H. Voelcker, *New J. Chem.*, 2006, **30**, 908.
17. J. Athinarayanan, V. S. Periasamy, M. Alhazmi, K. A. Alatiah and A. A. Alshatwi, *Ceram. Int.*, 2015, **41**, 275.
18. A. A. Alshatwi, J. Athinarayanan and V. S. Periasamy, *Mater. Sci. Eng., C*, 2015, **47**, 8.
19. H. Nsubuga, C. Basheer, M. B. Haider and R. Bakdash, *J. Chromatogr. A*, 2018, **1554**, 16.
20. G. Singh, H. B. Dizaji, H. Puttuswamy and S. Sharma, *Sustainability*, 2022, **14**, 539.
21. A. B. Corradi, F. Bondioli, A. M. Ferrari, B. Focher and C. Leonelli, *Powder Technol.*, 2006, **167**, 45.
22. D. Schneider, R. Kircheis, S. Wassersleben, W.-D. Einicke, R. Gläser and D. Enke, Low-Cost Microwave-Assisted Partial Pseudomorphic Transformation of Biogenic Silica, *Front. Chem.*, 2019, **7**, 1.
23. H. Chang, J.-H. Park and H. D. Jang, *Colloids Surf., A*, 2008, **313–314**, 140.
24. L. Rubio, G. Pyrgiotakis, J. Beltran-Huarac, Y. Zhang, J. Gaurav, G. Deloid, A. Spyrogianni, K. A. Sarosiek, D. Bello and P. Demokritou, *Part. Fibre Toxicol.*, 2019, **16**, 40.
25. M. S. Wooldridge, P. V. Torek, M. T. Donovan, D. L. Hall, T. A. Miller, T. R. Palmer and C. R. Schrock, *Combust. Flame*, 2002, **131**, 98.
26. H. Beidaghy Dizaji, T. Zeng, I. Hartmann, D. Enke, T. Schliermann, V. Lenz and M. Bidabadi, *Appl. Sci.*, 2019, **9**, 1083.
27. D. Bokov, A. Turki Jalil, S. Chupradit, W. Suksatan, M. Javed Ansari, I. H. Shewael, G. H. Valiev and E. Kianfar, *Adv. Mater. Sci. Eng.*, 2021, **2021**, 1.

28. M. C. Gonçalves, *Molecules*, 2018, **23**, 2021.
29. M. Zielecka, E. Bujnowska, K. Suwala and M. Wenda, in *Recent Applications in Sol-Gel Synthesis*, InTech, 2017.
30. L. Usgodaarachchi, C. Thambiliyagodage, R. Wijesekera and M. G. Bakker, *Curr. Res. Green Sustainable Chem.*, 2021, **4**, 100116.
31. S. Porrang, N. Rahemi, S. Davaran, M. Mahdavi and B. Hassanzadeh, *Eur. J. Pharm. Sci.*, 2021, **163**, 105866.
32. S. Horikoshi and N. Serpone, in *Microwaves in Nanoparticle Synthesis*, Wiley, 2013, pp. 1–24.
33. A. Franco, S. De, A. M. Balu, A. A. Romero and R. Luque, *ACS Sustainable Chem. Eng.*, 2018, **6**, 11555.
34. E. C. Peres, J. C. Slaviero, A. M. Cunha, A. Hosseini-Bandegharaei and G. L. Dotto, *J. Environ. Chem. Eng.*, 2018, **6**, 649.
35. G. A. G. Solero, *Nanosci. Nanotechnol.*, 2017, **7**, 21.
36. K. Cho, H. Chang, D. S. Kil, J. Park, H. D. Jang and H. Y. Sohn, *Aerosol Sci. Technol.*, 2009, **43**, 911.
37. J. M. Mäkelä, J. Haapanen, J. Harra, P. Juuti and S. Kujanpää, *KONA Powder Part. J.*, 2017, **34**, 141.
38. F. Cignoli, S. Maffi, R. Dondè, G. Zizak, G. Solero, I. Brescia and S. Alberti, *Combustion Colloquia*, 2009, p. 1.
39. S. Gu, J. Zhou, C. Yu, Z. Luo, Q. Wang and Z. Shi, *Ind. Crops Prod.*, 2015, **65**, 1.
40. P. G. Jeelani, P. Mulay, R. Venkat and C. Ramalingam, *Silicon*, 2020, **12**, 1337.
41. T. Suteewong, H. Sai, R. Cohen, S. Wang, M. Bradbury, B. Baird, S. M. Gruner and U. Wiesner, *J. Am. Chem. Soc.*, 2011, **133**, 172.
42. C. Fruijtier-Pölloth, *Toxicology*, 2012, **294**, 61.
43. V. Selvarajan, S. Obuobi and P. L. R. Ee, *Front. Chem.*, 2020 **8**, 602.
44. R. Huang, X. Hu, Y. Guo, J. Wang and B. Yang, *ACS Appl. Mater. Interfaces*, 2020, **12**, 35259.
45. N. N. Maseko, D. Schneider, S. Wassersleben, D. Enke, S. A. Iwarere, J. Pocock and A. Stark, *Sustainability*, 2021, **13**, 577.
46. V. Shirshahi and M. Soltani, *Contrast Media Mol. Imaging*, 2015, **10**, 1.
47. Y. Shen, *Renewable Sustainable Energy Rev.*, 2017, **80**, 453.
48. C. O. Prempeh, S. Formann, T. Schliermann, H. B. Dizaji and M. Nelles, *Appl. Sci.*, 2021, **11**, 10363.
49. S. J. So, I. S. Jang and C. S. Han, *J. Nanosci. Nanotechnol.*, 2008, **8**, 5367.
50. S. K. Sharma, A. R. Sharma, S. D. V. N. Pamidimarri, J. Gaur, B. P. Singh, S. Sekar, D. Y. Kim and S. S. Lee, *Nanomaterials*, 2019, **9**, 1440.
51. F. Chen, G. Hableel, E. R. Zhao and J. V. Jokerst, *J. Colloid Interface Sci.*, 2018, **521**, 261.
52. Y. Yang, M. Zhang, H. Song and C. Yu, *Acc. Chem. Res.*, 2020, **53**, 1545.
53. J. G. Croissant, K. S. Butler, J. I. Zink and C. J. Brinker, *Nat. Rev. Mater.*, 2020, **5**, 886.

54. F. Akhter, A. A. Rao, M. N. Abbasi, S. A. Wahocho, M. A. Mallah, H. Anees-ur-Rehman and Z. A. Chandio, *Silicon*, 2022, **14**, 8295.
55. V. Gaudin, *Electrochemical Biosensors*, Elsevier, 2019, p. 307.
56. C. Amador and L. Martin de Juan, *Comput.-Aided Chem. Eng.*, 2016, **39**, 509.
57. L. K. Harris and J. A. Theriot, *Trends Microbiol.*, 2018, **26**, 815.
58. I. Rea and L. De Stefano, *Sensors*, 2019, **19**, 5208.
59. V. S.-Y. Lin, K. Motesharei, K.-P. S. Dancil, M. J. Sailor and M. R. Ghadiri, *Science*, 1997, **278**, 840.
60. L. De Stefano, M. Rossi, M. Staiano, G. Mamone, A. Parracino, L. Rotiroti, I. Rendina, M. Rossi and S. D'Auria, *J. Proteome Res.*, 2006, **5**, 1241.
61. A. Mcheik, S. Cassaignon, J. Livage, A. Gibaud, S. Berthier and P. J. Lopez, *Front. Mar. Sci.*, 2018, **5**, 123.
62. J. Liu, J. Zhu, P. Zhang, L. Han, O. L. Reynolds, R. Zeng, J. Wu, Y. Shao, M. You and G. M. Gurr, *Front. Plant Sci.*, 2017, **8**, 1265.
63. V. Zarei, M. Mirzaasadi, A. Davarpanah, A. Nasiri, M. Valizadeh and M. J. S. Hosseini, *Processes*, 2021, **9**, 334.
64. D. Losic, J. G. Mitchell and N. H. Voelcker, *Adv. Mater.*, 2009, **21**, 2947.
65. P. Yang, S. Gai and J. Lin, *Chem. Soc. Rev.*, 2012, **41**, 3679.
66. F. Ren, J. Campbell, X. Wang, G. L. Rorrer and A. X. Wang, *Opt. Express*, 2013, **21**, 15308.
67. A. Kamińska, M. Sprynskyy, K. Winkler and T. Szymborski, *Anal. Bioanal. Chem.*, 2017, **409**, 6337.
68. M. Wanunu, T. Dadosh, V. Ray, J. Jin, L. McReynolds and M. Drndić, *Nat. Nanotechnol.*, 2010, **5**, 807.
69. S. Garaj, W. Hubbard, A. Reina, J. Kong, D. Branton and J. A. Golovchenko, *Nature*, 2010, **467**, 190.
70. M. J. Kim, M. Wanunu, D. C. Bell and A. Meller, *Adv. Mater.*, 2006, **18**, 3149.
71. K.-C. Lin, V. Kunduru, M. Bothara, K. Rege, S. Prasad and B. L. Ramakrishna, *Biosens. Bioelectron.*, 2010, **25**, 2336.
72. T. A. Saleh, *Environ. Technol. Innovation*, 2020, **20**, 101067.
73. J. A. Faucheaux, A. L. D. Stanton and P. K. Jain, *J. Phys. Chem. Lett.*, 2014, **5**, 976.
74. A. G. Cullis, L. T. Canham and P. D. J. Calcott, *J. Appl. Phys.*, 1997, **82**, 909.
75. P. Xiao, Y. Ke, J. Lu, Z. Huang, X. Zhu, B. Wei and L. Huang, *Biochem. Eng. J.*, 2018, **139**, 109.
76. H. Xie, E. Chen, Y. Ye, S. Xu and T. Guo, *J. Phys. Chem. Lett.*, 2020, **11**, 1428.
77. L. De Stefano, I. Rendina, M. De Stefano, A. Bismuto and P. Maddalena, *Appl. Phys. Lett.*, 2005, **87**, 233902.
78. F. F. E. Kohle, J. A. Hinckley and U. B. Wiesner, *J. Phys. Chem. C*, 2019, **123**, 9813.

79. R. Rajan, Y. Zakaria, S. Shamsuddin and N. F. Nik Hassan, *Egypt J. Forensic Sci.*, 2019, **9**, 50.
80. J. Romann, J.-C. Valmalette, M. S. Chauton, G. Tranell, M.-A. Einarsrud and O. Vadstein, *Sci. Rep.*, 2015, **5**, 17403.
81. K. Kieu, C. Li, Y. Fang, G. Cohoon, O. D. Herrera, M. Hildebrand, K. H. Sandhage and R. A. Norwood, *Opt. Express*, 2014, **22**, 15992.
82. T. Fuhrmann, S. Landwehr, M. El Rharbi-Kucki and M. Sumper, *Appl. Phys. B: Lasers Opt.*, 2004, **78**, 257.
83. C. Chiappini, E. De Rosa, J. O. Martinez, X. Liu, J. Steele, M. M. Stevens and E. Tasciotti, *Nat. Mater.*, 2015, **14**, 532.
84. J.-H. Park, L. Gu, G. von Maltzahn, E. Ruoslahti, S. N. Bhatia and M. J. Sailor, *Nat. Mater.*, 2009, **8**, 331.
85. L. Dalstein, E. Potapova and E. Tyrode, *Phys. Chem. Chem. Phys.*, 2017, **19**, 10343.
86. B. Qiao, T.-J. Wang, H. Gao and Y. Jin, *Appl. Surf. Sci.*, 2015, **351**, 646.
87. N. B. Arnfinnsdottir, C. A. Chapman, R. C. Bailey, A. Aksnes and B. T. Stokke, *Sensors*, 2020, **20**, 3163.
88. Y. Coffinier and R. Boukherroub, *Semiconducting Silicon Nanowires for Biomedical Applications*, Elsevier, 2022, p. 25.
89. S. Mariani, V. Robbiano, L. M. Strambini, A. Debrassi, G. Egri, L. Dähne and G. Barillaro, *Nat. Commun.*, 2018, **9**, 5256.
90. M. Rahimi-Mohseni, J. B. Raoof, R. Ojani, T. A. Aghajanzadeh and A. Bagheri Hashkavayi, *Int. J. Biol. Macromol.*, 2018, **113**, 648.
91. T. Zheng, *et al.*, Diatomite-Based Silica Nanoparticles for Sensitive Detection of Heavy Metal Ions, *ACS Appl. Mater. Interfaces*, 2019, **11**(3), 3167–3175.
92. Q. Li, *et al.*, Ultrasensitive detection of prostate-specific antigen based on TiO_2 nanorod arrays decorated with diatomite-derived silica nanoparticles, *Biosens. Bioelectron.*, 2017, **98**, 96–102.
93. X. Pang, *et al.*, Rapid and sensitive detection of Salmonella based on microsphere-labeled immuno-PCR assay using atomic force microscopy-based nanomechanical biosensor, *Sens. Actuators, B*, 2018, **258**, 1067–1074.
94. J. Guo, *et al.*, Rapid and label-free visual detection of glucose in saliva using microwave-assisted click chemistry, *Biosens. Bioelectron.*, 2016, **78**, 120–126.
95. Y. Zhang, *et al.*, DNA biosensor based on diatomite earth-supported screen-printed electrodes and Prussian blue labels, *Biosens. Bioelectron.*, 2018, **102**, 514–520.
96. N. Malik, T. Arfin and A. U. Khan, in *Nanomaterials for Drug Delivery and Therapy*, Elsevier, 2019, p. 373.
97. K. Lee, H.-Y. Moon, C. Park, O. R. Kim, E. Ahn, S. Y. Lee, H. E. Park, S.-H. Ihm, K.-B. Seung, K. Chang, T.-J. Yoon, C. Lee, C. Cheong and K. S. Hong, *Curr. Appl. Phys.*, 2009, **9**, S15.
98. S. Viarbitskaya, L. Ryderfors, T. Mikaelsson, E. Mukhtar and L. B.-Å. Johansson, *J. Fluoresc.*, 2011, **21**, 257.

99. N. Hagura, T. Ogi, T. Shirahama, F. Iskandar and K. Okuyama, *J. Lumin.*, 2011, **131**, 921.
100. M. Locritani, Y. Yu, G. Bergamini, M. Baroncini, J. K. Molloy, B. A. Korgel and P. Ceroni, *J. Phys. Chem. Lett.*, 2014, **5**, 3325.
101. D. Jurbergs, E. Rogojina, L. Mangolini and U. Kortshagen, *Appl. Phys. Lett.*, 2006, **88**, 233116.
102. C. M. Hessel, M. R. Rasch, J. L. Hueso, B. W. Goodfellow, V. A. Akhavan, P. Puvanakrishnan, J. W. Tunnel and B. A. Korgel, *Small*, 2010, **6**, 2026.
103. L. Gu, D. J. Hall, Z. Qin, E. Anglin, J. Joo, D. J. Mooney, S. B. Howell and M. J. Sailor, *Nat. Commun.*, 2013, **4**, 2326.
104. F. Erogbogbo, K.-T. Yong, I. Roy, G. Xu, P. N. Prasad and M. T. Swihart, *ACS Nano*, 2008, **2**, 873.
105. H. Ow, D. R. Larson, M. Srivastava, B. A. Baird, W. W. Webb and U. Wiesner, *Nano Lett.*, 2005, **5**, 113.
106. J.-T. Song, X.-Q. Yang, X.-S. Zhang, D.-M. Yan, M.-H. Yao, M.-Y. Qin and Y.-D. Zhao, *Dalton Trans.*, 2015, **44**, 11314.
107. R. Prieto-Montero, A. Katsumiti, M. P. Cajaraville, I. López-Arbeloa and V. Martínez-Martínez, *Sensors*, 2020, **20**, 5590.
108. P. Araichimani, K. M. Prabu, G. Suresh Kumar, G. Karunakaran, N. Van Minh, S. Karthi, E. K. Girija and E. Kolesnikov, *Ceram. Int.*, 2020, **46**, 18366.
109. V. Panwar and T. Dutta, *ACS Appl. Bio Mater.*, 2019, **2**, 2295.
110. I. Rea, M. Terracciano and L. De Stefano, *Adv. Healthcare Mater.*, 2017, **6**, 1601125.
111. J. Livage and P. J. Lopez, in *Encyclopedia of Glass Science, Technology, History, and Culture*, Wiley, 2021, p. 955.
112. P. J. Tréguer, J. N. Sutton, M. Brzezinski, M. A. Charette, T. Devries, S. Dutkiewicz, C. Ehlert, J. Hawkings, A. Leynaert, S. M. Liu, N. Llopis Monferrer, M. López-Acosta, M. Maldonado, S. Rahman, L. Ran and O. Rouxel, *Biogeosciences*, 2021, **18**, 1269.
113. S. Maher, T. Kumeria, M. S. Aw and D. Losic, *Adv. Healthcare Mater.*, 2018, **7**, 1800552.
114. J. E. N. Dolatabadi and M. de la Guardia, *TrAC, Trends Anal. Chem.*, 2011, **30**, 1538.
115. N. Rabiee, M. Khatami, G. Jamalipour Soufi, Y. Fatahi, S. Iravani and R. S. Varma, *ACS Biomater. Sci. Eng.*, 2021, **7**, 3053.
116. C. Tramontano, G. Chianese, M. Terracciano, L. de Stefano and I. Rea, *Appl. Sci.*, 2020, **10**, 6811.
117. B. Abdusatorov, J. Everaerts, A. I. Salimon and A. M. Korsunsky, *Proc. World Congr. Eng*, 2019.

CHAPTER 7

Biogenic Silica Composites for Biomedical Applications

PRABAKARAN SELVAKANI,[a] MUTHURAMAN RAGURAMAN[b] AND MARIAPPAN RAJAN*[b]

[a] Department of Chemistry, PSG College of Arts & Science, Coimbatore 641014, Tamil Nadu, India; [b] Biomaterials in Medicinal Chemistry Laboratory, Department of Natural Products Chemistry, School of Chemistry, Madurai Kamaraj University, Madurai 625021, Tamil Nadu, India
*Email: rajanm153@gmail.com

7.1 Introduction

7.1.1 Biomedical Industry and Research

The biomedical industry and research deals with products benefiting medical and healthcare biological applications which consists of the pharmaceutical industry, biotech industry, medical devices, diagnostic technology and healthcare information technology. Biotechnology is an essential part of the biomedical industry and emphasizes molecular biology, genomics, and new drug development and is a knowledge-concentrated industry that is frequently producing and adapting to new technologies. These industries deliver the equipment, knowledge and therapeutic products to the healthcare industry including hospitals, medical and dental clinics and other clinical services to precaution for and treat patients. The global competition and the difficulties in the distributing system of therapeutic products worldwide have risen from growth in world population and occurrence of chronic diseases, combined with mounting research costs, and are employing extraordinary cost pressures on the biomedical and healthcare industries. The challenges including globalization, social, health and disease

Biogenic Silica: Fundamentals and Applications
Edited by Ashok Kumar Nadda and Vijai Kumar Gupta

Published by the Royal Society of Chemistry, www.rsc.org

issues, innovations in science and advancements, digitalization of healthcare system are being involved in the biomedical industry.[1]

Currently, the urgent need for improvement in health care technologies is ever more conspicuous. The yearly expenditure for global healthcare processes has seen persistent growth and is estimated to decline by 0.1% to USD 8.3 trillion in 2020 before growing 5.8% to USD 8.8 trillion in 2021.[2] The avoidance, identification and management of disease as well as patient recovery are the main goals of biomedical engineering research by delivering various health technologies. For example, the NIH 2016–2020 strategic plan emphases on point of care and accuracy of various medicine technologies including genetic engineering, microfluidics, nanomedicine, imaging, digital/mobile-health and big data.[3] The notable biomedical applications that help human health systems include drug delivery,[4] tissue regeneration,[5] antibiotic treatment,[6] cancer treatment,[7–9] bio-sensing[10] and bio-imaging.[11]

7.1.2 Silica (Si) in Human Health

Si is a significant biologically beneficial element which is a non-metallic element with an atomic weight of 28. It is the second most abundant element in the Earth's crust at 28 wt%. Human exposure to Si communicates health benefits and is essentially acquired through food materials derived from plant resources. The bioavailability of Si in the human diet strengthens bones, improves immune response, and improves the health of neuronal and connective tissues.[12] Apart from humans, there is circumstantial proof for the necessity of Si in animals, most of the cells and in basic organisms such as bacteria, viruses, and fungi recommend that silica may have an appropriate function and even a vital role in all organisms. Universally, osteoporosis is an increasing health concern which is initiated by low mass and increased fragility of bones, and leads to severe incapacities and raises mortality rates. This has lead scientists to focus on the understanding of the relationship between dietary Si and bone health.[13] In humans, the Si is taken in as silicic acid with the daily intake ranges from 9 to 14 mg, however, bone health might be increased if the daily intakes of Si near 25 mg d^{-1}.[14] In 1979, Schiano *et al.* who did the initial investigations about nutritional importance of Si for human health, found that, by supplying external silica source it significantly increased the volume of trabecular bone by using monomethyl trisilanol.[15] Moreover, the biosynthesis of bone matrix forming molecules, collagen and glycosaminoglycan, are done by the presence of Si element.[16] The intake Si element is maintained at equilibrium within the body by various processes such as absorption, transport, retention, and excretion.[17] About 41% of the absorbed silica from food is excreted in urine while its concentration in the blood serum remains constant (10–31 μg dL^{-1}). Part of the intake Si content is engaged in connective tissues, including bone, skin, trachea, and tendon while another part is being transferred to the brain.[18] Aluminium induced Alzheimer's disease can be prevented, drinking a silica rich water with concentrations ≥11 mg L^{-1} is recommended.[19] Founding of a biological role of Si, will have beneficial consequences for nutrition as an anticipatory measure, or, for bone and connective tissue diseases by treating with Si containing supplements.

7.2 Biosilica

7.2.1 Structural Diversity and Composition

Various structural forms are exhibited in biosilica comprising sheet-like, globular, fibrillar, helical, tubular, and folded sheets like structures along with a diverse density and different composition. A cyanobacterial cell enveloped in silica colloids with a polysaccharide sheath and a layer of colloidal amorphous silica has been studied by V. R. Phoenix *et al.*, in 2006.[20] In 2020, Yang Pu *et al.*, have fabricated the C-phycocyanin (C-PC), a natural photosensitizer conjugated diatom biosilica for enhanced photodynamic effect towards tumor cells. The bare biosilica was obtained from diatom strain *Nanofrustulum shiloi*. Images for bare biosilica at different magnifications show a highly mesoporous surface given in Figure 7.1A (2k) and B (50k). The new biomaterials appeared mostly uniform and complete. The C-PC functionalized biosilica was significantly different from the pristine bare sample at 50k magnification given in Figure 7.1C which shows an uneven coating with surface protein molecules (see Figure 7.1).[21] Various examples of siliceous structures extracted from various plants, bacteria, yeast, fungi, protists, sponges, molluscs, ascidians, crustaceans, brachiopods, fishes, mammals, and humans are tabulated in Table 7.1.[22]

7.2.2 Micro Level Biosilica: Protozoa and Multiplicity of Siliceous Morphology

Dujardin in 1841, first discovered the siliceous environment of particular protozoan plates.[23] Amoeboid protists provide the cell surface for the deposition of siliceous products including various structures that are highly species-specific such as spicules, scales, solid plates, granules, mesh work frustules, and other elaborate geometric forms as showed in Table 7.1.

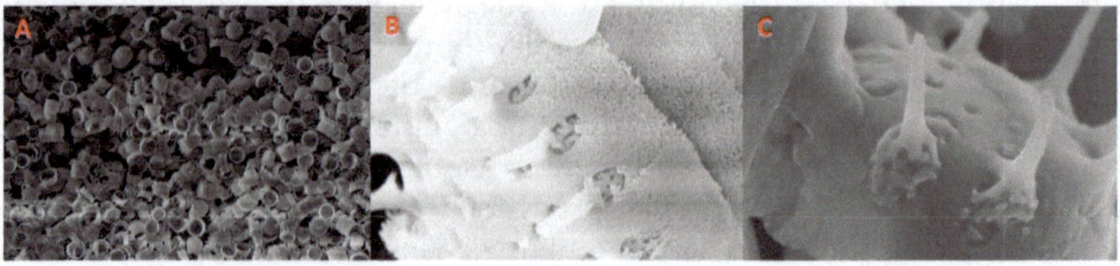

Figure 7.1 (A) FE-SEM image of bare biosilica (2k) (B) FE-SEM image of bare biosilica (50k) (C) FE-SEM image of biosilica conjugated with C-PC (50k). Reproduced from ref. 21 with permission from Elsevier, Copyright 2020.

Table 7.1 Obtained structures of biosilica from various organisms. Adapted from ref. 22 with permission from American Chemical Society, Copyright 2010.

Organism	Structure and form of biosilica
Cyanobacteria	Siliceous "geyser eggs"
Filamentous microorganisms	Siliceous shrubs
Hyperthermophilic microorganisms	Siliceous shrubs
Nostoc-type cyanobacteria	"Bio speleotherms" & shrubs
Biofilm community	Siliceous microstromatolite laminae
Thermus thermophilus	TMY silica scale from geothermal power plant
Phormidium, Fischerella	Lilypad stromatolites – up to 3 m long and 1.5 m wide
Cyanidium, Alicyclobacillus	Siliceous spicules – up to 3 cm high and 5 mm in diameter
Cyanidium caldarium	Granular silica spherules
Sarcodines	Protozoa
Gymnamoebae	Siliceous boat-shaped scales
Testate amoebae	Siliceous curved rods
Heliozoa	Siliceous needles, spines
Radiolaria	Siliceous Spicules
Flagellates	Stomatocysts, scales, spines, bristles
Dinoflagellates	Siliceous skeletons, cysts, granules, external skeleton, scales
Ebridians	Statosphores (siliceous resting cysts)
Silicoflagellates	Silicified cell walls (quartz-cellulose-calcite composite)
Thaumatomastigids	Silicified cell wall
Silicosigmoilina futabaensis	Siliceous cement, diatoms
Hexactinellida	Skeletal frameworks
Demospongiae	Spicule (macrosclerae) microsclerae
Patellacea spp. (Gastropoda)	Silica in radula teeth
Onchidiella celtica	Siliceous spicules and penial spines
Discinisca tenuis	Siliceous tablets
Psammobatis extenta (Rajidae)	Chalcedony in electrocytes and cholinergic nerves
Dog	Silica calculi and stones
Cattle	Silica calculi and stones
Monkey (*Macaca fuscata*)	Dental calculi
Rat liver mitochondria	Silica nanogranules
Chaetoceros gracilis (Bacillariophyceae)	Siliceous setae
Canary grass (*Phalaris canariensis*)	Silica fiber
Foxtail millet (*Setaria italica*)	Inflorescence bristles
Bambusoideae	Tabasheer
Higher plants	Siliceous nodular deposits
Pleioblastus chino (Poaceae, Bambusoidea)	Silica cells
Heath grass (*Sieglingia decumbens*)	Silica bodies (phytoliths)

Siliceous protist scales can persevere in the sediments of marine and fresh water and they leave a record of their presence once the organism has perished. In the sediments of freshwater and marine habitations, the siliceous scales can be conserved for a long time as sub-fossils or as micro fossils, which are then providing a space into ancient, modern and potential communities of protist.[24,25] It is very significant to consider the role of these unicellular organisms in the cycle of silica and its distribution in the oceans of the world. For example, the second major source of amorphous silica (suspended) in water, after diatoms are radiolarians. They are most abundant in the equatorial zone and come about from the Arctic to the Antarctic Oceans.[26]

7.2.3 Secretory Mechanism of Biosilica and Morphology

In some organisms like testate amoebae, heliozoa and heliozoa-like amoebae, and adiolarians, a general secretory mechanism of siliceous products has been reported previously.[27] In testate amoeba, heliozoans and relative organisms, the site of silicification is placed in the cell cytoplasm or within a widened portion of the peripheral cytoplasm known as a cytokalymma where the silica deposition vesicles (SDVs) can be deposited. By increasing the membrane surface area, it is possible in the surface of the SDV, to incorporate the additional silica secreting membrane with the help of Golgi-derived vesicles (GDVs) which are, in some testate amoebae, coupled with the membrane neighboring silica deposition sites. By utilizing exocytosis, silica products of testate amoebae and heliozoa are formed on the cell surface. However, the cytokalymma of radiolaria contain a silica-secreting vacuolar space, they are different in terms of activity and form from the intracellular secretory spaces of testate amoebae and heliozoa. The cytoplasmic flowing activity is exhibited by cytokalymma which has adynamic structure and the species-specific shape of the skeleton is achieved in a mold-like manner. Subsequently, we can say that the formed or deposited silicate product of the radiolarian is an endoskeleton and is not released on the surface by exocytosis. However, there is some evidence of silica deposition on the cell surface, till now, no significant SDVs have been sequestered. The siliceous scale of the *R. ambigua* (Protista, Centrolheliozoa) shows the rounded body form that lengthens a small number of stiff radiating axopodia used for food capture. Periplast lies around the cell body of loosely bounding boat-shaped scales.[22,28] Within the single species *Polyplacocystis ambigua* (Protista, Centrohelida), the diversity of siliceous scale morphologies has been reported by Gaponova and Dovga.[29] A SEM image in Figure 7.2, indicated the valid evidence of the existence of the fusiform, naviculoid, and oval forms of the *P. ambigua* periplast scales. The longitudinal and transverse (or helical) arrays of patterned siliceous costae in the basic structure of the choanoflagellate lorica have been shown in Figure 7.2.[30]

7.2.4 Marine Glass Sponges: Macro Level Biosilica

Sponges (Porifera) are different from other living beings including animals earlier in the history of evolution which are the simplest and eldest

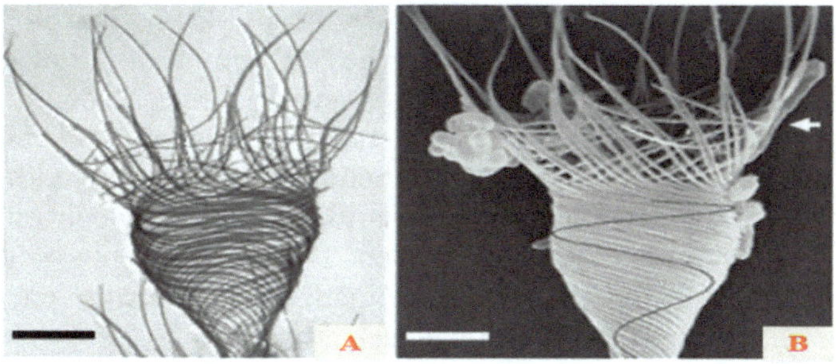

Figure 7.2 TEM image (A) of *Acanthoeca spectabilis* siliceous lorica showing the left-handed orientation of costae comprising the lorica chamber and the continuity of individual helical costae with the costal strips subtending the spines (bar) 2 μm. SEM image (B) of a lorica showing the arrangement of helical costae forming the chamber in continuity with bases of the spines (bar) 1 μm. The inner layer of costal strips within the lantern of spines is also apparent (arrow). Reproduced from ref. 30 with permission from Elsevier, Copyright 2008.

multicellular animals on Earth and live on the seabed or another substratum. The three types of phylum Porifera are, Hexactinellida, Demospongiae and the Calcarea. The first two comprise a siliceous skeleton and the later one is a calcareous skeletal network.[31] Hexactinellida Schmidt (Porifera) is primarily deep-water marine sponges and they are identified from their synthesis of siliceous spicules of hexactinic, triaxonic (cubic) symmetry or shapes. In addition, the silicious spicules of hexactinellids become a vital ecological factor.[32] The silica demineralization–bio-mineralization–re-mineralization cycle is maintained by sponges. One of the important parts of the mound-forming process is catching fine-grained siliciclastic sediments present in suspension in near-bottom currents, because it precludes the reframe work from disintegrating due to its own aggregate weight.[32]

7.3 Bio-silicification in Diatoms

For the construction of frustules, a silica source is needed for the diatoms which they can get from their surrounding environment. In aquatic environments, an existing form of soluble silicon is orthosilicic acid ($Si(OH)_4$)[33] which comprises 97% of total dissolved silicon. Mostly, diatoms utilize this silicic acid, but the utilization of other silicon forms by diatoms is unknown. At first, Lewin reported the uptake process of silicates from the diatoms' surroundings and absorption into cells.[34] Then, in 1999, Del Amo and Brzezinski reported that silicic acid ($Si(OH)_4$) is the most common silicon species which is taken into diatoms and those are highly cumulative in seawater at pH 8.0.[35] This mechanism of silica uptake is actively processed by various pathways, since the exact mechanism is not fully known. The concentration of silicic acid naturally ranges from <1 to 70 mM in oceanic seawater, however the higher concentration is found inside the diatoms cell body from 19 to 340 mM.[36] This enormously large quantity clearly indicates that simple passive diffusion is not the sole reason of silica formation across the cell membrane,[37] but somewhat a function of energetic transport. There are several theories, including protein mediated uptake by specialized silicon transfer proteins (STPs).[38] The three proposed modes are found based on chemostat studies including (i) surge uptake, (ii) externally controlled uptake, and (iii) internally controlled uptake which are highly environment dependent.[39] Then, diatoms' silicified wall components are controlled by two interactive mechanisms, called macro-morphogenesis and micro-morphogenesis involving SDVs.[40] The detailed process of silicon bio-mineralization was reviewed by W. E. G. Muller in 2003.[41]

7.3.1 Templates for Silicification: Self-assembled Biomolecules

Apart from diatoms, some self-assembled biomolecules including peptides, DNA, bacteria, virus and collagen have the ability of silicification with themselves by electrostatic forces by the way of direct interaction of positively charged biomolecules with negatively charged silicates to form nano-hybrid composites with highly complex and ordered structures with distinctive morphologies for various applications.[42] For example, in tissue regeneration applications, proteins and peptides (self-assembled from amino acids) are used as the powerful scaffolds since they mimic the extracellular matrix (ECM) of bone and soft tissues.[43] In particular, a network of silica nanospheres was generated with silaffins by utilizing its distinct ability of biosilification.[44] Similarly, the peptide NH_2-NSGAITIG-$CONH_2$ acts as a template for the bio-silicification by the mechanism of the attraction of proton with N-terminal amino group from the hydroxyl group of the peptide, resulting in the increase the nucleophilicity of serine oxygen, which attacks the tetra ethyl orthosilicate (TEOS) silicon atom.[45] Such bio self-assembled silicified molecules could be used for drug delivery and controlled release applications.[46] Viruses also have unique structural parts. In virion, *i.e.* viruses coated on protein, the nucleic acid is transported *via* a self-assembled cage structure consisting of hundreds of subunits.[47] These nano materials can be used for drug delivery applications by coating the guest drug molecules on them. These virion proteins and lipids can have the ability to attract the silica precursors molecule towards them.[48] Moreover, bacterial species like bacterial flagellate are designed as bio-templates for silicification. From the rod-like type-1 bacterial pili, a highly well-organized 3D lattice structure has been built by Cao *et al.* in the presence of inducer molecules for molecular recognition-based self-assembly. Then, this structure was poly-condensed with TEOS to form pili-silica hybrids for encompassing foreign peptide molecule for bio engineering and nanomedicine applications.[49] Apart from these biomolecules, by programming DNA sequences, nucleic acids can also act as bio-templates for silicification on their self-assembled structure. Wu *et al.* have reported Si nanoparticles assembled

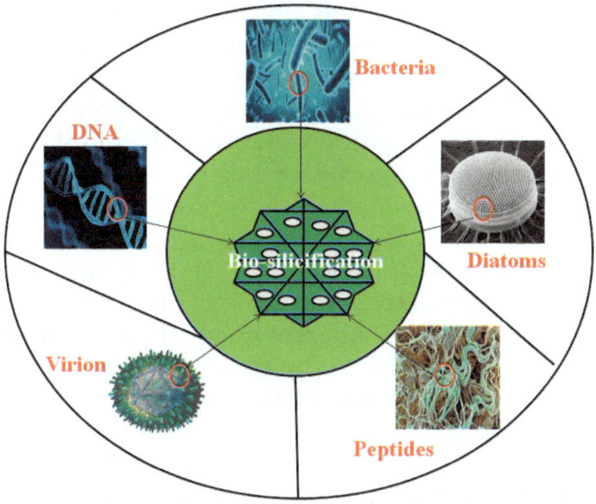

Figure 7.3 Graphical representation of bio-silicification process in different self-assembled biomolecules.

on DNA by grafting of DNA sequences with Si nanoparticles of two variations such as 120 SiNP-DNA1 and 290 SiNP-DNA2. Among these two, 290 SiNP-DNA2 acted as aptamer, the central sequence which is more sensitive to adenosine triphosphate (ATP). This structure could be used for reversible bio-sensing devices in the future since they will disassemble in the presence of ATP.[50,51] Figure 7.3 shows the graphical representation of the bio-silicification process in self-assembled biomolecules (see Figure 7.3).

7.4 Biosilica Composites in Biomedical Applications

7.4.1 Diatom Biosilica in Drug Delivery Applications: Surface Functionalization

7.4.1.1 For Anticancer Treatment: Delivery of Organic Drugs

Initially, Kröger and Voelcker's group used the diatom biosilica for delivering anticancer drugs. They utilized the diatom *Thalassiosira pseudonana* for obtain the biosilica genetically engineered with immunoglobulin G (IgG)-binding domain of protein G (GB1-bearing fusion protein) to deliver the drugs camptothecin and its derivative 7-ethyl-10-hydroxycamptothecin to the site of Burkitt's lymphoma HR1K cells and neuroblastoma. They treated the neuroblastoma mouse model with their fabricated drug-loaded biosilica *via* intraperitoneal injection and they found that their drug loaded biosilica lead to tumor growth regression in the mice model.[52] Apart from this unique genetically modified diatom biosilica, Terracciano *et al.* prepared the polyethylene glycol (PEG) and a cell-penetrating peptide functionalized diatomite nanoparticles and found their low toxicity, hemocompatibility, optimal intracellular uptake in cancer cells and their capability to distribute the poorly water-soluble anticancer drug, sorafenib to the breast cancer MCF-7 and MDA-MB-231 cells.[53] In another work, one of the potent anticancer drugs doxorubicin (DOX) was encapsulated on the porous diatom *Amphora subtropica* (AS) which was chemically altered with chitosan. The composite exhibited the capacity of high drug loading and drug delivery, resilient luminescence, biodegradability and biocompatibility nature and it had lower toxicity related to free DOX.[54] In another work, diatomaceous earth micro-particles (DEMPs) was dually loaded with two dissimilar anticancer drugs, namely curcumin or paclitaxel (hydrophobic) and DOX (hydrophilic), in a single drug delivery carrier at different compartments. The molar ratio of these two drugs was maintained as the same for anticancer treatment[55] (see Figure 7.4). However, apart from these surfaces chemically altered composites of diatoms, the diatoms like diatom frustules are used as templates or scaffolds in some of the drug delivery research works. For example, Vona *et al.* have used the unchanged diatomaceous earth for the loading and delivery of the Ophiobolin A (OphA) one of the fungal phytotoxic-derived chemotherapeutic agents active against lung, melanoma, brain and ovarian cancer cell lines. However, this work deals with the activation of diatomite by acid-oxidative treatment for better loading and releasing behavior of OphA.[56]

7.4.1.2 For Anticancer Treatment: Delivery of Inorganic Drugs

The discovery of cisplatin was a revolution in metal based anticancer inorganic complex drugs. At first, Delasoie *et al.* altered the surface of the diatomaceous earth micro-particles by functionalizing them with vitamin B12 (VB12). Then they loaded some anticancer drugs such as cisplatin (CP), 5-fluorouracil (5-FU), and a polypyridyl ruthenium(II) complex on the VB12 modified diatoms. They have found that the releasing profile was higher for CP and 5-FU than for the Rh(II) complex. In lipophilic environments like the cell membrane, the drugs were discharged quickly. On the other hand, in aqueous media, they remained in the diatom surface for up to 5 days. This scaffold was active against colorectal micro cancer cells (HT-29) of targeted tissues *via* the activation of transcobalamin II (TC(II)) and transcobalamin II receptor (TC(II)-R) expression.[57] The same group, later activated the surface functionalization of diatom microalgae by photochemistry which encouraged the increased cytotoxicity of particular anticancer rhenium complexes against microbes and colorectal carcinomas *in vivo*.[58–61] In another work, polypyridyl ruthenium(II) complexes were delivered from the drug carrier made up of diatom-like silica-protein (human serum albumin) nanocomposites to the cancer cells. The authors exhibited that the ROS generation through light irradiation is the cause of the activation of the composite material, which then achieved an outstanding photodynamic therapy efficiency, so apart from its sustained drug delivery property, this carrier is also used for bio-imaging and photodynamic therapy.[62]

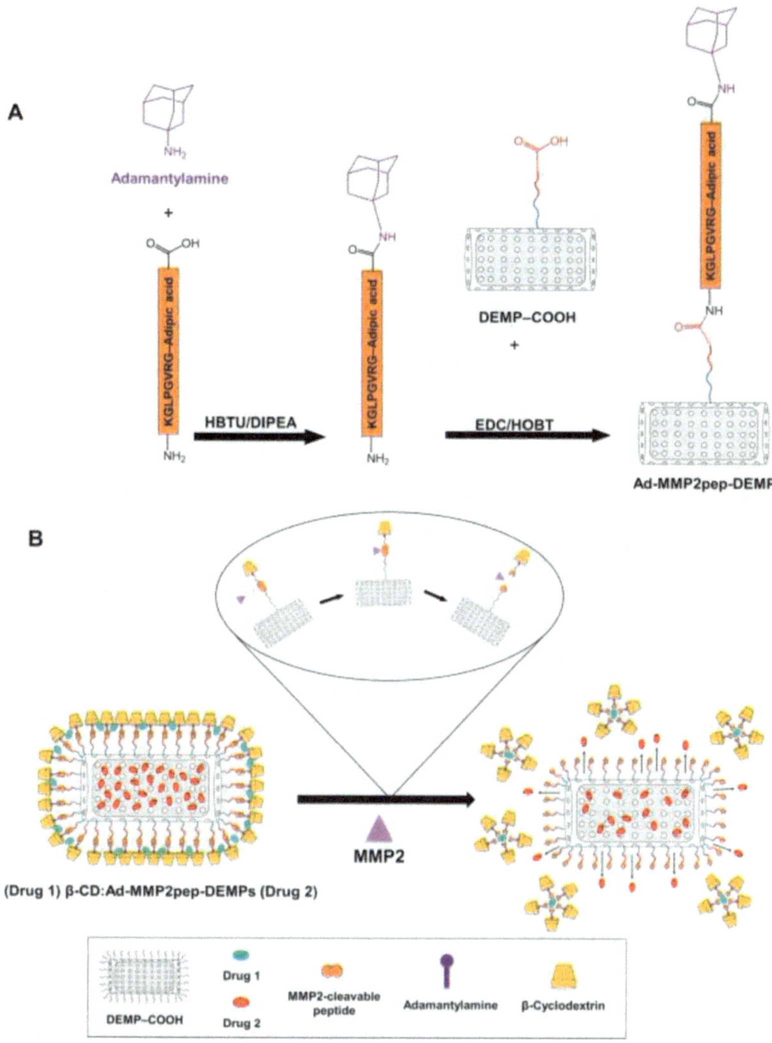

Figure 7.4 Schematics depicting the (A) synthesis of metallomatrix proteinase-2 (MMP2) substrate peptide modified β-cyclodextrin capped DEMPs, and (B) release of two drugs (PTX and DOX) from the core of DEMPs and from the corona of β-cyclodextrin and adamantane, upon cleaving of MMP2 specific peptide in the presence of MMP2 enzyme. Reproduced from ref. 55 with permission from Elsevier, Copyright 2020.

7.4.1.3 *For Antibiotic Drug Delivery*

Unlike cancer drug delivery, studies about diatoms for antibiotic drug delivery are limited. Using silanization chemistry, Vasani *et al.* in 2015 have prepared the thermo responsive *Aulacoseira* sp. diatoms biosilica microcapsules by embedding oligo (ethylene glycol) ethacrylate copolymers loaded with an antimicrobial drug levofloxacin. Authors found that the antimicrobial drug had controlled release on the infected site by an influence of temperature. This levofloxacin drug loaded diatom shows activity against *Staphylococcus aureus* and *Pseudomonas aeruginosa* mycobacterium.[63] In another study, Cicco *et al.* have fabricated the chemically modified *Thalassiosira weissflogii* diatoms. They loaded two drugs for the integrated delivery of the antibiotic ciprofloxacin and the concomitant anti-inflammatory agent against the confrontational side effects of the antibiotic ciprofloxacin.[64] In order to deliver the antituberculosis drug isoniazid (INH), by using iron(II) chloride tetrahydrate in the presence of the diatom biosilica followed by trimesic acid treatment, Uthappa *et al.* have fabricated a porous diatom biosilica and metal–organic frameworks (MOFs) hybrid material. They found the drug-loading capacity of the composite was $9.6 \pm 1.6\%$ and organized drug discharge process was observed over 23 days.[65] Using this type of prolonged release carrier is more beneficial for curing the disease over direct use of the drug. Another antibiotic drug gentamycin was loaded on the *Aulacoseria genus* diatoms with gold nanoparticles (AuNPs) in simulated body fluid conditions. The AuNPs were loaded on the diatom using both *in situ* and *ex situ* methods. Among these two methods, authors found that the composite prepared using an *ex situ* method exhibited 93% of gentamicin release in 10 days and a lower drug-loading efficiency. On the other hand, the composite prepared using an *in situ* method exhibited longer and controlled drug release behavior.[66] In another study, the antimicrobial activity of metal coated diatoms were reported by Cong *et al.* They prepared copper coated *Cyclotella cryptica* diatoms for bacterial infected wound therapy. The presence of Cu metal on *C. cryptica* diatoms considerably enhanced the photothermal and photodynamic behaviour of the material by producing ROS under the irradiation of infrared light. As a result, an enhanced effect of sterilization and about 100% antibacterial efficiency were observed against *S. aureus*- and *E. coli*-infected wounds. Further *in vivo* studies confirmed the antibacterial activity and wound healing activity of the prepared material *via* resisting inflammation, enhancing angiogenesis and collagen deposition, and moderating the ratio of type-III to type-I collagen.[67] Sherief *et al.* coated the diatomite particles with AgNPs using a parallel approach, and exhibited that the prepared material is active against five various pathogenic organisms.[68]

7.4.2 Diatom Biosilica Composites in Bio-sensing Applications

7.4.2.1 Fluorescence and Surface-enhanced Raman Scattering (SERS) Bio-sensing

An immobilization of various chemicals like antibodies, proteins like enzymes, medications, and DNA aptamers on diatom based biosilica scaffold are widely used as biosensors for bio-sensing applications. Fluorescence biosensing is the most general and well evaluated bio-sensing method, then surface-enhanced Raman scattering (SERS). Using silanization chemistry, the antibodies are cross-linked on the surface of the micro-particles to form modified diatom biosilica for immunoassays for bio-sensing. Using a magnesiothermic process, in 2016, Rea *et al.* created silicon replicas *via* the transformation of *Aulacoseira* sp. diatomite and they further, as biological probes, immobilized proteins and antibodies on their surface.[69] The bio-sensing ability of the modified diatom-based silica material was indicated by an increased intensity in photoluminescence of the same biomolecular interactions with the analyte.[69] In order to detect the mouse immunoglobulin G (IgG) protein, Squire *et al.* in 2018, have fabricated the photonic crystals using pristine *Pinnularia* sp. diatoms. The authors have taken rhodamine 6G (R6G) and the R6G-labeled mouse IgG immuno-complex and inspected the fluorescence improvement ability and its effect of the diatom photonic biosilica with these proteins on the diatom frustule using a standard antibody–antigen-labeled antibody immunoassay protocol. The authors concluded that the fabricated photonic crystals of diatom biosilica have a great ability of fluorescence immunoassay to achieve the high sensitivity and to enhance fluorescence. The authors came to this conclusion by examining various concentrations of antigen and found a 100-fold increase in the recognition limit of fluorescence spectroscopy, with a limit of detection of the mouse IgG of 14 fg mL^{-1}.[70] The same group further detected another biomarker corresponding to cardiovascular disease by using the same diatoms *Pinnularia* sp.,[71] (see Figure 7.5). In 2018, Selvaraj *et al.* have prepared the glutaraldehyde crosslinked *Amphora* sp. diatoms which are functionalized with amine molecule and immobilized with *Salmonella typhi* antibody for the recognition of typhoid fever. For this process, the authors used photonic crystals as optical biosensors. Finally, the authors concluded that the diatom-based biosensors had excellent bio-sensitivity owing to their large surface to volume ratio which provided enough space for immobilizing

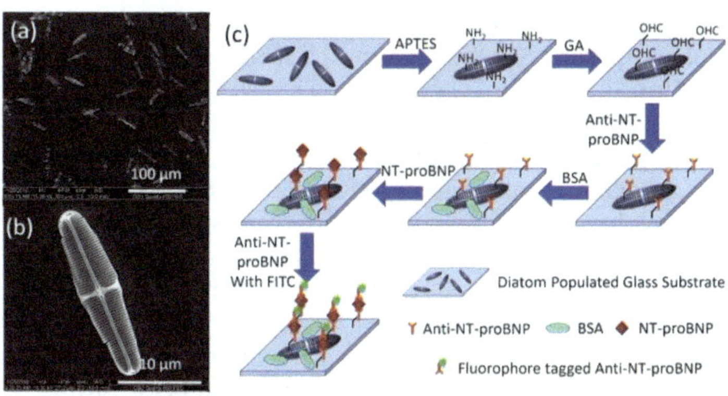

Figure 7.5 (a) SEM image of diatom frustule (*Pinnularia* sp.) presented on a glass slide, (b) zoomed in on a single frustule. (c) Schematic view of diatom-based immunoassay for NT-pro BNP detection. Reproduced from ref. 71 with permission from Elsevier, Copyright 2019.

diatoms.[72] Another application regarding bio-sensing ability of diatom based biosilica is SERS, the phenomenon of inelastic scattering of photons during the interaction of the elementary particles with matter. Fortunately, diatoms have the ability of increasing the localized surface-plasmon resonance (LSPR) due to their unique crystal-like, hierarchical, micro- and nanoscale periodicity feature which leads to the near-field optical amplification of signal.[73,74] Various chemical and biological analytes were prepared using photonic crystals as nano-plasmonic sensors with the sensing mechanism of SERS which was fabricated from the diatoms from *Pinnularia* sp. The *in situ* grown silver or self-assembled gold nanoparticles on the large surface area of the diatom frustules is responsible for the sensing SERS mechanism, whose plasmons optically combine with the guided-mode resonance of the diatom frustules.[75] Moreover, fluidic sensing system applications of diatom frustules were studied by various authors. For example, the AuNPs-decorated fluidic properties of diatomite biosilica immobilized on a chip for SERS biosensing, was reported by Kong *et al.*[76] The fabricated devices have the ability of splitting and sensing tiny molecules from mixed samples with an ultrahigh detection sensitivity, down to 1 ppm.[76]

7.4.3 Diatom Biosilica Composites in Bone Tissue Engineering Applications

For effective bone regeneration and reconstitution, the interplay between proteins such as bone morphogenetic protein, alkaline phosphatase, organic components such as organophosphate and inorganic minerals like calcium, hydroxyapatite, silica, *etc.* is necessary.[77,78] In mammals, along with calcium phosphate-based ceramics and polymeric–ceramic based composites, silicon also has a pivotal role in communicating strategies and the mineralization process of bone regeneration, and also various stages of calcification phenomenon can be controlled with the presence of minerals like Si, Mg and Zn. Among these minerals, Si plays a significant role in the initial stages of the bio-calcification process of bone.[79,80] The *in vitro* osteo-inductive and bioactive properties of composites made up of silica-collagen materials in simulated body fluid (SBF) solution have been further enhanced by the incorporation of calcium phosphate ceramic as a third constituent.[81,82] The mechanism involved in the bio-calcification process of Si is due to the ability of the binding nature of Si with glycosaminoglycans and controls the crosslink interactions between collagen and proteoglycans (extracellular matrix (ECM) components) and increases the quality of ECM to enable bone mineralization.[83,84] Likewise, silicon restricts the bone destroying process of osteoclast formation and bone resorption, straightforwardly.[82] However, various synthetic Si materials are employed for this bone regeneration process, the synthetic procedures contain hazardous chemicals and surfactants for crosslinking which may delaying the process of bone regeneration. To overcome this drawback of synthetic silica materials, biogenic silica from natural diatom has gained huge attention for bone reconstruction. Only the size, shape and surface properties like surface roughness are deciding factors in the bone construction remodeling of naturally occurring diatom Si, since it is non-toxic and highly bioactive.[85] In addition, the micro and nanoparticles based on diatom biosilica has the tendency to release Si ions which has beneficial properties for bone regeneration.[86] The biomimetic approach utilizes polymer incorporated inert matrices for the connection of histidine incorporated silicate for the nanoparticulate biosilica synthesis.[87] The suspension of Si based particles derived from diatoms release free Si ions in deionized water which was confirmed using inductively coupled plasma-optical emission spectrometry (ICP-OES). Its kinetics are affected only by its method of purification of diatomite and its particle size without showing any cytotoxicity *in vitro*.[86] To enhance the adhesion and proliferation of *L. murine* fibroblasts and human osteosarcoma cell lines MG63, *Thalassiosira weissflogii* frustules were utilized by Cicco, S. R. *et al.* in 2015. In order to avoid any inflammation in biological systems, they have covalently functionalized the *T. weissflogii* diatom with well-organized scavenger cyclic nitroxide 2,2,6,6-tetramethylpiperidine-oxyl (TEMPO). Moreover, to avoid any bacterial infection, the antibiotic drug ciprofloxacin was also physiosorbed on the TEMPO-biosilica to be utilized in teeth and bone. The authors identified that biosilica based diatom derived material is suitable for fibroblasts and osteoblast-like cells growth with anti-inflammatory and antibiotic effects.[87] The hydroxyapatite facilitation and amelogenin, ameloblastin, and enamel inexpression ability of biosilica in human osteoblast-like SaOS-2 cells were reported also.[88,89] Moreover, the osteoclast differentiation function of these biosilica material was eradicated by osteo-protegerin (OPG) and NF-kB ligand (RANKL) expression of SaOS-2 cells on biosilica materials[90] (see Figure 7.6).

7.4.4 Diatom Biosilica Composites in Hemorrhage Control

In unfortunate war situations, many hospital surgeries like cardiovascular, orthopedic and hepatic surgeries deal with severe frequent bleeding which results in hypovolemic shock. If the wound is not treated and bleeding is not stopped, this ultimately can lead to death.[91,92] However, many efforts have been made to treat hemorrhage bleeding with very potent hemostatic agents, but these agents are still lacking in efficiency to control hemorrhage bleeding. For example, QuikClot® zeolite is a generally accepted blood-coagulating agent. But while applying this agent into tissues, it will generate heat increasing from 45 °C to 95 °C which destroys the surrounding tissues.[93] To overcome this drawback of QuikClot® zeolite, HemCon® bandage containing a chitosan derivative has been introduced but its shape restricts its usage in larger and deeper wounds.[94] So, the ideal hemostatic agents should contain the properties for quick blood clotting in a variety of conditions, to reduce the damage at the neighboring site, be non-toxic, non-immunogenic and cost effective.

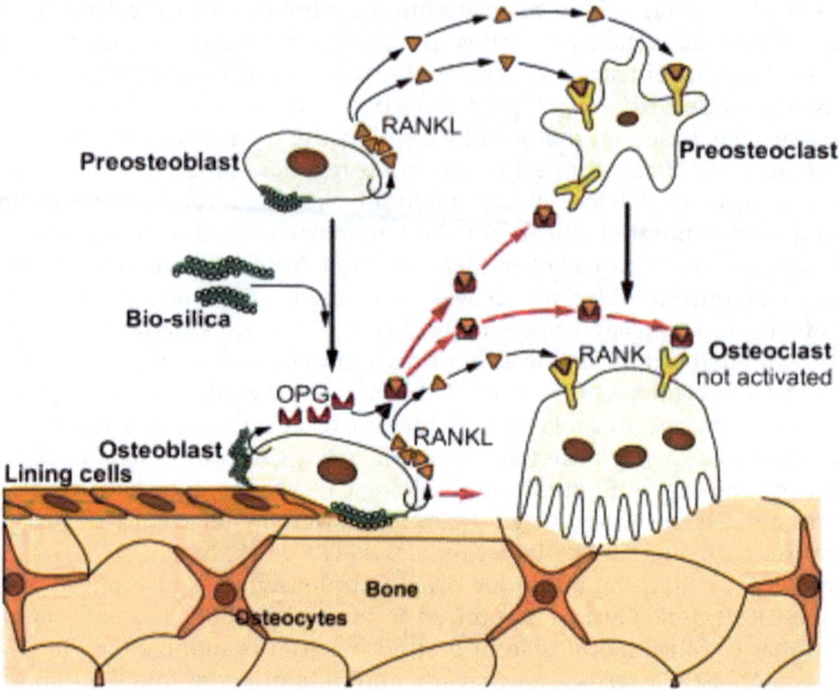

Figure 7.6 Anticipated influence of biosilica on osteoblasts, osteoclasts, and their progenitors. Biosilica facilitates OPG expression in osteoblasts. Osteoblasts have the potential to differentiate to osteocytes and lining cells. OPG counteracts various effects of RANKL, a cytokine that induces preosteoclast maturation and osteoclast activation. Reproduced from ref. 90 with permission from Elsevier, Copyright 2010.

Due to quick plasma absorption ability, temperature control in surrounding tissues, and potent exchange triggered homeostasis, materials made up of porous silica have attained extraordinary attention to fabricate hemostatic materials. Biosilica, one of the low price mesoporous silica materials, is subject to research for their hemorrhage control ability because of some drawbacks of synthetic silica materials like the high cost of production, elongated requisite of time for the production, containing toxic chemicals, *etc.* Diatomite biosilica coated with a biocompatible and biodegradable derivative of chitosan has a potent hemostatic quality and has effective plasma absorption ability, blood compatibility and importantly zero cytotoxicity, which eventually is mirrored in the enhanced efficiency in the control of hemorrhage bleeding.[92] Wang *et al.*, in 2018, have stated the application of chitosan, dopamine and diatom silica composite beads for quick blood coagulation.[95] Dopamine can act as a bio-glue in this composite. The authors found the composite beads acted as a potent biocompatible material with a hemolytic rate of below 5% and cell viability rate of above 80%.[95]

7.4.5 Sponge Biosilica Composites in Tissue Engineering Applications

The biodegradable, biocompatible and bioactive natures of silica-based bio-glasses are making silica a potential bone repairing biomaterial. Its bioactive property is achieved by the formation of bonding between the biosilica and bone tissues *via* the formation of hydroxyapatite (HAP) matrix. On the other hand, the requirement of a high temperature calcination process and the brittle nature of the synthesized biosilica are some drawbacks of these materials, which is making researchers move to some nature inspired marine based biomaterials.[96] Owing to marine sponges' divergent structures with pores, they can act as tissue scaffolds. In sponges, the biosilicification process, happening at the crossing point between the organic and inorganic phases, has directed researchers to improve their innovative hierarchical materials based on the "Darwinian approach".[97] Bio-inspired silicification is a favorable method for the design and improvement of bone-renovating biomaterials. In 2014, the osteo-inductive activity of biosilica based materials for curing bone disease was confirmed by Wang *et al.* in human osteoblast-like cells (SaOS-2) *in vitro*.[98] Amorphous glassy silica (SiO_2) can be formed in the sponges under the natural conditions of pH 8.2, 30–35 PSU salinity and −2 to 30 °C temperature.[99] The spicules are present in sponges in a very high amount up to 95% and also the sizes are varied between 0.01 and 1 mm. For example, the underwater *Monorhaphis chuni*, secretes 3 m long and 1 cm wide spicules.[100,101] The non-toxic biocompatible nature of the biosilica to mammalian cells was confirmed already.[102]

Some works are available that prove the bone mineralization ability of sponges derived biosilica materials. In 2007, Muller *et al.* have established the higher concentration of the enamel matrix such as amelogenin and enamelin structural molecules' levels in the presence of *S. domuncula* derived silica-based components that cause HAP crystallite formation. Moreover, when the SaOS-2 cells are exposed to biosilica derived from *S. domuncula*, it demonstrated the increasing process of bio-mineralization in *in vitro* study. From these results, we can conclude that this enzymatically amalgamated biosilica *via* silicatein (silica containing enzymes present in various sponges) may be a hopeful way for tooth and bone reconstruction *in vivo*.[89] The anabolic effect on bone forming osteoblast cells and destroying effects on bone resorbing osteoclast cells of silicatein prepared biosilica were evaluated. This was confirmed in 2010, by Wiens *et al.* They have demonstrated the expression of osteo-protegerin (OPG – inhibitory factor to osteoclastogenesis) was favored more than the osteoclastogenic NF-kB ligand (RANKL), in SaOS-2 cells when they are exposed to biosilica.[90] From this study, we can easily understand the influence of biosilica on the bone forming and bone resorbing mechanism. In addition, the same authors demonstrated silicatein/biosilica-modified materials from *S. domuncula* have an ability of osteogenic activity of SaOS-2 cells *via* enhanced

Table 7.2 The osteogenic ability of marine sponges derived biosilica. Adapted from ref. 106 with permission from John Wiley & Sons, Copyright © 2016 Wiley Periodicals, Inc.

Name of the sponges	Cells treated (*in vitro*)	Results
Suberites domuncula	SaOS-2 cells	Enhanced deposition of calcium phosphate ceramic in the surfaces with biosilica (silicatein)-modification.
Suberites domuncula	SaOS-2 cells	Improved expression of osteoclasts inhibitory factors OPG in biosilica exposed cells while osteoclasts development RANKL expression continued unaffected
Suberites domuncula	SaOS-2 cells	HAP formation and expression of bone morphogenetic protein 2 (BMP-2).
Suberites domuncula	Human multi-potent stromal cells (hMSC)	Stimulated viability and differentiation of hMSCs, enhanced mineralization in osteogenic cells, expression of BMP-2 and alkaline phosphatase (ALP)

proliferation and HAP formation.[103] Barros *et al.* in 2014, extracted the biosilica from *Petrosia ficiformis* for developing tissue engineered biomedical devices. They demonstrated that the bioactive crystals nucleation and cell colonization was more favored for the acidic treated biosilica than other treatments which reveals that the miscellany of pivotal biomedical applications of marine sponges in bone regenerative approaches.[104] Moreover, the tissue regenerating ability of biosilica based composite was also evaluated. In 2014, Wang *et al.* evaluated the tissue regenerating ability of β-tricalcium phosphate (β-TCP)/biosilica/biodegradable poly(D,L-lactide-*co*-glycolide) (PLGA) composite using SaOS-2 cells by MTT assay.[105] Table 7.2 displays the osteogenic ability of marine sponges derived biosilica.[106]

7.4.6 Sponge Biosilica Composite in Drug Delivery Applications

Despite the drug delivery applications based on biogenic silica from sponges having been little studied, they have benefits over synthetic silica materials in drug delivery applications.[107] In 2020, Kaya *et al.* extracted the multifunctional, open-channeled and hollow bioactive micro silica spheres from the *Geodia macandrewii* marine sponge. They used hydrogen fluoride for peeling the outer surface of sterrasters to get the micro biosilica with homogeneously dispersed pores, a cavity in the center of the sphere, and channels opening from each pore into the central cavity. They found that the bioactivity of the material is only valid for micro silica spheres. In case of untreated sterrasters, there was no bioactivity recorded in simulated body fluid (SBF). Due to their porous network, the obtained spheres exhibited great loading and releasing behavior for an anti-cancer drug, carboplatin, in the human HeLa cancer cell line.[108] This study proves that the biosilica from marine sponges also has the ability to deliver therapeutic drugs onto a targeted site with a controlled manner.

7.4.7 Plant Derived Biosilica Composites for Biomedical Applications

Preparation of Si nanoparticles (SiNPs) and Si composites involves tetra ethyl orthosilicate (TEOS) and tetra methyl orthosilicate organic precursors.[109] However, in biomedical and materials fields, biosilica derived from plant natural sources is utilized due to their low cost, eco-friendliness and their great abundance. Some of the silica can be effectively derived from various biomasses such as coffee husk, rice husk, sugarcane bagasse, wheat husk and corn cob. Rice husk comprises cellulose (75–90%), hemicellulose, lignin and ash components (17–20%) and contains a plentiful amount of silicon. The ash consists of greater than 90% of silica with some metallic impurities.[110] In the case of sugarcane bagasse industry; a main side-product *i.e.* sugarcane ash consists of about 40–50% of silica. This quantity can be increased up to 80% by acid pretreatment.[111] Moreover, corn cobs extracted from maize corn, wheat husk ash, bamboo leaf and raw teff straw contains more than 60%, 90%, 75.90–82.86 wt% and 52% of silica contents, respectively.[112–115] The Si content in raw teff straw can be enhanced up to 97% by thermal treatment. The SiNPs and composites derived from these resources using various methods and treatments like sol–gel method, acid-thermal treatment, sintering, alkali extraction and calcination can be used in various biomedical fields including bio-sensing and drug delivery along with super capacitors and bio-imaging.[116] The existence of silanol groups at the surface of the Si matrix, makes the SiNPs easily functionalized to load and discharge therapeutic drugs.[117] Functionalization of mesoporous SiNPs with amines are broadly used for delivering drugs to the targeted site.[118] Various drugs including heparin, hydrocortisone, metoprolol, salidroside, venlafaxine, vancomycin, methotrexate, sophoridine, insulin, cyclic adenosine monophosphate, doxorubicin and 5-fluorouracil were loaded on the Si surface with high loading capacity and controlled releasing behavior.[116] Mesoporous silica composited with magnetic carbon nanotubes exhibits a high loading tendency for therapeutics such as protein cytochrome *C* and gentamicin.[119] The drug allantoin was loaded in composites of SiNPs derived from rice husks and polycaprolactone nanofibers and released in a controlled manner.[120] In 2020, Dhinasekaran *et al.* prepared 5-fluorouracil (5-FU) loaded chitosan-coated biomass SiNPs in order to diminish the cost and enhance the biocompatibility, and the results exhibited that this composite has high potential in affecting the cancer cell lines MCF-7 and A549[121] (see Figure 7.7). The various natural sources of biosilica have been showed in Figure 7.8.

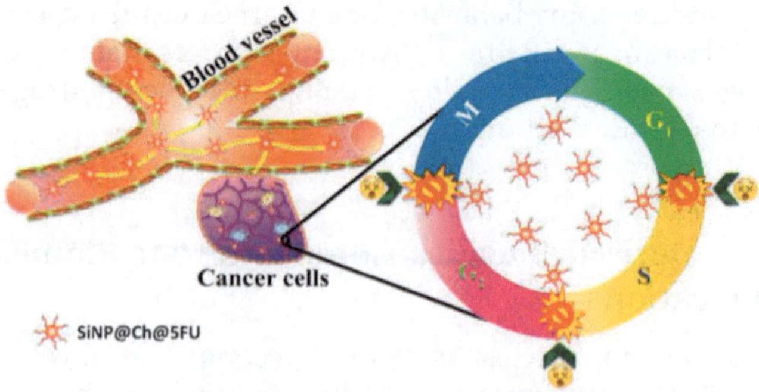

Figure 7.7 Proposed working mechanism of anticancer activity of 5-FU (5-fluorouracil) functionalized SiNPs extracted from rice husk. Reproduced from ref. 121 with permission from Elsevier, Copyright 2020.

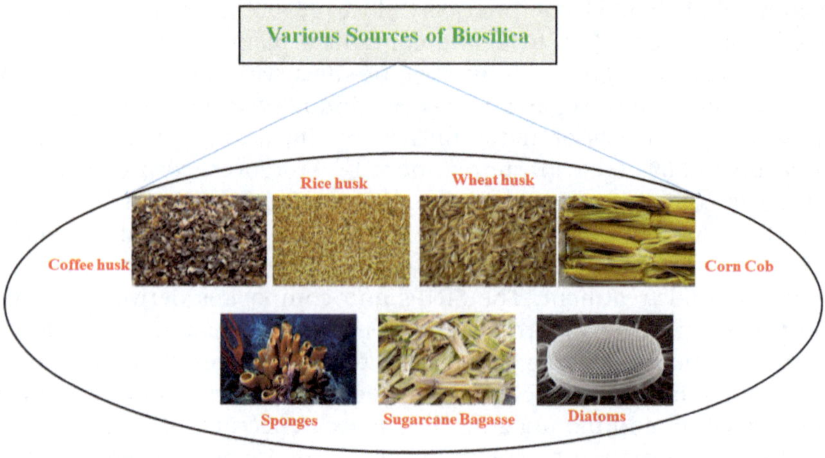

Figure 7.8 Various natural sources (marine as well as plants) of biosilica.

7.4.8 Influence of Minerals on Biomedical Properties of Diatom Biosilica

Essential elements or metals have a pivotal role in human biological organisms to implement their metabolic function. The same thing is also the same for diatoms. They are getting their trace metals and vitamins from their natural surroundings. Same as magnesium – a central metal cofactor in chlorophyll – these trace metals are also required for enzyme production.[122] Moreover, like nitrogen and phosphorus, iron is also an important element that is utilized by the respiratory electron transport system by the production of protein.[123] The naturally occurring raw frustules contain aluminium, iron, and titanium at about 1.600 wt%, 0.282 wt% and 0.125 wt%, respectively, in both of their organic cell phase and inorganic silica frustule.[124] Apart from these metals, substitution of some essential

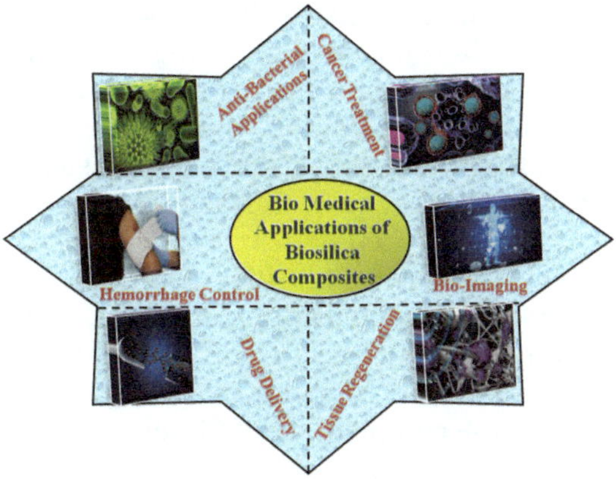

Figure 7.9 Overall biomedical applications of biosilica composites.

metals like germanium, titanium and calcium *via in vitro* and *in vivo* feeding also helps to regulate bone formation. The germanium ion was added into diatom frustules in the form of germanium dioxide and oscinod acid. The added germanium ion can act as an inhibitor for the growth of diatoms contaminating other algal cultures and also interrupt the construction of frustules in the diatoms.[43,125] By using the *in vivo* nurturing of algae with calcium chloride salts, Leone *et al.* in 2017, have prepared calcium-doped biosilica for bone repair. In this work, the authors matched nanostructured biosilica obtained from diatom frustules with and without Ca ion. They found that the Ca ion doped biosilica frustules have good viability as a result of osteoblast like cells, confirmed by MTT assay. They conclude that the Ca doped biosilica will be served as a best platform for regenerative medicine.[46] In 2018, Li *et al.* have prepared the Ca doped biosilica extracted from the *Oscinodiscus* sp. diatoms. They found that due to their unique structural feature and optimum composition with Ca ion, the Ca doped biosilica exhibited various hemostatic performances which promote blood clotting and *in vivo* hemorrhage control.[126] The various biomedical applications of biosilica composites have been shown in Figure 7.9.

7.5 Conclusions with Future Directions

As our knowledge of various natural species including marine sponges, diatoms and plant materials increases, we are providing novel and innovative paths to consume their biological benign properties for biomedical applications. In this chapter, along with different biomedical applications of biosilica based composites, we discussed the general structural features of biosilica with its formation mechanism in various bio-silicification sources. In diatoms, for the construction of frustules, a Si source is necessary and is obtained from their surrounding aqueous environments in the form of ortho silicic acid. However, other Si sources for diatoms and the exact mechanism of various pathways of silica uptake still needs to be investigated. Using diatoms-based Si composites; the delivery of single as well as duel anticancer and antibiotic organic and inorganic drugs done in a controlled manner was highlighted in this chapter. In addition, the benefits of using diatom and other biosilica over synthetic silica in bone tissue regeneration and hemorrhage control are highlighted. Various natural plants having silica contents are discussed with their potent applications. In the future, the delivery of antibiotic drugs using biosilica composites requires more detailed analysis. Moreover, the applications of plants derived Si based composites in drug delivery applications, are still lacking in research. Therefore, it is anticipated that numerous research groups will remain focused on Si based composites from the complete set of Si rich natural plants (*e.g.*, coffee husk, rice husk, sugarcane bagasse, wheat husk and corn cob) and examine their drug delivery potential. The molecular dynamic simulation and molecular biology techniques could provide us the knowledge about the biomolecule-silica affinity interactions which will help to predict the biosilicification mechanism exactly. This gained information will increase our current knowledge in the field of interactions-based loading and releasing of therapeutic proteins and drugs on and from Si composites. Therefore, we anticipate that the upcoming era of research could convey the bio-mineralization mechanism of plant materials, diatom cells and siliceous sponges toward the development of biomaterials and biomedical applications.

References

1. J. Shih Lee, *Being a Biomedical Entrepreneur*, Global Monitors, Inc., USA & University of Virginia, USA, 2018, p. 1.
2. Global healthcare spend to remain stable, Pricing and Market Access, 2020.
3. National Institutes of Health, The NIH-wide strategic plan, fiscal years 2016–2020: turning discovery into health, 2015.
4. K. McNamara and S. A. M. Tofail, Nanoparticles in biomedical applications, *Adv. Phys.*, 2016, **2**, 54–88.
5. M. Raguraman and M. Rajan, Nanoengineering/technology for tissue engineering and organ printing, *Emerging nanotechnologies for medical applications*, 2023, pp. 35–54.
6. A. Moretta, *et al.*, Antimicrobial Peptides: A New Hope in Biomedical and Pharmaceutical Fields, *Front. Cell. Infect. Microbiol.*, 2021, **11**, 668632.
7. M. Bencina *et al.*, *Advances in Biomembranes and Lipid Self-Assembly*, 1st edn, 2016, vol. 24.
8. N. Phan, Biological therapy: a new age of cancer treatment, *Biomed. Res. Ther.*, 2014, 6.
9. K. Vinothini, S. S. Kumar, H. Abrahamse and M. Rajan, Synergistic effect of polymer functionalized graphene oxide system for breast cancer treatment, *Int. J. Pharm.*, 2023, **632**, 122556.

10. M. Holzinger, A. Le Goff and S. Cosnier, Nanomaterials for biosensing applications: a review, *Front. Chem.*, 2014, **2**, 63.
11. S. Mehmet Erturk, C. Johnston, C. Tempany-Afdhal and A. D. Van den Abbeele, Imaging Tools in Human Research, *Clin. Transl. Sci.*, 2009, 87–104.
12. R. Jugdaohsingh, Silicon and bone health, *J. Nutr., Health Aging*, 2007, **11**, 99–110.
13. R. K. Iler, *The Chemistry of Silica: Solubility, Polymerization, Colloidal and Surface Properties, and Biochemistry*, John Wiley & Sons, NY, Newyork, 1979.
14. M. A. Farooq and K. J. Dietz, Silicon as Versatile Player in Plant and Human Biology: Overlooked and Poorly Understood, *Front. Plant Sci.*, 2015, **6**, 994.
15. A. Schiano, F. Eisinger, P. Detolle, A. M. Laponche, B. Brisou and J. Eisinger, Silicon, bone tissue and immunity, *J. Rev. Rhum.*, 1979, **46**, 483–486.
16. M. Hott, C. Pollak, D. Modrowski and P. J. Marie, Short-term effects of organic silicon on trabecular bone in mature ovariectomized rats, *Calcif. Tissue Int.*, 1993, **53**, 174–179.
17. R. Jugdaohsingh, S. H. Anderson, K. L. Tucker, H. Elliott, D. P. Kiel, R. P. Thompson and J. J. Powell, Dietary silicon intake and absorption, *Am. J. Clin. Nutr.*, 2002, **75**, 887–893.
18. H. Robberecht, R. Van-Cauwenbergh, V. Van-Vlaslaer and N. Hermans, Dietary silicon intake in Belgium: Sources, availability from foods, and human serum levels, *Sci. Total Environ.*, 2009, **407**, 4777–4782.
19. S. Gillette-Guyonnet, S. Andrieu, F. Nourhashemi, V. de La Guéronnière, H. Grandjean and B. Vellas, Cognitive impairment and composition of drinking water in women: findings of the EPIDOS Study, *Am. J. Clin. Nutr.*, 2005, **81**, 897–902.
20. V. R. Phoenix, P. C. Bennett, A. S. Engel, S. W. Tyler and F. G. Ferris, Chilean high-altitude hot-spring sinters: a model system for UV screening mechanisms by early Precambrian cyanobacteria, *Geobiology*, 2006, **4**, 15.
21. P. Yang, W. Mengjiao, W. Andrzej, K. Marta, W. Yinchu and L. Wenjun, A hybrid biomaterial of biosilica and C-phycocyanin for enhanced photodynamic effect towards tumor cells, *Biochem. Biophys. Res. Commun.*, 2020, **533**, 573–579.
22. H. Ehrlich, K. D. Demadis, O. S. Pokrovsky and P. G. Koutsoukos, Modern views on desilicification: biosilica and abiotic silica dissolution in natural and artificial environments, *Chem. Rev.*, 2010, **110**, 4656–4689.
23. E. C. Bovee, Distribution and Forms of Siliceous Structures Among Protozoa, *Silicon and Siliceous Structures in Biological Systems*, New York, 1981.
24. A. J. Gooday, G. F. Esteban and K. J. Clarke, Organic and siliceous protistan scales in north-east Atlantic abyssal sediments, *J. Mar. Biol. Assoc. U. K.*, 2006, **86**(4), 679–688.

25. G. F. Esteban, A. J. Gooday and K. J. Clarke, Siliceous scales of filose-amoebae (Pompholyxophryidae, Rotosphaerida) from deep Southern Ocean sediments, including first records for the Southern Hemisphere, *Polar Biol.*, 2007, **30**, 945–950.
26. E. Fourcade, J. Mendez, J. Azema, P. Cros, P. De Wever, J. Duthou, J. E. Romero and F. Michaud, *C. R. Acad. Sci., Ser. IIa: Sci. Terre Planetes*, 1994, **318**, 527.
27. O. R. Anderson, C. Perry Carole and N. P. Hughes, Transmission and scanning electron microscopic evidence for cytoplasmic deposition of strontium sulphate crystals in colonial radiolarian, *Philos. Trans. R. Soc. London, Ser. B*, 1990, **329**, 8–86.
28. D. J. Patterson and M. Dürrschmidt, The formation of siliceous scales by Raphidiophrys ambigua (Protista, Centroheliozoa), *J. Cell Sci.*, 1988, **91**, 33.
29. L. P. Gaponova and I. P. Dovgal, The Variability of Exoskeleton Elements In Poly placocystis Ambigua (Protista, Centrohelida), *Vestn. Zool.*, 2008, **42**, 79–83.
30. B. S. C. Leadbeater, M. Henouil and N. Berovic, Choanoflagellate Lorica Construction and Assembly: The Nudiform Condition. II. Acanthoeca spectabilis Ellis, *Protist*, 2008, **159**, 495–505.
31. P. R. Bergquist, *Sponges*, Hutchinson & Co., London, 1978, p. 268.
32. K. R. Tabachnick and H. M. Reiswig, Dictionary of Hexactinellida, in *Systema Porifera*, ed. J. N. A. Hooper, R. W. M. Van Soest and P. Willenz, Springer, Boston, MA, 2002, p. 1224.
33. E. Paasche, *The physiological ecology of phytoplankton*, Blackwell Scientific, 1980, p. 259.
34. J. C. Lewin, Silicon metabolism in diatoms. III. Respiration and silicon uptake in Navicula pelliculosa, *J. Gen. Physiol.*, 1955, **39**, 1.
35. Y. D. Amo and M. A. Brzezinski, The chemical form of dissolved Si taken up by marine diatoms, *J. Phycol.*, 1999, **35**, 1162.
36. V. Martin-Jézéquel, M. Hildebrand and M. A. Brzezinski, Silicon metabolism in diatoms: implications for growth, *J. Phycol.*, 2000, **36**, 821.
37. K. Thamatrakoln and A. B. Kustka, When to say when: can excessive drinking explain silicon uptake in diatoms?, *BioEssays*, 2009, **31**, 322.
38. E. Paasche, Silicon and the ecology of marine plankton diatoms. II. Silicate-uptake kinetics in five diatom species, *Mar. Biol.*, 1973, **19**, 262.
39. P. J. Harrison, H. L. Conway, R. W. Holmes and C. O. Davis, Marine diatoms grown in chemostats under silicate or ammonium limitation. III. Cellular chemical composition and morphology of Chaetoceros debilis, Skeletonema costatum, and Thalassiosira gravida, *Mar. Biol.*, 1977, **43**, 19.
40. J. D. Pickett-Heaps, A. M. M. Schmid and L. A. Edgar, The cell biology of diatom valve formation, *Prog. Phycol. Res.*, 1990, 7, 1–168.
41. W. E. G. Muller, *Silicon Biomineralization*, Springer Berlin Heidelberg, Berlin, Heidelberg, 2003, p. 33.

42. K. Albert, X. C. Huang and H. Y. Hsu, Bio-templated silica composites for next-generation biomedical applications, *Adv. Colloid Interface Sci.*, 2017, **249**, 272.
43. Y. Loo, M. Goktas, A. B. Tekinay, M. O. Guler, C. A. E. Hauser and A. Mitraki, Self-Assembled Proteins and Peptides as Scaffolds for Tissue Regeneration, *Adv. Healthcare Mater.*, 2015, **4**, 2557.
44. N. Kroger, R. Deutzmann and M. Sumper, Polycationic Peptides from Diatom Biosilica That Direct Silica Nanosphere Formation, *Science*, 1999, **286**, 1129.
45. E. Kasotakis and A. Mitraki, Silica bio-templating by self-assembling peptides *via* serine residues activated by the peptide amino terminal group, *Biopolymers*, 2012, **98**, 501.
46. S. V. Patwardhan, Biomimetic and bioinspired silica: recent developments and applications, *Chem. Commun.*, 2011, **47**, 7567.
47. J. A. Speir, S. Munshi, G. Wang, T. S. Baker and J. E. Johnson, Structures of the native and swollen forms of cowpea chlorotic mottle virus determined by X-ray crystallography and cryo-electron microscopy, *Structure*, 1995, **3**, 63.
48. F. M. Fernandes, T. Coradin and C. Aimé, Self-Assembly in Bio silicification and Bio templated Silica Materials, *Nanomaterials*, 2014, **4**, 792.
49. B. Cao, H. Xu and C. Mao, Controlled Self-Assembly of Rodlike Bacterial Pili Particles into Ordered Lattices, *Angew. Chem. Int. Ed.*, 2011, **50**, 6264.
50. J. Wu, T. Coradin and C. Aime, Reversible bio responsive aptamer-based nanocomposites: ATP binding and removal from DNA-grafted silica nano particles, *J. Mater. Chem. B*, 2013, **1**, 5353.
51. J. Wu, J. Silvent, T. Coradin and C. Aimé, Biochemical Investigation of the Formation of Three-Dimensional Networks from DNA-Grafted Large Silica Particles, *Langmuir*, 2012, **28**, 2156.
52. B. Delalat, V. C. Sheppard, S. R. Ghaemi, S. Rao, C. A. Prestidge, G. McPhee, M. L. Rogers, J. F. Donoghue, V. Pillay, T. G. Johns, N. Kröger and N. H. Voelcker, Targeted drug delivery using genetically engineered diatom biosilica, *Nat. Commun.*, 2015, **6**, 8791.
53. M. Terracciano, M. A. Shahbazi, A. Correia, I. Rea, A. Lamberti, L. D. Stefano and H. A. Santos, Surface bioengineering of diatomite based nano vectors for efficient intracellular uptake and drug delivery, *Nanoscale*, 2015, 7, 20063.
54. R. Sasirekha, T. S. Sheena, M. S. Deepika, P. Santhanam, H. E. Townley, K. Jeganathan, S. Dinesh Kumar and K. Premkumar, Surface engineered Amphora sub tropica frustules using chitosan as a drug delivery platform for anticancer therapy, *Mater. Sci. Eng., C*, 2019, **94**, 56.
55. A. Kabir, N. Nazeer, R. Bissessur and M. Ahmed, Diatoms embedded, self-assembled carriers for dual delivery of chemotherapeutics in cancer cell lines, *Int. J. Pharm.*, 2020, **573**, 118887.
56. D. Vona, G. Leone, R. Ragni, F. Palumbo, A. Evidente, M. Vurro, G. M. Farinola and S. R. Cicco, Diatoms Biosilica as Efficient Drug-Delivery System, *MRS Adv.*, 2016, **1**, 3825.

57. J. Delasoie, J. Rossier, L. Haeni, B. Rothen-Rutishauser and F. Zobi, Slow-targeted release of a ruthenium anticancer agent from vitamin B 12 functionalized marine diatom microalgae, *Dalton Trans.*, 2018, **47**, 17221.
58. S. N. Sovari, S. Vojnovic, S. S. Bogojevic, A. Crochet, A. Pavic, J. Nikodinovic-Runic and F. Zobi, Design, synthesis and *in vivo* evaluation of 3-arylcoumarin derivatives of rhenium(I) tricarbonyl complexes as potent antibacterial agents against methicillin-resistant *Staphylococcus aureus* (MRSA), *Eur. J. Med. Chem.*, 2020, **205**, 112533.
59. J. Delasoie, A. Pavic, N. Voutier, S. Vojnovic, A. Crochet, J. Nikodinovic-Runic and F. Zobi, Identification of novel potent and non-toxic anticancer, anti-angiogenic and antimetastatic rhenium complexes against colorectal carcinoma, *Eur. J. Med. Chem.*, 2020, **204**, 112583.
60. J. Delasoie, N. Radakovic, A. Pavic and F. Zobi, Neovascularization Effects of Carbon Monoxide Releasing Drugs Chemisorbed on Coscinodiscus Diatoms Carriers Characterized by Spectro microscopy Imaging, *Appl. Sci.*, 2020, **10**, 7380.
61. H. Shi, J. Lou, S. Lin, Y. Wang, Y. Hu, P. Zhang, Y. Liu and Q. Zhang, Diatom-like silica–protein nanocomposites for sustained drug delivery of ruthenium polypyridyl complexes, *J. Inorg. Biochem.*, 2021, **221**, 111489.
62. M. Kaksonen and A. Roux, Mechanisms of clathrin-mediated endocytosis, *Nat. Rev. Mol. Cell Biol.*, 2018, **19**, 313.
63. R. B. Vasani, D. Losic, A. Cavallaro and N. H. Voelcker, Fabrication of stimulus-responsive diatom biosilica microcapsules for antibiotic drug delivery, *J. Mater. Chem. B*, 2015, **3**, 4325.
64. S. R. Cicco, D. Vona, E. D. Giglio, S. Cometa, M. Mattioli-Belmonte, F. Palumbo, R. Ragni and G. M. Farinola, Chemically Modified Diatoms Biosilica for Bone Cell Growth with Combined Drug-Delivery and Antioxidant Properties, *ChemPlusChem*, 2015, **80**, 1104.
65. U. T. Uthappa, G. Sriram, O. R. Arvind, S. Kumar, J. Ho Young, G. M. Neelgund, D. Losic and M. D. Kurkuri, Engineering MIL-100(Fe) on 3D porous natural diatoms as a versatile high performing platform for controlled isoniazid drug release, Fenton's catalysis for malachite green dye degradation and environmental adsorbents for Pb^{2+} removal and dyes, *Appl. Surf. Sci.*, 2020, **528**, 146974.
66. S. Briceño, E. A. Chavez-Chico and G. González, Diatoms decorated with gold nanoparticles by In-situ and Ex-situ methods for *in vitro* gentamicin release, *Mater. Sci. Eng., C*, 2021, **123**, 112018.
67. X. Cong, Y. Mu, D. Qin, X. Sun, C. Su, T. Chen, X. Wang, X. Chen and C. Feng, Copper deposited diatom-biosilica with enhanced photothermal and photodynamic performance for infected wound therapy, *New J. Chem.*, 2022, **46**, 2140.
68. M. A. Sherief, G. T. El-Bassyouni, A. A. Gamal and M. A. Esawy, Modification of diatom using silver nanoparticle to improve antimicrobial activity, *Mater. Today: Proc.*, 2021, **43**, 3369.

69. I. Rea, M. Terracciano, S. Chandrasekaran, N. H. Voelcker, P. Dardano, N. M. Martucci, A. Lamberti and L. De Stefano, Bioengineered Silicon Diatoms: Adding Photonic Features to a Nanostructured Semiconductive Material for Biomolecular Sensing, *Nanoscale Res. Lett.*, 2016, **11**, 405.
70. K. Squire, X. Kong, P. LeDuff, G. L. Rorrer and A. X. Wang, Photonic crystal enhanced fluorescence immunoassay on diatom biosilica, *J. Biophotonics*, 2018, **11**, e201800009.
71. K. J. Squire, Y. Zhao, A. Tan, K. Sivashanmugan, J. A. Kraai, G. L. Rorrer and A. X. Wang, Photonic crystal-enhanced fluorescence imaging immunoassay for cardiovascular disease biomarker screening with machine learning analysis, *Sens. Actuators, B*, 2019, **290**, 118.
72. V. Selvaraj, A. Muthukumar, P. Nagamony and V. Chinnuswamy, Detection of typhoid fever by diatom-based optical biosensor, *Environ. Sci. Pollut. Res.*, 2018, **25**, 20385.
73. E. De Tommasi and A. C. De Luca, Diatom biosilica in plasmonics: applications in sensing, diagnostics and therapeutics, *Biomed. Opt. Express*, 2022, **13**, 3080.
74. G. Chen, Y. Wang, M. Yang, J. Xu, S. J. Goh, M. Pan and H. Chen, Measuring Ensemble-Averaged Surface-Enhanced Raman Scattering in the Hotspots of Colloidal Nanoparticle Dimers and Trimers, *J. Am. Chem. Soc.*, 2010, **132**, 3644.
75. X. Kong, K. Squire, E. Li, P. Le Duff, G. L. Rorrer, S. Tang, B. Chen, C. P. McKay, R. Navarro-Gonzalez and A. X. Wang, Chemical and Biological Sensing Using Diatom Photonic Crystal Biosilica With In-Situ Growth Plasmonic Nanoparticles, *IEEE Trans. Nanobiosci.*, 2016, **15**, 828.
76. X. Kong, E. Li, K. Squire, Y. Liu, B. Wu, L. J. Cheng and A. X. Wang, Plasmonic nanoparticles-decorated diatomite biosilica: extending the horizon of on-chip chromatography and label-free biosensing, *J. Biophotonics*, 2017, **10**, 1473.
77. X. Wang, H. C. Schro, M. Wiens, U. SchloBmacher and W. E. G. Muller, Chapter five - Biosilica: Molecular Biology, Biochemistry and Function in Demosponges as well as its Applied Aspects for Tissue Engineering, *Adv. Mar. Biol.*, 2012, **62**, 231.
78. R. A. Praphakar, M. Sumathra, R. S. Ebenezer, S. Vignesh, H. Shakila and M. Rajan, Fabrication of bioactive rifampicin loaded κ-Car-MA-INH/Nano hydroxyapatite composite for tuberculosis osteomyelitis infected tissue regeneration, *Int. J. Pharm.*, 2019, **565**, 543–556.
79. S. Prabakaran, M. Rajan, Z. Geng and Y. Liu, Fabrication of substituted hydroxyapatite-starch-clay bio-composite coated titanium implant for new bone formation, *Carbohydr. Polym.*, 2021, **271**, 118432.
80. N. B. Matsko, N. Znidaršič, I. Letofsky-Papst, M. Dittrich, W. Grogger, J. Strus and F. Hofer, Silicon: The key element in early stages of biocalcification, *J. Struct. Biol.*, 2011, **174**, 180.
81. D. Eglin, S. Maalheem, J. Livage and T. Coradin, In vitro apatite forming ability of type I collagen hydrogels containing bioactive glass and silica sol-gel particles, *J. Mater. Sci.: Mater. Med.*, 2006, **17**, 161.

82. S. Heinemann, C. Heinemann, R. Bernhardt, A. Reinstorf, B. Nies, M. Meyer, H. Worch and T. Hanke, Bioactive silica–collagen composite xerogels modified by calcium phosphate phases with adjustable mechanical properties for bone replacement, *Acta Biomater.*, 2009, **5**, 1979.
83. S. Heinemann, T. Coradin and M. F. Desimone, Bio-inspired silica–collagen materials: applications and perspectives in the medical field, *Biomater. Sci.*, 2013, **1**, 688.
84. C. T. Price, K. J. Koval and J. R. Langford, Silicon: a review of its potential role in the prevention and treatment of postmenopausal osteoporosis, *Int. J. Endocrinol.*, 2013, **2013**, 1.
85. D. M. Brown, N. Kanase, B. Gaiser, H. Johnston and V. Stone, Inflammation and gene expression in the rat lung after instillation of silica nanoparticles: Effect of size, dispersion medium and particle surface charge, *Toxicol. Lett.*, 2014, **224**, 147.
86. T. D. H. Le, W. Bonani, G. Speranza, V. Sglavo, R. Ceccato, D. Maniglio, A. Motta and C. Migliaresi, Processing and characterization of diatom nanoparticles and microparticles as potential source of silicon for bone tissue engineering, *Mater. Sci. Eng., C*, 2016, **59**, 471.
87. S. R. Cicco, D. Vona, E. D. Giglio, S. Cometa, M. Mattiolo-Belmonte, F. Palumbo, R. Ragni and G. M. Farinola, Chemically Modified Diatoms Biosilica for Bone Cell Growth with Combined Drug-Delivery and Antioxidant Properties, *ChemPlusChem*, 2015, **80**, 1104.
88. H. C. Schröder, O. Boreiko, A. Krasko, A. Reiber, H. Schwertner and W. E. G. Müller, Mineralization of SaOS-2 cells on enzymatically (silicatein) modified bioactive osteoblast-stimulating surfaces, *J. Biomed. Mater. Res., Part B*, 2005, **75**, 387.
89. W. E. G. Müller, A. Boreiko, X. Wang, A. Krasko, W. Geurtsen, M. R. Custódio, T. Winkler, L. Lukic-Bilela, T. Link and H. C. Schroder, Morphogenetic Activity of Silica and Bio-silica on the Expression of Genes Controlling Biomineralization Using SaOS-2 Cells, *Calcif. Tissue Int.*, 2007, **81**, 382.
90. M. Wiens, X. Wang, H. C. Schroder, U. Kolb, U. Schlo Bmacher, H. Ushijima and W. E. G. Muller, The role of biosilica in the osteoprotegerin/RANKL ratio in human osteoblast-like cells, *Biomaterials.*, 2010, **31**, 7716.
91. A. M. Behrens, M. J. Sikorski and P. Kofinas, Hemostatic strategies for traumatic and surgical bleeding, *J. Biomed. Mater. Res., Part A*, 2014, **102**, 4182.
92. C. Fenc, J. Li, G. S. Wu, Y. Z. Mu, M. Kong, C. Q. Jiang, X. J. Cheng, Y. Liu and X. C. Chen, Chitosan-Coated Diatom Silica as Hemostatic Agent for Hemorrhage Control, *ACS Appl. Mater. Interfaces*, 2016, **8**, 34234.
93. R. Peter, B. Carlos, M. Matthew, S. Ali, P. Dave, G. Donald, C. Lowell, D. Demetrios, V. George and A. Hassan, Quik Clot Use in Trauma for Hemorrhage Control: Case Series of 103 Documented Uses, *J. Trauma*, 2008, **64**, 1093.

94. C. Dai, C. Liu, J. Wei, H. Hong and Q. Zhao, Molecular imprinted macroporous chitosan coated mesoporous silica xerogels for hemorrhage control, *Biomaterials*, 2010, **31**, 7620.
95. Y. Wang, Y. Fu, J. Li, Y. Mu, X. Zhang, K. Zhang, M. Liang, C. Feng and X. Chen, Multifunctional chitosan/dopamine/diatom-biosilica composite beads for rapid blood coagulation, *Carbohydr. Polym.*, 2018, **200**, 6.
96. S. K. Kim, *Marine Biomaterials: Characterization, Isolation and Applications*, CRC Press, 2013.
97. M. Wiens, X. Wang, U. Schlossmacher, I. Lieberwirth, G. Glasser, H. Ushijima, H. C. Schröder and W. E. Müller, Osteogenic Potential of Biosilica on Human Osteoblast-Like (SaOS-2) Cells, *Calcif. Tissue Int.*, 2010, **87**, 513.
98. S. Wang, X. Wang, F. G. Draenert, O. Albert, H. C. Schröder, V. Mailänder, G. Mitov and W. E. G. Müller, Bioactive and biodegradable silica biomaterial for bone regeneration, *Bone*, 2014, **67**, 292.
99. W. E. G. Muller and I. M. Muller, Porifera: An Enigmatic Taxon Disclosed by Molecular Biology/Cell Biology, *Porifera Research: Biodiversity, Innovation and Sustainability*, 2007, p. 89.
100. D. Barthel, Tissue composition of sponges from the Weddell Sea, Antarctica: not much meat on the bones, *Mar. Ecol.: Prog. Ser.*, 1995, **123**, 149.
101. X. Wang, L. Gan, K. P. Jochum, H. C. Schröder and W. E. G. Müller, The Largest Bio-Silica Structure on Earth: The Giant Basal Spicule from the Deep-Sea Glass Sponge Monorhaphis chuni, *Evidence-Based Complementary Altern. Med.*, 2011, **2011**, 1.
102. H. C. Schröder, X. Wang, U. Schloßmacher, M. Wiens and W. E. G. Müller, Biogenic Inorganic Polysilicates (Biosilica): Formation and Biomedical Applications, *Prog. Mol. Subcell. Biol.*, 2013, **54**, 197.
103. M. Wiens, X. Wang, U. Schlossmacher, I. Lieberwirth, G. Glasser, H. Ushijima, H. C. Schroder and W. E. Muller, Osteogenic Potential of Biosilica on Human Osteoblast-Like (SaOS-2) Cells, *Calcif. Tissue Int.*, 2010, **87**, 513.
104. A. A. Barros, I. M. Aroso, T. H. Silva, J. F. Mano, A. R. C. Duarte and R. L. Reis, Surface Modification of Silica-Based Marine Sponge Bioceramics Induce Hydroxyapatite Formation, *Cryst. Growth Des.*, 2014, **14**, 4545.
105. S. Wang, X. Wang, F. G. Draenert, O. Albert, H. C. Schroder, V. Mailander, G. Mitov and W. E. Muller, Bioactive and biodegradable silica biomaterial for bone regeneration, *Bone*, 2014, **67**, 292.
106. R. N. Granito, M. R. Custodio and A. C. M. Renno, Natural marine sponges for bone tissue engineering: The state of art and future perspectives, *J. Biomed. Mater. Res., Part B*, 2017, **105**, 1717–1727.
107. J. Venkatesan, A. Sukumaranl, E. P. Chalisserry, S. Kwon Kim, R. Pallela and H. Ehrlich, Marine Sponges as Future Biomedical Models, *Marine Sponges: Chemico biological and Biomedical Applications*, 2016, p. 349.
108. M. Kaya, I. Bilican, M. Mujtaba, I. Sargin, M. E. Haskoylu, E. T. Oner, K. Zheng, A. R. Boccaccini, D. Cansaran-Duman, M. Serdar Onses,

I. Torun, L. Akyuz, C. Elbuken and M. V. Sørensen, Sponge-derived natural bioactive glass microspheres with self-assembled surface channel arrays opening into a hollow core for bone tissue and controlled drug release applications, *Chem. Eng. J.*, 2021, **407**, 126667.
109. A. Bosio, N. Rodella, A. Gianoncelli, A. Zacco, L. Borgese, L. E. Depero, P. A. Bingham and E. Bontempi, A new method to inertize incinerator toxic fly ash with silica from rice husk ash, *Environ. Chem. Lett.*, 2013, **11**, 329.
110. M. A. Azmi, N. A. A. A. Ismail, M. Rizamarhaiza, W. M. Hasif and H. Taib, Characterisation of silica derived from rice husk (Muar, Johor, Malaysia) decomposition at different temperatures, *AIP Conference Proceedings*, AIP Publishing, College Park, 2016.
111. S. K. L. Singh and S. All Jawald, *IOSR J. Mech. Civil Eng.*, 2015, **12**, 12, e-ISSN: 2278-1684.
112. S. S. Owoeye, B. Oji and J. Aderiye, Effect of Temperature, Time and Atmospheric Condition on Active Silica Extraction from Corn Cob Ash, *Int. J. Eng. Technol. Innovat.*, 2015, **2**, 2348.
113. H. J. Naqvi, A. Saeed, A. Umair and S. Tajwar, Precipitated Silica from Wheat Husk, *J. Pakistan Instit. Chem. Eng.*, 2011, **39**, 51.
114. S. Silviana and W. J. Bayu, Silicon Conversion From Bamboo Leaf Silica By Magnesiothermic Reduction for Development of Li-ion Baterry Anode, MATEC Web of Conferences', EDP Sciences, 2018.
115. A. B. Wassie and V. C. Srivastava, Synthesis and Characterization of Nano-Silica from Teff Straw, *J. Nano Res.*, 2017, **46**, 64.
116. S. Prabha, D. Durgalakshmi, S. Rajendran and E. Lichtfouse, Plant-derived silica nanoparticles and composites for biosensors, bioimaging, drug delivery and supercapacitors: a review, *Environ. Chem. Lett.*, 2021, **19**, 1667.
117. M. Vallet-Regí, F. Balas and D. Arcos, Mesoporous Materials for Drug Delivery, *Angew. Chem. Int. Ed.*, 2007, **46**, 7548.
118. M. Vallet-Regí, M. Colilla and B. González, Medical applications of organic–inorganic hybrid materials within the field of silica-based bioceramics, *Chem. Soc. Rev.*, 2011, **40**, 596.
119. R. K. Singh, K. D. Patel, J. J. Kim, T. H. Kim, J. H. Kim, U. S. Shin, E. J. Lee, J. C. Knowles and H. W. Kim, Multifunctional Hybrid Nanocarrier: Magnetic CNTs Ensheathed with Mesoporous Silica for Drug Delivery and Imaging System, *ACS Appl. Mater. Interfaces*, 2014, **6**, 2201.
120. M. Ke, J. A. Wahab, B. Hyunsik, K. H. Song, J. S. Lee, M. Gopiraman and I. S. Kim, Allantoin-loaded porous silica nanoparticles/polycaprolactone nanofiber composites: fabrication, characterization, and drug release properties, *RSC Adv.*, 2016, **6**, 4593.
121. D. Dhinasekaran, R. Raj, A. R. Rajendran, B. Purushothaman, B. Subramanian, A. Prakasarao and G. Singaravelu, Chitosan mediated 5-Fluorouracil functionalized silica nanoparticle from rice husk for anticancer activity, *Int. J. Biol. Macromol.*, 2020, **156**, 969.

122. R. E. Blankenship, *Molecular Mechanisms of Photosynthesis*, Blackwell Science Ltd, Oxford, 2002.
123. J. A. Raven, M. C. W. Evans and R. E. Korb, The role of trace metals in photosynthetic electron transport in O2-evolving organisms, *Photosynth. Res.*, 1999, **60**, 111.
124. J. H. Martin and G. A. Knauer, The elemental composition of plankton, *Geochim. Cosmochim. Acta*, 1973, **37**, 1639.
125. F. Azam, B. B. Hemmingsen and B. E. Volcani, Germanium incorporation into the silica of diatom cell walls, *Arch. Mikrobiol.*, 1973, **92**, 11.
126. J. Li, J. Han, Q. Sun, Y. Wang, Y. Mu, K. Zhang, X. Dou, M. Kong, X. Chen and C. Feng, Biosynthetic calcium-doped biosilica with multiple hemostatic properties for hemorrhage control, *J. Mater. Chem. B*, 2018, **6**, 7834.

CHAPTER 8

Biogenic Silica for Drug Delivery and Gene Therapy

RUBAI AHMED,[a] SWARNALI DAS,[a] SOVAN SAMANTA,[a] JHIMLI BANERJEE,[a] ARINDAM PRAMANIK*[b] AND SANDEEP KUMAR DASH*[a]

[a] Department of Physiology, University of Gour Banga, Malda-732103, West Bengal, India; [b] Amity Institute of Biotechnology, Amity University, Noida-201301, India
*Emails: arindampramanik87@gmail.com; deep.vu@gmail.com

8.1 Introduction

In the rapidly evolving biomedical field of drug and gene delivery, the necessity of safer and more effective carriers has driven researchers to investigate a wide range of materials. Silicon, the second most prevalent element on Earth remains as silicon dioxide in both crystalline and amorphous form.[1,2] The silica nanoparticle was synthesized by Stöber and his coworkers in 1968 by using a sol–gel method where tetraethyl orthosilicate (TEOS) was used as a precursor of silica.[3] The qualities of silica-based nano systems, which include low physical stability, variable size, and ease of surface modification, as well as their capacity to be generated on a large scale with excellent biocompatibility, have made them stand out among other inorganic materials.[4] This nanoparticle emerged to discover numerous applications in various fields, such as in the food industry, technical and medical fields with their wide surface area, controlled particle size, and excellent biocompatibility.[5] Even though silica nanoparticles can be synthesized using many processes with many potential and diverse structures, densities, and content, biogenic silica (BS) is a great substitute for fabricated

Biogenic Silica: Fundamentals and Applications
Edited by Ashok Kumar Nadda and Vijai Kumar Gupta

Published by the Royal Society of Chemistry, www.rsc.org

silica.[6] Side by side, due to the low biocompatibility of the synthetic silica, and as compared to them, the more availability, density, low cost, and eco-friendliness along with the more efficacy to modulate the toxicological effect of the biogenic forms of silica draw more attention as an alternative especially in the biomedical field.[2,6] BS, a type of silica also defined as biosilica, is derived from involving living aquatic and terrestrial organisms such as diatoms, sponges, plant extracts or metabolites, bacteria, algae, and fungi, along with numerous essential activities in biological functions.[7,8] Most of the literature is based on the synthesis of silica nanoparticles depending upon chemical synthesis for various uses. However, synthesizing them also can be possible through eco-friendly methods like green synthesis, done by using plants, agricultural waste, worms, bacteria, fungi, yeast, *etc.*[9–11] The BS emerged as a promising candidate for advancing the frontiers of numerous therapeutic and medicinal applications due to their unique set of properties which make them an attractive choice in the research field. This bioconjugate form of silica has been studied to show therapeutic potential like antibacterial[12,13] and anticancer activities,[14] along with some others like wound healing,[15] control of haemorrhaging,[16] bone tissue repairing[17] *etc.* However, the BS nanoparticles also have been investigated as potential drug and gene delivery systems, and their bioconjugation also helps to enhance the activity inside the body.[18,19] The mesoporous silica allows for encapsulation of drugs, making them suitable for controlled release along with enhancing the therapeutic activity of drugs by providing sustained release and targeted delivery to specific tissues or cells along with allowing to minimize the side effect.[20,21] Additionally, due to the degradation by endonucleases, the naked genetic materials are unable to undergo active site-specific targeting, to overcome these drawbacks with improved transfection specificity and efficiency the silica nanoparticle is also being used for gene delivery.[6,22] Surface functionalization of silica nanoparticles has also been done to alter the outer layer or surface of silica nanoparticles with various chemical groups or molecules to tailor their properties for specific applications.[18] Being a naturally occurring component, this biogenic form of silica is less toxic than crystalline silica.[23,24] In this chapter, the significant recent advancements in biomedical applications of BS are discussed. Mainly, this chapter contains information about the synthesis of biogenic forms of silica from different biological sources and their various applications in the biomedical field, like as a carrier for drug or gene delivery, contrast agents for imaging, and therapeutic agents.

8.2 Different Types of BS Synthesis

Several ways such as the sol–gel method, chemical vapor deposition method, hydrothermal method, and leaching method along with numerous living organisms are employed to prepare a biogenic form of silica.[25] Green synthesis of silica is an attractive method to prepare BS to avoid chemical synthesis and a huge number of literature have been reported on it.

8.2.1 Extracted BS from Some Agriculture Waste Products

Most studies mentioned the synthesis procedure of BS by using different agricultural waste products which also emerge as precursors of BS with the advantages of availability at that time when the harvesting season comes to an end and low cost. As in a prior study, it was noted that groundnut shell, coconut husk, orange peel, banana peel, and walnut shell had been chosen by the researchers as a precursor of silica.[26] Rice plant is a great source of amorphous silica. In a previously reported study researchers used rice straw to extract highly pure amorphous silica *via* a base dissolution-acid precipitation method and synthesized silica nano-discs with an average diameter of 172 (±70) nm and a thickness of 3.09 (±0.74) nm.[27] The palm ash and rice husk utilized an useful sustainable resource of silica which was extracted by using the leaching method.[28] Another study also represents the formation of silica nanoparticles by chemical and biomass methods where silica was extracted from various agricultural waste such as sugarcane bagasse, coffee husk, corn cob, wheat husk, rice husk, *etc.*[29] A way to synthesize mesoporous silica nanoparticles, utilizing rice husk as a raw material through the sol–gel method has been reported in another study.[30] Rice hull was used to fabricate silica micro- and nanoparticles by using the precipitation method same as commercial silica.[31] Similarly, Athinarayanan *et al.*, also synthesized silica in high quantity with 10–30 nm size and spherical shape by utilizing acid pretreated rice husk.[6] Following the calcination process some other researchers collect amorphous BS from rice husk.[2] It was also reported the use of sugarcane bagasse developed less toxic spherical silica nanoparticles with 30 nm size through an extraction and precipitation process.[32] Similarly, Falk *et al.* successfully produced nano silicon with large surface area, pore volume and more than 96% purity level through magnesia thermal reduction.[33] By using the thermo-chemical treatment method utilizing African sugarcane and maize residue, the fabrication of silica nanoparticles also has been demonstrated in a study.[34] In east America, teff straw, an agriculture waste was efficiently used to extract the silica in nanoform by using heat and acid treatment.[35] Similarly, another study also noted down that *Eragrostis tef* straw also can be used in production of silica nanoparticles *via* thermal methods.[36] Fatima *et al.* showed in their work that from the coconut husk, silica was extracted by using a chemical extraction method.[37] Following the sol–gel polymeric method Prempeh *et al.* used corn husk to synthesize porous silica[38] (Figure 8.1).

8.2.2 Plant Derived BS

Some studies also indicate that some parts of the plant were also used in the synthesis procedure of BS by different methods. Following the sol–gel method a previously reported study demonstrated the silica nanoparticle was developed by using *Bambusa vulgaris* leaves ash.[39] Babu and his coworker successfully synthesized silica nanoparticles with 70 to 80 nm size

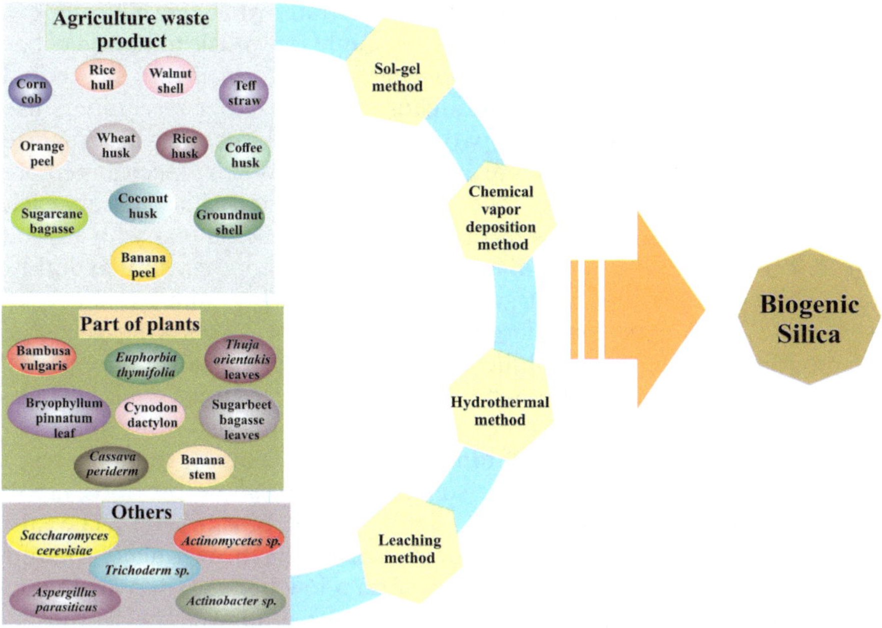

Figure 8.1 Diagrammatic representation of different synthesis procedures for biogenic silica from agricultural waste products, part of plants, fungus, and bacteria.

and spherical shape from *Cynodon dactylon* using a green approach method.[40] Banana stem was being used to extract silica by an acid leaching method.[41] There are also some studies that reported the fabrication of silica nanoparticles by using bamboo leaves.[42,43] Other studies figured out that using of sugarbeet bagasse leaves[44] and *Thuja orientalis* leaves[45] are efficient to synthesize silica induced nanomaterial. The sol–gel method has been used to synthesize BS-induced nanomaterial from Cassava periderm,[46] and silica from brown teff straw.[47] Production of biosynthesized silica-mediated nanoparticles was noted in another study, following a green chemistry approach, using an aqueous extract of *Euphorbia thymifolia*, which acts as a stabilizing and capping agent of TEOS.[9] Similarly, silica nanoparticles have been fabricated using an aqueous extract of the *Bryophyllum pinnatum* leaf (Figure 8.1).[48]

8.2.3 Bacteria, Fugus and Worm Mediated Synthesis of BS

As mentioned above not only plants but also bacteria and fungus are involved in synthesized silica. Singh *et al.* reported the bacterium-mediated synthesis of silicon or silica nanocomposite through *Actinobacter* sp. where K_2SiF_6 has been used as a precursor of silica and bacteria exposed together under ambient conditions.[49] From another study, it was also established that *Saccharomyces cerevisiae* yeast was being used for synthesizing silica nanoparticles.[50] Kaur and colleagues demonstrated that synthesizing crystalline structure silica nanocomposite through rice husk and wheat barn using microorganisms such as *Actinomycetes* and *Trichoderma* sp. took 48 and 24 hours, respectively, for biotransformation.[51] Other researchers used rice husk to synthesize silica nanoparticles along with that they also reported on their study about the fungal *Aspergillus parasiticus* species-mediated biotransformation.[52] Similarly, another study reported silica nanoparticles fabrication with 55–250 nm size range from rice husk along with the biodigestion through the Californian red worm.[53] In another similar kind of study Rohtagi *et al.* also fabricate silica nanoparticles from rice husk and used white rot fungus for microbial fermentation.[54] Following the green technique and using thermophilic bacterium (BKH1) the silica nanoparticle fabrication was noted down in a previous study.[55] Vetchinka *et al.* also used bacterial culture in the development of silica nanospheres with a small (5 to 15 nm) to respectively large (250 nm) size range.[56] Diatom, single cell algae also have been used as a natural source of silica. Some previous reports suggested different applications of diatom based biosilica.[57,58] The silica exoskeletons were able to form through unicellular brown algae diatoms, with a small amount of natural precursor on a low budget and within a few hours at room temperature.[18,59] A prior study also employed the synthesis of silica nanoparticles following the green synthesis method using coal fly ash.[60] Diatomaceous earth also called BS or biosilica holds a pivotal role in various contemporary applications, including drug delivery and nucleic-acid delivery systems in the field of modern medicine. Furthermore, it has garnered recognition for its vital role in diverse commercial and industrial sectors, such as the synthesis of carbon-neutral fuels, the production of health foods, the development of biomolecules, and the creation of materials associated with nanotechnology. Additionally, BS has gained recent prominence in highly specialized industries like drug delivery systems, electronic devices, biomolecule diagnostic tools, chemical sensors, and energy applications. Conversely, diatoms contribute significantly to the preservation of water quality (Figure 8.1).

8.3 BS-based Drug Delivery Systems

The primary hurdle facing the drug and pharmaceutical industry revolves around delivering therapeutic substances to specific areas within the human body, namely diseased tissues while minimizing any adverse effects on healthy tissues.[61] Over the past several decades, a diverse range of innovative drug delivery systems has been developed with the aim of shielding drugs from degradation and rapid elimination, enhancing their physicochemical properties, and improving their uptake by cells. These advanced delivery systems of drugs have demonstrated their ability to enhance drugs distribution and increase therapeutic effectiveness in targeted tissues while minimizing undesired side effects.[62] Due to its unique physicochemical properties, synthetic porous silica, whether microporous, mesoscopic, or macroscopic has been the subject of substantial research as a drug carrier.[63,64]

However, its benefits are outweighed by expensive, manufacturing procedures, that use hazardous solvents and could leave residues in the finished items and it is time-consuming. BS, a naturally occurring substance produced by unicellular algae known as diatoms, emerges as a promising candidate for an advanced drug delivery system that could pave the way for safer and more efficient therapeutics in the future. BS has slow degradation compared to synthetic silica particles (Figure 8.2).

The concept of harnessing the biological structure of diatoms for creating novel materials was initially introduced by Morse in 1999.[66] He highlighted that diatom's genetically controlled nanoscale silica architecture surpassed current human engineering capabilities. Rosi *et al.* in 2004 was the first to show the utilization of BS as a delivery system.[67] They successfully loaded and released gold nanoparticles from DNA-functionalized diatom surfaces in a regulated manner.

Using *in vitro* tests, Aw *et al.* investigated the potential of BS microcapsules for encapsulating medicines for oral delivery.[68,69] The results confirmed the effectiveness of diatoms in delivering water-insoluble drugs like indomethacin, achieving a drug loading capacity of approximately 22% and sustaining the release of the drug over 2 weeks. In this study, two stages of drug release were seen: an initial burst release over the period of 6 hours, followed by a gradual, protracted release over the course of 14 days, with zero-order kinetics, which was explained as the emission from the BS pores (Figure 8.3). In a similar work, Zhang *et al.* confirmed sustained drug release by examining the oral administration of mesalamine and prednisone encapsulated BS. Moreover, *in vitro* toxicity assessments demonstrated minimal toxicity of BS even at concentrations up to 1 mg mL^{-1}, when tested on HT-29/Caco-2 cocultured monolayers.[70] Furthermore, both drugs exhibited

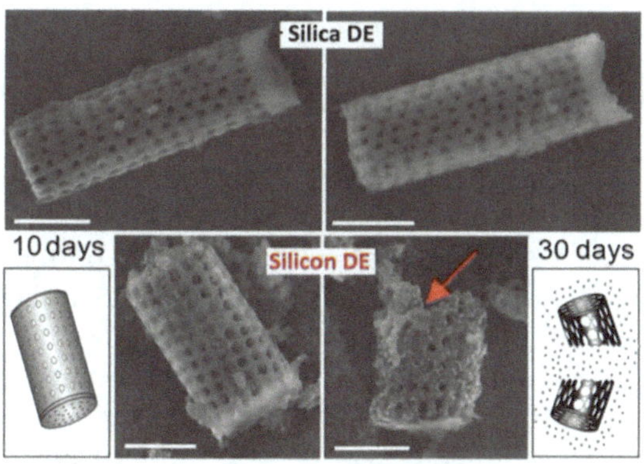

Figure 8.2 Degradation rate of BS particles (Silica DE) compared to synthetic silica (Silicon DE). BS showing enhanced stability even after 30 days. Reproduced from ref. 65 with permission from John Wiley & Sons, Copyright © 2015 Wiley-VCH Verlag GmbH & Co. KGaA, Weinheim.

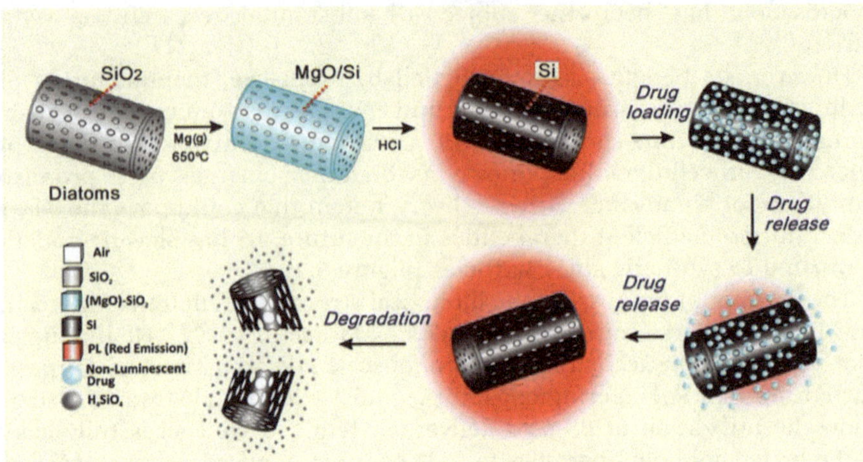

Figure 8.3 Mechanism of drug loading and release in BS particles. Reproduced from ref. 65 with permission from John Wiley & Sons, Copyright © 2015 Wiley-VCH Verlag GmbH & Co. KGaA, Weinheim.

enhanced permeation when tested using these monolayers. The presence of surface hydroxyl groups on the BS provides an opportunity to enhance surface characteristics and exercise greater control over the encapsulating and release patterns of drugs. In this context, Aw *et al.* examined two different organic modifications to examine the effects of surface functionalization on BS.[71] These changes gave the microcapsules hydrophobic and hydrophilic qualities. Results indicated that with the proper surface functionalization, drug loading and release may be adjusted from sluggish to sustained release for almost more than two weeks. Notably, hydrophobic alteration led to decreased drug loading and quicker drug release while hydrophilic functionalization enhanced drug efficiency and sustained release of drug. The drug encapsulation and release characteristics of both hydrophobic drug (indomethacin) and hydrophilic drug (gentamicin) were similarly controlled by Bariana *et al.* by using various functional groups.[71] In these studies, the functional modification of the BS was confirmed by using Fourier-transform infrared spectroscopy. Indomethacin released more slowly from hydrophilic surfaces containing hydrolyzed epoxy groups or polar carboxyl amines, but gentamicin released slowly and steadily from hydrophobic diatom surfaces treated with long-chain hydrocarbons.[71,72] These BS microcarriers demonstrated the potential for sustained drug release over a two-week period, particularly for poorly water-soluble drugs like indomethacin.

Another advancement in BS based drug carriers has been achieved by Kumeria *et al.*, who developed a hybrid diatom microcapsule decorated with 2D graphene oxide (GO) nanosheets, providing pH-dependent triggered drug release capabilities.[73] This was achieved by modifying the surface of BS structures to introduce amine groups, allowing for the attachment of GO nanosheets either electrostatically or covalently. This approach significantly

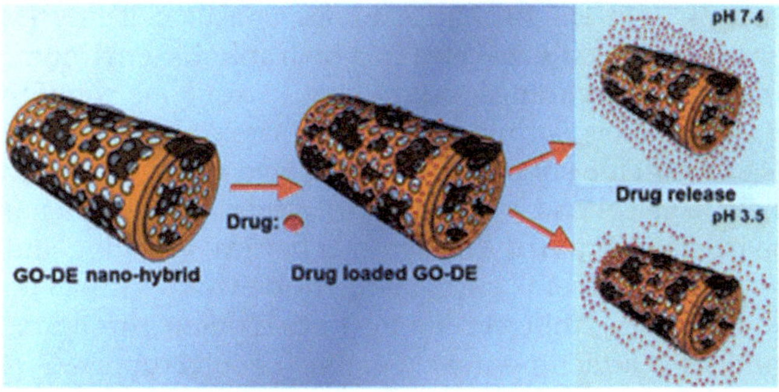

Figure 8.4 Schematic representation of covalent functionalization of diatoms frustule surface for pH dependent release. Reproduced from ref. 73 with permission from the Royal Society of Chemistry.

influenced the release of drugs (*e.g.* indomethacin), under different pH conditions (Figure 8.4).

Vasani *et al.* developed a BS based delivery system where polymer was adsorbed on the surface rather than chemical conjugation as a way to control medication release from diatom surfaces.[74] They grafted thermo-responsive copolymers onto the surface of BS microcapsules *via* surface-initiated atom transfer radical polymerization, demonstrating their efficacy in thermo-responsive drug delivery. The grafted copolymer's reduced critical solution temperature caused strong temperature dependence release of levofloxacin from these BS microcapsules.

Diatom species are known to produce biological molecules with demonstrated anti-cancer, antibacterial, anti-inflammatory, and antibiofilm properties.[75] In particular, the BS produced from marine diatom *Thalassiosira weissflogii* has been termed a "Smart Drug Delivery System" (DDS), capable of loading curcumin (CUR) at an impressive capacity of nearly 79%. It has been observed that under an acidic environment, CUR is released more rapidly compared to physiological conditions. It appears that this phenomenon is due to the adsorption of CUR into the porous structure of BS, which is influenced by a surface charge of approximately −3.05. Human renal adenocarcinoma cells (ACHN) are cytotoxic to the BS encapsulated with CUR, whereas human embryonic kidney (HEK293) cells are not affected. Additionally, it exhibits a wide range of antibacterial efficacy against Aeromonas species, *Staphylococcus aureus*, *Bacillus subtilis*, *Escherichia coli* and *Streptococcus pneumoniae*.[76] In addition, BS obtained from *Chaetoceros* sp. diatoms and conjugated with an antibody (Trastuzumab) showed fluorescence properties and the ability to selectively target breast cancer cell lines (SKBR3 cells) while sparing non-cancerous cells. This BS from *Chaetoceros* sp. is safe for the environment and has the potential for application of targeted medication delivery and biosensing chips.[77] Additionally, various drugs such as 5-fluorouracil (5-FU), ruthenium(II) complex and cisplatin, have been loaded into BS produced from marine microalgae for targeted delivery to a colon cancer cell line (HT-29) and a breast cancer cell line (MCF-7). The ruthenium complex is easily released in a hydrophobic environment, such as that present in cell membranes, but is kept in aqueous solutions for up to 5 days while 5-FU and cisplatin demonstrate faster release from the carrier.[78] The anticancer medicine sorafenib was delivered using BS fragments coated with polyethylene glycol and cell-penetrating peptide (CPP) as nanoparticles. These nanoparticles, obtained through mechanical crushing and sonication of diatom frustules, were surface-functionalized with 3-aminopropyltriethoxysilane (APTES), followed by PEG attachment through amide bond formation and CPP attachment to enhance cellular uptake[79] (Figure 8.5). These nanoparticles showed a two-fold increase in drug loading compared to uncoated nanoparticles in aqueous solutions and were also biocompatible with breast cancer cells and red blood cells. The porous BS produced through genetic modification of diatoms has been formulated for the purpose of delivering drugs specifically to B-lymphocytes and neuroblastomas, as well as to a mouse xenograft model.[80] Mesoporous silica materials have been extensively utilized in drug delivery contexts,[71] but BS particles derived from diatoms presents a better economical alternative as a drug delivery carrier.[81]

According to a prior report tetrahydro-1-benzopyrans was synthesised by using amino functionalized silica gel (AP-SiO_2) and used to treat numerous neurodegenerative disorders.[82] Synthesis of silica nanoparticles with the help of olive residue ash (ORA) shows anticancer activity against the breast cancer cell.[83] Bamboo (tabasheer) plant which is a silicon accumulator used for synthesizing porous silicon nanoparticles has the capability of inhibiting the activity of inflammatory signalling protein NF-κB, and was tested on a human embryonic kidney fibroblast HEK 293 cell line.[84] BS derived from sugarcane ash has the ability to heal wounds without being toxic to skin keratinocytes at concentrations up to 5 wt%.[85] Nanosilica which was derived from sugarcane bagasse ash has skin tissue regeneration ability, wound healing properties were examined by *in vivo* experimental study on male and female Wistar albino rats.[86] Another study also reported wound-healing properties of mucoadhesive films which are entrapped with BS through releasing of sialic acid and also show good antibacterial properties against Gram-negative bacteria.[87] Highly porous combinations, *i.e.* biosilica which is

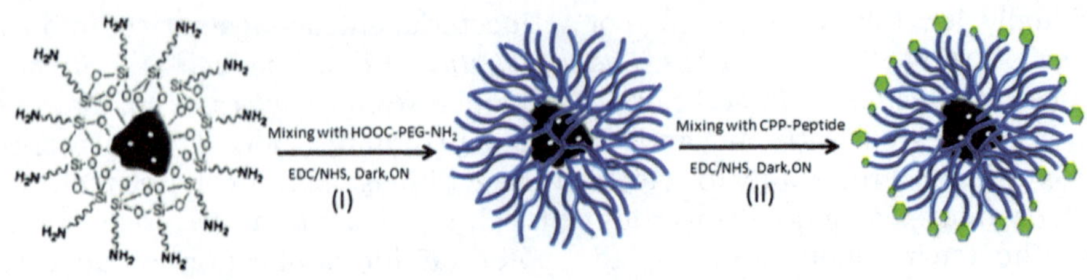

Figure 8.5 Schematic representation of the BS functionalization with cell-penetrating peptide for delivery of sorafenib. Reproduced from ref. 79 with permission from the Royal Society of Chemistry.

derived from diatomite and loaded with chitosan, show effective results on different osteoblast cell lines for bone tissue engineering.[88] Silica based bioactive glasses derived from rice hull ash can act as an alternative to commercial silica for bone tissue repairing.[89]

8.4 BS in Therapeutics

8.4.1 Antibacterial Application

A waste of agriculture, rice husk ash was used to fabricate the silica nanoparticles which showed antibacterial potency by decreasing 94.12% bacterial growth at 50 $gm\,L^{-1}$ concentration. This study claimed that the bacterial aggregation around the nanoparticle was enriched because of the association of Fe^{3+}/Al^{3+} species, electrostatic contact, and reactive oxygen species (ROS) production leading to bacterial membrane degradation and cellular content leakage.[90] A prior study also reported that silver-decorated mesoporous silica nanoparticles synthesized from rutin extract had significant antimicrobial properties against *Staphylococcus aureus* and *Escherichia coli* with low cytotoxicity.[91] *Cynodon dactylon* synthesized silica nanoparticles have better antimicrobial activity than chemically synthesized silica nanoparticles against *E. coli*, *P. aeruginosa*, *B. subtilis* and *S. aureus* and which was observed *via* the disc diffusion method.[92] The leaf extract of *Punica granatum* fabricated silica nanoparticles that showed good antibacterial efficacy against two multi-drug resistant Gram-negative bacterial strain of *E. coli* and *Salmonella* sp.[93] *Camellia sinensis* (green tea) synthesized silica-coated silver nanoparticles have antibacterial potency against *Streptococcus mutans* with a non-toxic effect on human dental cells.[94] Nanohybrid silica synthesized from rice husk also acts as a filler for dental composites.[95] Biogenic porous silicon fabricated form tabasheer includes extraction also shows antibacterial potency against *E. coli* and *S. aureus* as well as *Candida albicans* fungus[96] (Figure 8.6).

8.4.2 Anticancer Application

Iron oxide nanoparticles encapsulated within BS and associated with small molecules have the potential for magnetic targeting at the tumor site.[97] This concept holds promise for magnetically guided targeted drug delivery to improve cancer treatment, providing benefits over current drug delivery methods. The utilization of magnetic fields offers several benefits, including diverse manipulative capabilities and a broad spectrum of achievable outcomes with BS nanoparticles, achieved through magnetic alignment under a constant magnetic field or elevation in temperature when subjected to an alternating magnetic field.[98] Thus, this complex substance can be useful for theranostic applications.[99]

Previous research indicated that the silica matrix's gadolinium and europium ions have the potential for T_1-weighted magnetic resonance

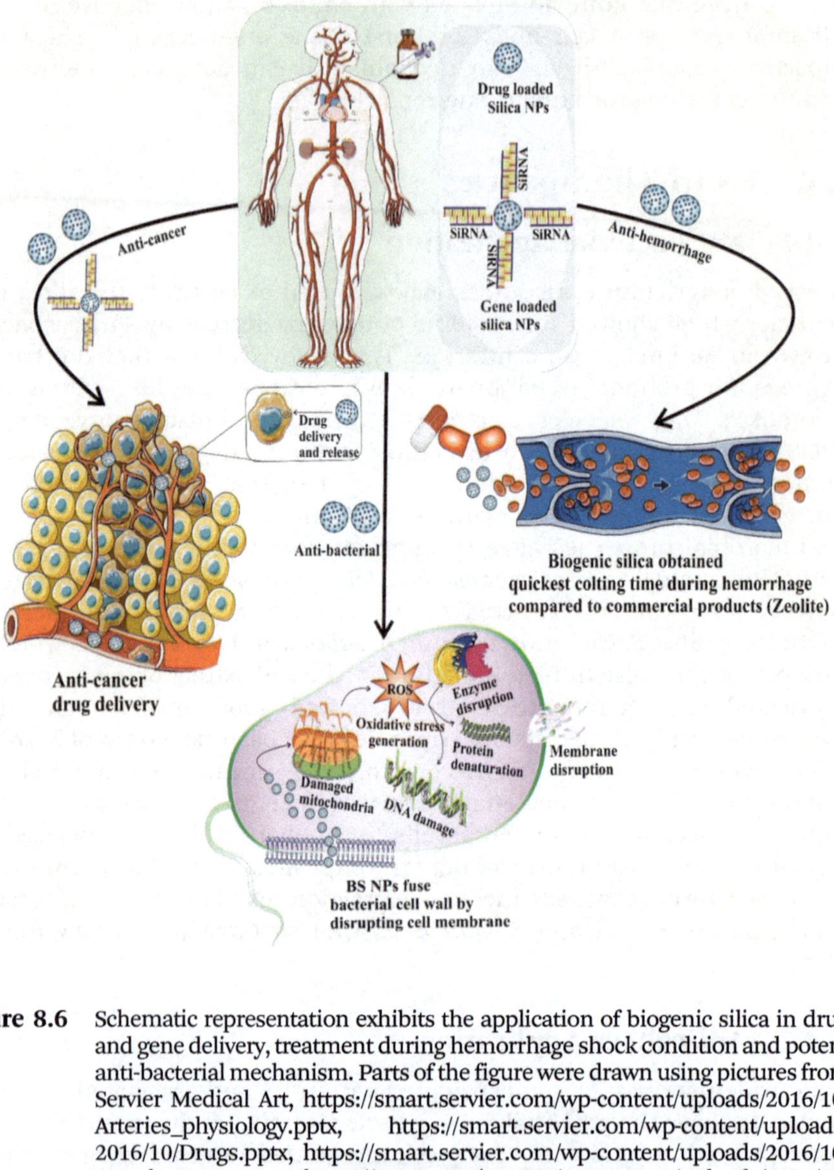

Figure 8.6 Schematic representation exhibits the application of biogenic silica in drug and gene delivery, treatment during hemorrhage shock condition and potent anti-bacterial mechanism. Parts of the figure were drawn using pictures from Servier Medical Art, https://smart.servier.com/wp-content/uploads/2016/10/Arteries_physiology.pptx, https://smart.servier.com/wp-content/uploads/2016/10/Drugs.pptx, https://smart.servier.com/wp-content/uploads/2016/10/Oncology.pptx, https://smart.servier.com/wp-content/uploads/2016/10/Drugs.pptx, https://smart.servier.com/wp-content/uploads/2016/10/Veins.pptx, https://smart.servier.com/wp-content/uploads/2016/10/Intracellular_components.pptx, https://smart.servier.com/wp-content/uploads/2016/10/Cell_membrane.pptx, https://smart.servier.com/wp-content/uploads/2016/10/Nucleic_acids.pptx, https://smart.servier.com/wp-content/uploads/2016/10/Intracellular_components.pptx, https://smart.servier.com/wp-content/uploads/2016/10/Nucleic_acids.pptx, https://smart.servier.com/wp-content/uploads/2016/10/Infectiology.pptx, under the terms of the CC BY 4.0 license https://creativecommons.org/licenses/by/4.0/.

imaging and fluorescence imaging, respectively. Surface functionalization allows for the modification of diatomaceous BS with various biological or synthetic materials, imparting additional properties such as magnetic manipulation and controlled drug release. In order to create magnetically guided drug microcarriers, Losic *et al.* surface functionalized BS with iron oxide nanoparticles (IONPs) that had been dopamine-modified.[100] Iron oxide nanoparticles have been incorporated into the BS and directed towards the tumor site (*in vivo*), through the influence of an external magnetic field. BS with dimensions of 10 nm is employed for both the iron oxide nanoparticles and molecule loading. The versatile array of functionalities offered by diatoms, coupled with the low levels of toxicity and biodegradability associated with IONP loaded BS, indicates substantial promise for their utilization in multimodal imaging and as therapeutic carriers within combination therapy approaches.[97] When a magnetic field was provided, there was noticeably 6-fold higher BS accumulation at the tumor location compared to the control condition. Javalkote *et al.* developed magnetized BS loaded with curcumin (CUR), which demonstrated notably enhanced toxicity against the human cervical cancer cell line (HeLa) in contrast to pure CUR[101] (Figure 8.6). Cervical cancer has also been treated *via* scavenger receptor class B type I antibody conjugated silica-coated gold nanoparticles. This result was reported by showing an effective result on human cervical epithelial (H8) and cervical cancer-derived (Ca Ski) cell lines.[102] Furthermore, these magnetic BS nanoparticles can be efficiently incorporated in the cells and adapted to the lipid-rich cellular membrane without causing detrimental effects on cellular morphology and viability.[103] However, it is worth noting that these dimensions may potentially result in obstructions within the narrow pulmonary capillaries and considerable cargo losses due to the relatively large pore size. Consequently, there is a compelling need to optimize the sizes of these magnetic BS nanoparticles.

8.4.3 Application in Treatment of Hemorrhage

Uncontrolled haemorrhage can cause hypovolemic shock, which can be deadly if it is not treated immediately by replacing lost fluid and stopping the bleeding. Present hemostatic agents, such as QuikClot® zeolite and HemCon® chitosan bandages have several numbers of limitations. For instance, QuikClot® zeolite can generate excessive heat, reaching temperatures of up to 95 °C, potentially causing tissue burns. BS have emerged as an alternative hemostatic agent that is noncytotoxic, nonimmunogenic, and cost-effective, effectively addressing the limitations of commonly used agents.[104] BS exhibit a better capacity for absorbing plasma without generating heat, unlike zeolite.

Diatomite (bought commercially) and diatoms grown in a lab were two types of pure diatom frustules that were coated with chitosan at various concentrations. Blood clotting activity experiments *in vivo* and *in vitro* were then carried out. The results demonstrated that chitosan coating resulted in minimal hemolytic activity when compared to BS. In addition, while measuring their capacity to cease bleeding following rat tail amputation, chitosan-coated BS obtained the quickest clotting time in comparison to gauze and the commercial QuikClot® zeolite. Observations of the rat tail also revealed that chitosan-coated BS did not yield large clots, unlike gauze and QuikClot® zeolite. This can be explained by the chitosan-coated BS's quick and improved fluid absorption capabilities as compared to commercial QuikClot® zeolite or gauze[104] (Figure 8.6).

8.4.4 BS for Gene Therapy

A revolutionary approach to treating and curing illnesses that are typically resistant to traditional medicines is gene therapy. Recently, there has been exploration into the use of BS from diatoms for applications in gene therapy. One effective approach in cancer treatment involves the delivery of small interfering RNA (siRNA), which can selectively inhibit the expression of mutant genes in cancer cells, leading to their controlled growth or demise. However, the limitation of direct siRNA delivery includes, once it enters the cytoplasm of the cell, it is degraded by nucleases. To overcome this hurdle, various nanocarriers, including polymeric nanoparticles and liposomes have been developed[105,106] (Figure 8.6).

The qualities of silica-based nano systems, which include low physical stability, variable size, and easy surface modification, as well as their capacity to be generated on a large scale with excellent biocompatibility, have made them stand out among other inorganic materials.[107–109] Despite the fact that silica nanoparticles can be produced by using a variety of ways but with many potential and diverse structures, densities, and contents, BS is a great substitute for fabricated silica.[6] A previous study demonstrated that a plant, especially Bambuseae (tabasheer) and *E. telmateia,* derived porous silica nanoparticle, surface functionalized with PEG, loaded with APTES-DNA was able to slowly diffuse the DNA into HEK 293 cells and within 48 hours the cell was able to express eGFP. This study indicates the possibility of gene delivery with porous silica nanoparticles derived from plants.[110] Another study also reported that the silica nanoparticle can be found throughout living nature from unicellular organisms to higher-order plants and is useful for a variety of processes. A mesoporous form of it has been discovered that can be used as a gene (plasmid DNA) delivery carrier to human cells as a critical therapeutic alternative for the treatment of disorders that are genetically based, like cancer.[111] There was also reported in previous studies the synthesis of BS nanoparticles from rice husk with excellent biocompatibility for human mesenchymal stem cells and human lung fibroblast cells, and it was also claimed that these fabricated nanoparticles are able to be used in biomedical applications.[6,112] A previous study also suggested that biodegradable porous silicon nanoparticles were used to encapsulate an anti-miR21 locked nucleic acid to treat ovarian cancer.[113] BS nanoparticles have been engineered as a nanocarrier for the conveyance of siRNA molecules into cancerous cells. This process entails the comminution, sonication, and filtration of microporous frustules to achieve particles of approximately 450 nm in size, followed by surface modification through siRNA conjugation. It has been empirically demonstrated that these frustule carriers exhibit non-toxic characteristics when utilized to deliver the siRNA into the tumor cells.[114] The impact of small interfering RNA (siRNA) should be directed specifically toward the disease site, minimizing any harm to healthy cells. In one pioneering effort to utilize diatoms for siRNA delivery, a covalent linkage was established between a poly-D-arginine peptide/siRNA complex and diatom surfaces modified with APTES.[114] *In vitro* experimentation has demonstrated that BS nanoparticles at concentrations up to 300 $\mu g\, mL^{-1}$ over a 72 hour period exhibit negligible toxicity. However, when siRNA is conjugated with BS NPs, it leads to cytoplasmic localization and it subsequently induces gene silencing in human epidermoid cancer cells (H1355)[114] (Figure 8.7). Importantly, it also demonstrated effective siRNA delivery into the cytoplasm with successful gene silencing outcomes.

Several research groups have successfully harnessed this concept to regulate gene inhibition or upregulation using biogenic silica NPs.[115,116] For instance, Delalat and colleagues engineered BS frustules that displayed the Immunoglobulin G (IgG)-binding domain of protein G (GB1-biosilica) on their surfaces.[117] Facilitating the delivery of cytotoxic agents that are insoluble in water to tumor cells was the objective. Anticancer drugs such as camptothecin (CPT) and SN38, were used to be delivered in cancer cells specifically. For this, IgG antibodies were conjugated to GB1-biogenic silica, forming anti-p75NTR-GB1-biosilica. Subsequently, the chemotherapeutic agents were loaded onto these engineered BS nanoparticles through electrostatic interactions. As a result, these modified biogenic nanoparticles demonstrated the ability to selectively deliver the drugs in B-lymphoma and neuroblastoma cells. When injected into neuroblastoma tumor model in

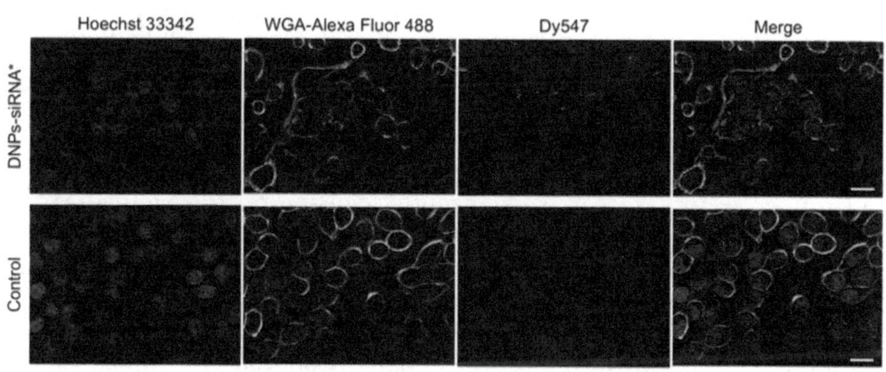

Figure 8.7 Dy547 dye was conjugated with the siRNA. BS loaded with siRNA (DNPs-siRNA) delivery shows localization of siRNA in the cell nucleus as observed by the red fluorescence (Dy547).[114] Reproduced from ref. 114 with permission from Elsevier, Copyright 2014.

mice, the above-modified biosilica encapsulated with SN38 led to a significant inhibition in the growth of the tumor.[117] The capacity to preferentially target B-lymphoma and neuroblastoma cells was therefore established by these genetically modified biosilica nanoparticles. Anti-p75NTR-GB1-biosilica nanoparticles containing SN38 were given to mice with SH-SY5Y neuroblastoma tumors, and this resulted in a substantial decrease in tumor development.[117]

8.4.5 Photonic Application

The photonic and optical characteristics of BS in biosensing applications are widely recognized, often likening diatoms to "jewels of the sea" or "living opals". Diatoms primary component, silica, is also the primary constituent of glass, essentially associating diatoms as organisms living within a glass structure.[118] The nanostructured three-dimensional arrangement of BS exhibits strong light interaction, resulting in pronounced scattering effects, leading to distinct photonic patterns as initially discovered by Fuhrmann *et al.*[119] The photonic characteristics of BS vary depending on the diatom species, offering unique characteristics in terms of light guidance, transmission, and diffraction, making it a promising avenue for the application in sensing devices or optics.[120] As an alternative, while naturally occurring silica sources including corn cob, rice husk, coffee husk, sugarcane bagasse and wheat husk wastes are frequently inexpensive and plentiful, they also include contaminants. These biomass wastes, which are typically dumped in lands, ponds, and rivers, can be used to create homogeneous silica nanoparticles.[29]

BS, when conjugated with specialized light-emitting and luminescent agents, can be used for dual purposes *i.e.*, for carrying payload along with the real-time monitoring of its localization using the luminescent characteristics. During an animal trial, engineered BS derived from diatoms *Thalassiosira weissflogii* has been successfully achieved by incorporating red-emitting organic dye molecules. The resulting biosilica exhibits high luminescence, which is a product of the metabolic conversion of these dye molecules within the frustules.[121] Subsequently, the stained biosilica is isolated using an acidic-oxidative method. Although silica nanoparticles are employed in various industries, including the textile, electrical, biomedical, and rubber sectors, research is currently focusing on the biomedical sector.[29] Through the use of hot injection, meso-silica nanoparticles made from rice husk were combined with green, fluorescent carbon dots. Compared to carbon dots, this complex exhibits great drug-loading efficiency and brighter fluorescence.

Another research team utilized BS isolated from *Aulacoseria* sp., functionalized with gold nanoparticles and encapsulated with the antibiotic 'gentamicin' (GMC).[122] UV-Vis spectrophotometry was used to study the *in vitro* release of the drug in simulated body fluid. The fluorescence intensity of this material correlates with the quantity of GMC loaded into it.

The preparation method, whether *ex situ* or *in situ*, influences the loading and release dynamics of GMC. In particular, the *in situ* approach produced slow and sustained release of gentamicin, whereas the *ex situ* strategy reduced loading of gentamicin but increased release rate. This finding could lead to a more methodical design and management of the optical characteristics and decorated structure of diatom BS for delivery systems of drugs.[122]

Noyes *et al.* conducted a comprehensive investigation into how light interacts with the internal structure of BS frustules produced by *Coscinodiscus wailesii*. Their findings revealed that light interacts with the valves as well as the entire diatom structure, resulting in sixfold symmetric diffraction patterns during transmission.[123] This effect exhibited strong wavelength dependence, with the red spectral region diffracting the most (approximately 50%), followed by the green (approximately 30%) and blue spectral regions (approximately 20%). These results shed light on the role of periodic nanostructures in diatom's survival and growth under various conditions. Chloroplasts within these BS move as per the light intensity to capture it and turn it into energy.[124] Digital holography was used by Di Caprio *et al.* to investigate how the diatom shell's photonic character assist in chloroplast migration in the biosilica, particularly produced by *C. wailesii*.[125] Their findings confirmed that the silica shell's photonic characteristics enhance light harvesting through chloroplast migration, using the inherent photoluminescence of the silica shell and the autofluorescence of the chloroplasts to validate plastid migration.

The distinctive optical characteristics possessed by diatoms serve not only for their survival but also find practical utility, primarily in sensing applications.[126] Several reports have highlighted diatoms' capability to detect the presence of nitrogen dioxide (NO_2) through a reduction in their photoluminescence.[127–129] Generally, when NO_2 adheres to the diatomaceous earth (DE) silica shell, it leads to the quenching of surface-oxygen defects responsible for luminescence emission. Lettieri *et al.* conducted an investigation into how temperature and NO_2 concentration affect the photoluminescence intensity of biosilica shells and successfully detected NO_2 levels as low as 50 parts per billion (ppb).[130]

In addition, the BS's three-dimensional structure has been used in a number of investigations to construct plasmonic metal nanoparticles for surface-enhanced Raman spectroscopy (SERS) based sensing.[131,132] Recently, a technique for producing plasmonic silver nanoparticles on BS surfaces was devised by Kong *et al.* This technique makes the particles SERS-active and useful for identifying pollutants in soil, water, and air.[133] Similar sensing applications have also been reported by other researchers using self-assembled plasmonic nanoparticles onto BS. In another study, silver nanoparticles were grown *in situ* to create nanoplasmonic sensors based on SERS using inexpensive and eco-friendly diatomaceous earth biosilica. These sensors can be investigated for a variety of uses, including food sensing, monitoring air and water quality, medical and scientific research.[80]

Biosilica shells have also found application in electrochemical sensing, where they act as filters to isolate target analytes from samples. Biosilica has also been found to be capable of detecting cardiac biomarkers at ultralow concentrations, as low as approximately 1 pg mL^{-1}.[134] Biosilica is also used for detecting food dyes using optical probes and separating them by microfluidic techniques.[135]

The three-dimensional porous structure of biosilica could be utilised as an invaluable template for growing other complex microstructures, which are difficult to fabricate using conventional methods. Losic *et al.* utilized diatom templates, specifically *C.* sp. and *Thalassiosira eccentric*, to create positive and negative replicas of metallic and polymeric structures.[136] They demonstrated the unique optical properties of these fabricated structures and their potential in biosensing applications. These structures created using diatom templates are promising and are expected to be investigated further in the future to create affordable biosensing devices for a range of biomedical applications.

Biosilica is loaded with enzymes, antibodies, drugs and DNA aptamers able to act as a biosensor. To detect interleukin 8 (IL-8) in blood plasma the development of a surface-enhanced Raman scattering (SERS) immunoassay based on diatomic biosilica integrated with gold nanoparticle was reported in a previous study.[137] Additionally, another study also developed a biosensor APTES-functionalized diatom frustules which was derived from *Amphora* sp. to detect bovine serum albumin (BSA).[138] Using chemical bonding with FA (folic acid) and embedded molybdenum disulfide (MoS_2) nanosheets, researchers grafted PhENH2, an aggregation-induced emission (AIE) dye, onto SiNPs, which displayed sustained fluorescence for photothermal therapy in breast cancer.[139] Additionally, Li and his team discovered that photosensitizer 5,10,15,20-tetrakis(1-methyl-4-pyridinio)porphyrin tetra (*p*-toluenesulfonate) was embedded in FA-modified SiNPs to enable targeted two-photon excitation fluorescence imaging in human breast cancer cells.[140] Managò and team exhibited that diatomite NPs coupled with a nontargeting siRNA could be in H1355 lung cancerous cells for more than 72 hours by applying Raman imaging. These NPs showed low systemic toxicity, cost-effectiveness, chemical and thermal stability which makes them ideal therapeutic agents and nano vectors against cancer.[80,103]

8.5 Surface Functionalization and Targeting

Numerous studies revealed that the biofunctionalized nanoparticles have both drug-loading and drug-releasing capabilities, as well as low toxicity and optimum cellular absorption.[141,142] Diatomite is a sedimentary natural porous form of silica substance with extraordinary characteristics, which can also be used as a legitimate substitute for synthetic porous silica in nano-based drug delivery.[79] Amorphous silica is a substance that diatoms produce to surround their cell walls. The advantages of frustules (empty silica shells) are their biocompatibility, non-toxic, biodegradability and abundance of hydroxyl groups on their surface.[143] For delivering the poorly water-soluble anticancer drug sorafenib to the MCF-7 and MDA-MB-231 breast cancer cells, Terracciano *et al.* functionalized diatomite nanoparticles with cell-penetrating peptide and polyethylene glycol and examined the hemocompatibility, construct's toxicity and intracellular uptake in cancer cells. This work demonstrates that the increased cellular uptake, biocompatibility, stability and drug release profiles of diatomite NPs after biofunctionalization turn into a possible platform for applications involving delivery of drugs for cancer therapy.[79] Chitosan was utilized by Sasirekha *et al.* to chemically alter the Amphora subtropical diatoms' surface, indicating that the chitosan-grafted diatoms were used for encapsulating the chemotherapy medication (drug) doxorubicin (DOX) in their porous compartments.[141] To modify and functionalize the surface of diatomaceous earth microparticles researchers created core shell materials by using a chemical process *via* encapsulating two chemotherapeutic drugs (paclitaxel or hydrophobic curcumin and hydrophilic DOX) at constant molar ratios, in different compartments of the single drug delivery vehicle.[144] To improve the loading capacity of the drug, the gold-coated, gelatin-capped diatomite nanoparticles were used in cancer treatment, and the results were effective.[145] Uthappa *et al.* developed curcumin (CUR) loaded diatoms (DE) which was surface-modified with polydopamine (PDA) and investigated the mechanism of action of the drug-loading and drug-releasing capacities of it. Later, folic acid (FA) was labeled with DE-CUR-PDA for targeted drug administration and the results showed that the PDA layer coated on DE, serves as a barrier or gatekeeper which helps to regulate the drug release rate, and was responsible for the decrease in drug release. The outcomes indicated that the DE-CUR-PDA-FA might be exploited as a promising targeted drug delivery method in cancer as well as tumor therapy.[146] An experimental demonstration of using the diatom as a delivery vehicle for a specific goal was made by Rosi *et al.* in 2004, by which they were able to extract gold nanoparticles from the diatom's surface which was functionalized with DNA.[67] In order to create magnetically guided drug microcarriers, Losic and team have successfully functionalized the diatom skeleton with dopamine-modified iron oxide nanoparticles using the Aulacoseira species which showed excellent efficacy.[100] Researchers previously investigated the functionalization of diatom microcapsules' surfaces and their effect on water-insoluble medicines' drug loading and release characteristics. In other studies investigators decided to model poorly soluble drugs after indomethacin. Two organosilanes, 3-aminopropyltriethoxy silane, *N*-(3-(trimethoxy silyl)propyl)ethylene diamine, and phosphonic acids (2-carboxyethyl-phosphonic acid and 16-phosphono-hexadecanoic acid), which provide organic surface hydrophilic and hydrophobic properties, were used to modify the surface on diatoms. In this study, results demonstrated the diatom porous silica as a potentially effective drug carrier that may further improve its performance by modifying the surface functionalities to achieve the desired drug loading and releasing capacity against numerous therapeutics.[71] Using vitamin B12 (cyanocobalamin) as a tumor-targeting substance and cisplatin, 5-fluorouracil (5-FU), and a tris-tetraethyl[2,2′-bipyridine]-4,4′-diamine-ruthenium(II) complex as anti-cancer agents, Delasoie and team coated diatomaceous earth microparticles made from marine microalgae. *In vitro* testing showed that the B12-coated diatoms adhered more strongly to the colorectal cancer cell line HT-29 and the breast cancer cell line MCF-7. The expression of both the transcobalamin II (TC(II)) and transcobalamin II receptor (TC(II)-R) on targeted tissue increased along with the adhesion of the B12-modified diatoms which was at least three times more than that of the unmodified ones. By acting as a micro-shuttle and interacting with the target areas prior to administering the medication in close contact with the tumor tissue, the result implies that the group of B12-modified diatoms may offer an efficient approach for delivering water-insoluble inorganic complexes to tumor tissues specifically.[78] Research has emphasised the improved loading and release capacity of drug of acid-treated diatoms in comparison to bare diatoms. The improved drug transport characteristics of diatoms are explained by the well-hydroxylated surfaces produced by the acid treatment.[147] Following acid and peroxide treatments to enhance the silanol groups at the surface of diatoms and raise the percentage drug loading efficacy from 0.47 to 9.47 wt%, researchers described the enhanced drug loading and release features of diatoms.[147] To acquire a high biocide payload, researchers investigated the self-similar nano-organized architecture of BS particles. The interfaces were adjusted to control the release rate and responsiveness, utilizing either the carrier's natural hydroxyl surfaces or systems modified with amines or carboxylic acids in high density. Thymol nanostructuring in BS-enabled systems promote controlled and responsive release, remarkable, high and localized biocidal activity.[148]

Camptothecin and its more effective derivative of it named 7-ethyl-10-hydroxycamptothecin were precisely delivered to neuroblastoma and Burkitt's lymphoma HR1K cells using genetically modified biosilica produced from the diatom *Thalassiosira pseudonana*. *Thalassiosira pseudonana* was genetically altered by incorporating the protein G (GB1)-bearing fusion protein's immunoglobulin G (IgG)-binding domain into the diatoms' genome. The resultant design contained a stable GB1 fusion protein that was available for IgG binding and could connect to the surface of the biosilica, allowing the attachment of cell-targeting antibodies. In a xenograft mouse model of neuroblastoma, treatment with intraperitoneal injection of the same drug-loaded biosilica resulted in tumor growth regression.[117] Bare silica materials obtain a negative surface charge in aqueous environments because of the silanol groups present on their outermost layers. Therefore to accomplish effective gene delivery, surfaces are typically functionalized with the cationic groups which allows the systems to condense and load DNA utilizing the electrostatic interactions with the negative charges on the phosphate backbone of nucleic acids. Amination modification with trialkoxysilanes like (3 aminopropyl)triethoxysilane (APTES) or (3-aminopropyl)-trimethoxysilane (APTMS) is an easy and popular method of improving the gene loading capacity of silica materials.[149] Rea and coworkers demonstrated, using confocal imaging and fluorescence spectrometry, the effective internalization of siRNA into the cytoplasm of H1355 cancer cells and the continuous release of siRNA *in vitro* through the functionalization of diatomite with poly-D-arginine peptide/siRNA combination. Additionally, western blot analysis showed that specifically targeted mRNA was silenced by the siRNA, demonstrating the effectiveness of their newly created siRNA delivery mechanism.[147]

8.6 Biocompatibility and Toxicity Assessment

In contrast to its biogenic counterparts, which have a variety of positive impacts in modern toxicological applications, BS is more biocompatible than synthetic versions of silica.[112] Diatomaceous earth, an amorphous silica that occurs naturally, is regarded as less dangerous.[150] The elevated level of biocompatibility of silica nanoparticles (SNPs) (amorphous SNPs originate from bamboo leaf wastes) is supported by the dose-dependent *in vitro* cytotoxicity against A549 lung cancer cell lines, which is assessed based on LDH release. The teratogenic impact of amorphous SNPs was investigated on the embryos of zebrafish in both *in vivo* and *in vitro* environments, which illustrated the morphological abnormalities and non-significant histology deformations during the developmental phases. Even at larger concentrations of SNPs, genotoxicity experiments showed the least DNA strand breakage, which also highlights the need for a better understanding of molecular basis of nanotoxicity. Based on the teratogenic characteristics and the genotoxic potential, the current analysis validates the high concentration of SNPs' non-toxic behaviour. As a result, profitable processing of amorphous SNPs with high purity promotes the fabrication of numerous biomedical nanocomposites containing important inorganic–organic hybrids.[151] Researchers conducted a viability assay to examine the biocompatibility of the biogenic SNPs with human lung fibroblast cells (hLFCs), evaluating cellular morphological alterations, mitochondrial transmembrane potential, intracellular ROS production and oxidative stress-related gene expression. The outcomes showed that the biogenic SNPs were compatible with these *in vitro* cell-based methods without any substantial incompatibility. According to these preliminary results, biogenic SNPs are biocompatible, may be used as a best substitute for synthetic silica and also may able to be used in biomedical applications as well as food additives.[112] In a different study, the biocompatibility of biogenic SNPs was assessed by examining the cell survival, mitochondrial activity, morphology, reactive oxygen species and gene expression of human lung fibroblast cells. The results of these assays and microscopy experiments revealed that biogenic SNPs have a minor effect on the mitochondrial membrane potential at high doses but have no significant effect on cell viability or morphology.[152] Numerous difficulties exist in figuring out how hazardous nanomaterials affect people, so it is important to carefully evaluate the toxicity of nanoparticles.[139] These studies showed that the biogenic SNPs were less toxic and biocompatible, suggesting that they might be applicable in several biomedical sectors.

8.7 Conclusion and Future Perspectives

During the last decade, BS has revealed many attractive traits which can be synergistically applied in the development of drug delivery, gene therapy and multiple therapeutics. Due to high dispersibility and substantial surface area, BS can increase the rate of drug dissolution and bioavailability of drugs that are not water-soluble. Additionally, surface chemistry and pore size are a couple of other variables that may affect the drug release rate from BS. In the treatment of cancer, BS has demonstrated several benefits over other nanocarriers for the transport of chemotherapeutic drugs, including effective drug-loading capacity and endocytosis behaviour. In this book chapter, we discussed about different types of surface modifications of BS with various stimuli-responsive molecules and tumor-recognition molecules to boost the therapeutic effect of various diseased conditions, photonic applications, and antitumor agents. Through diverse surface modifications, BS can successfully complex with genes and transfect a variety of cells and demonstrates excellent potential for regulated drug delivery. Numerous difficulties exist in figuring out the hazards of nanomaterials and how they affect people. Since there are so many variables in toxicity studies, there isn't yet a single uniform standard. Studies on the toxicity of BS should consider real applications and improve safety evaluation metrics both *in vitro* and *in vivo*. Despite the recent intensive study into the development of BS carriers for drug/gene delivery and various therapies, there are several difficulties that must be resolved to support their further development. So, a thorough investigation should be done into these materials' degradability, biocompatibility, and pharmacokinetics. For better understanding a thorough and rigorous demonstration of the *in vivo* and *in vitro* therapeutic advantages of BS-based systems is required. The needed information concerning the circulation properties in blood, clearance time in the body, accumulation in tissues and possible immunogenicity should be investigated before the clinical trial of BS. Given the appropriate resolution of these problems, BS-based formulations may result in promising advances in the treatment of a variety of significant disorders.

Abbreviations

BS	Biogenic silica
SNPs	Silica nanoparticles
TEOS	Tetraethyl orthosilicate
GO	Graphene oxide
DDS	Drug delivery system
CUR	Curcumin
ROS	Reactive oxygen species
IONPs	Iron oxide nanoparticles
DE	Diatomaceous earth
SERS	Surface-enhanced Raman spectroscopy

FA	Folic acid
MoS_2	Molybdenum disulfide
AIE	Aggregation-induced emission
hLFCs	Human lung fibroblast cells

Acknowledgements

Authors are highly grateful to the Department of Physiology, University of Gour Banga and Amity Institute of Biotechnology, Amity University for providing the necessary facilities.

References

1. G. J. Gascho, in *Studies in Plant Science*, ed. L. E. Datnoff, G. H. Snyder and G. H. Korndörfer, Elsevier, 2001, vol. 8, pp. 197–207.
2. A. A. Alshatwi, J. Athinarayanan and V. S. Periasamy, *Mater. Sci. Eng., C*, 2015, **47**, 8–16.
3. W. Stöber, A. Fink and E. Bohn, *J. Colloid Interface Sci.*, 1968, **26**, 62–69.
4. F. D. Bruckmann, F. B. Nunes, T. D. Salles, C. Franco, F. C. Cadoná and C. R. Bohn Rhoden, *Magnetochemistry*, 2022, **8**, 131.
5. Y. Huang, P. Li, R. Zhao, L. Zhao, J. Liu, S. Peng, X. Fu, X. Wang, R. Luo, R. Wang and Z. Zhang, *Biomed. Pharmacother.*, 2022, **151**, 113053.
6. J. Athinarayanan, V. S. Periasamy, M. Alhazmi, K. A. Alatiah and A. A. Alshatwi, *Ceram. Int.*, 2015, **41**, 275–281.
7. H. C. Schröder, X. Wang, W. Tremel, H. Ushijima and W. Müller, *Nat. Prod. Rep.*, 2008, **25**, 455–474.
8. F. Akhter, A. A. Rao, M. N. Abbasi, S. A. Wahocho, M. A. Mallah, H. Anees-ur-Rehman and Z. A. Chandio, *Silicon*, 2022, **14**, 8295–8310.
9. R. Periakaruppan, N. Renuga Devi, S. A. Abed, P. Vanathi and J. S. Kumar, *Silicon*, 2023, **15**, 4309–4316.
10. Y. A. Hasanien, M. A. Mosleh, A. S. Abdel-Razek, G. S. El-Sayyad, E. H. El-Hakim and E. H. Borai, *Biomass Convers. Biorefin.*, 2023, 1–4.
11. S. Singh, U. Bhatta, P. Satyam, A. Dhawan, M. Sastry and P. Blv, *J. Mater. Chem.*, 2008, **18**, 2601–2606.
12. R. Periakaruppan, S. Manju Praveena, C. Priya, P. Ranjitha, S. Gokul Raj and J. Danaraj, *Appl. Biochem. Biotechnol.*, 2022, **194**, 5594–5605.
13. R. Babu, P. Yugandhar and N. Savithramma, *Bull. Mater. Sci.*, 2018, **41**, 1–8.
14. H. Yu, R. Zheng, F. Lei, W. Wang, W. Guo, L. Zhang, Y. Liu, X. Chen and Y. Wang, *Am. J. Transl. Res.*, 2022, **14**, 1518–1534.
15. H. Ramphul, F. Gimié, J. Andries, D. Jhurry and A. Bhaw-Luximon, *Int. J. Biol. Macromol.*, 2020, **157**, 296–310.
16. J. Li, J. Han, Q. Sun, Y. Wang, Y. Mu, K. Zhang, X. Dou, M. Kong, X. Chen and C. Feng, *J. Mater. Chem. B*, 2018, **6**, 7834–7841.
17. A. C. Özarslan and S. Yücel, *Mater. Sci. Eng., C*, 2016, **68**, 350–357.

18. F. Zobi, *Micro*, 2022, **2**, 342–360.
19. A. Pieła, E. Żymańczyk-Duda, M. Brzezińska-Rodak, M. Duda, J. Grzesiak, A. Saeid, M. Mironiuk and M. Klimek-Ochab, *Bioorg. Chem.*, 2020, **99**, 103773.
20. C. Salazar-Hernández, M. Salazar-Hernández, R. Lona-Ramos, E. Elorza-Rodríguez and H. Rocha-Ramírez Agustín, in *Rice*, ed. Amanullah and F. Shah, IntechOpen, Rijeka, 2017, ch. 5.
21. I. Ruggiero, M. Terracciano, N. M. Martucci, L. De Stefano, N. Migliaccio, R. Tatè, I. Rendina, P. Arcari, A. Lamberti and I. Rea, *Nanoscale Res. Lett.*, 2014, **9**, 329.
22. X. Pan, H. Veroniaina, N. Su, K. Sha, F. Jiang, Z. Wu and X. Qi, *Asian J. Pharm. Sci.*, 2021, **16**, 687–703.
23. J. Y. Liu and C. M. Sayes, *Toxicol. Res.*, 2022, **11**, 565–582.
24. J. Rabovsky, *Scand. J. Work, Environ. Health*, 1995, 108–110.
25. B. Tessema, G. Gonfa, S. Mekuria Hailegiorgis and S. Venkatesa Prabhu, *Adv. Mater. Sci. Eng.*, 2023, **2023**, 4865273.
26. J. G. Peerzada and R. Chidambaram, *Silicon*, 2021, **13**, 2089–2101.
27. P. Lu and Y.-L. Hsieh, *Powder Technol.*, 2012, **225**, 149–155.
28. C. P. Faizul, C. Abdullah and B. Fazlul, *Adv. Environ. Biol.*, 2013, 7, 3690–3695.
29. S. Prabha, D. Durgalakshmi, S. Rajendran and E. Lichtfouse, *Environ. Chem. Lett.*, 2021, **19**, 1667–1691.
30. L. Usgodaarachchi, C. Thambiliyagodage, R. Wijesekera and M. G. Bakker, *Curr. Res. Green Sustainable Chem.*, 2021, **4**, 100116.
31. W. Jansomboon, K. Boonmaloet, S. Sukaros and P. Prapainainar, *Key Eng. Mater.*, 2017, **718**, 77–80.
32. N. K. Mohd, N. N. A. N. Wee and A. A. Azmi, *AIP Conf. Proc.*, 2017, 020123.
33. G. Falk, G. P. Shinhe, L. B. Teixeira, E. G. Moraes and A. P. N. de Oliveira, *Ceram. Int.*, 2019, **45**, 21618–21624.
34. N. N. Maseko, D. Schneider, S. Wassersleben, D. Enke, S. A. Iwarere, J. Pocock and A. Stark, *Sustainability*, 2021, **13**, 577–590.
35. A. B. Wassie and V. C. Srivastava, *J. Nano Res.*, 2017, **46**, 64–72.
36. A. B. Bageru and V. C. Srivastava, *Mater. Lett.*, 2017, **206**, 13–17.
37. M. F. Anuar, Y. W. Fen, M. H. M. Zaid, K. A. Matori and R. E. M. Khaidir, *Results Phys.*, 2018, **11**, 1–4.
38. C. O. Prempeh, S. Formann, I. Hartmann and M. Nelles, *Biomass Convers. Biorefin.*, 2022, 1–11.
39. K. Durairaj, P. Senthilkumar, P. Velmurugan, K. Dhamodaran, K. Kadirvelu and S. Kumaran, *J. Sol–Gel Sci. Technol.*, 2019, **90**, 653–664.
40. R. H. Babu, P. Yugandhar and N. Savithramma, *Bull. Mater. Sci.*, 2018, **41**, 65.
41. N. Yusaidi, S. Abdullah and N. Zarib, *Int. J. Eng. Adv. Technol.*, 2019, **9**, 5858–5862.
42. N. K. Sethy, Z. Arif, P. K. Mishra and P. Kumar, *J. Polym. Eng.*, 2019, **39**, 679–687.

43. Irzaman, N. Oktaviani and Irmansyah, *IOP Conf. Ser.: Earth Environ. Sci.*, 2018, **141**, 012014.
44. N. O. San, C. Kurşungöz, Y. Tümtaş, Ö. Yaşa, B. Ortac and T. Tekinay, *Particuology*, 2014, **17**, 29–35.
45. N. Jabeen, Q. Maqbool, S. Sajjad, A. Minhas, U. Younas, S. Anwaar, M. Nazar, R. Kausar and S. Z. Hussain, *IET Nanobiotechnol.*, 2017, **11**, 557–561.
46. J. O. Agunsoye, J. A. Adebisi, S. A. Bello, M. Haris, J. B. Agboola and S. B. Hassan, *Mater. Sci. Technol.*, 2018, 701–709.
47. T. A. Amibo, S. M. Beyan and T. M. Damite, *J. Chem.*, 2022, **2022**, 9770520.
48. M. Sankareswaran, R. Periakaruppan, M. Sasivarnam, J. Danaraj, S. Dhanasekaran and M. M. Abomughaid, *Appl. Biochem. Biotechnol.*, 2022, **194**, 4266–4277.
49. S. Singh, U. M. Bhatta, P. V. Satyam, A. Dhawan, M. Sastry and B. L. V. Prasad, *J. Mater. Chem.*, 2008, **18**, 2601–2606.
50. H. Zamani, A. Jafari, S. M. Mousavi and E. Darezereshki, *J. Pet. Sci. Eng.*, 2020, **190**, 107002.
51. T. Kaur, G. P. Singh, G. Kaur, S. Kaur and P. K. Gill, *Mater. Res. Express*, 2016, **3**, 085026.
52. A. Zielonka, E. Żymańczyk-Duda, M. Brzezińska-Rodak, M. Duda, J. Grzesiak and M. Klimek-Ochab, *Fungal Biol.*, 2018, **122**, 333–344.
53. M. Estevez, S. Vargas, V. M. Castaño and R. Rodriguez, *J. Non-Cryst. Solids*, 2009, **355**, 844–850.
54. K. Rohatgi, S. V. Prasad and P. K. Rohatgi, *J. Mater. Sci. Lett.*, 1987, **6**, 829–831.
55. S. Show, A. Tamang, T. Chowdhury, D. Mandal and B. Chattopadhyay, *Colloids Surf., B*, 2015, **126**, 245–250.
56. E. Vetchinkina, E. Loshchinina, M. Kupryashina, A. Burov and V. Nikitina, *Ind. Eng. Chem. Res.*, 2019, **58**, 17207–17218.
57. M. Terracciano, L. De Stefano and I. Rea, *Pharmaceutics*, 2018, **10**, 242–256.
58. F. Zobi, Diatom biosilica in targeted drug delivery and biosensing applications: Recent studies, *Micro*, 2022, **2**, 342–360.
59. P. J. Lopez, C. Gautier, J. Livage and T. Coradin, *Curr. Nanosci.*, 2005, **1**, 73–83.
60. F. Yan, J. Jiang, K. Li, N. Liu, X. Chen, Y. Gao and S. Tian, *Environ. Sci. Technol.*, 2017, **51**, 7606–7615.
61. V. Wagner, A. Dullaart, A.-K. Bock and A. Zweck, *Nat. Biotechnol.*, 2006, **24**, 1211–1217.
62. J. F. Coelho, P. C. Ferreira, P. Alves, R. Cordeiro, A. C. Fonseca, J. R. Góis and M. H. Gil, *EPMA J.*, 2010, **1**, 164–209.
63. J. Ezzati Nazhad Dolatabadi, Y. Omidi and D. Losic, *Curr. Nanosci.*, 2011, 7, 297–314.
64. M.-A. Shahbazi, B. Herranz and H. A. Santos, *Biomatter*, 2012, **2**, 296–312.
65. S. Maher, M. Alsawat, T. Kumeria, D. Fathalla, G. Fetih, A. Santos, F. Habib and D. Losic, *Adv. Funct. Mater.*, 2015, **25**, 5107–5116.
66. D. E. Morse, *Trends Biotechnol.*, 1999, **17**, 230–232.

67. N. L. Rosi, C. S. Thaxton and C. A. Mirkin, *Angew. Chem.*, 2004, **116**, 5616–5619.
68. M. S. Aw, S. Simovic, Y. Yu, J. Addai-Mensah and D. Losic, *Powder Technol.*, 2012, **223**, 52–58.
69. M. S. Aw, S. Simovic, J. Addai-Mensah and D. Losic, *Nanomedicine*, 2011, **6**, 1159–1173.
70. H. Zhang, M.-A. Shahbazi, E. M. Mäkilä, T. H. da Silva, R. L. Reis, J. J. Salonen, J. T. Hirvonen and H. A. Santos, *Biomaterials*, 2013, **34**, 9210–9219.
71. M. S. Aw, M. Bariana, Y. Yu, J. Addai-Mensah and D. Losic, *J. Biomater. Appl.*, 2013, **28**, 163–174.
72. M. Bariana, M. S. Aw, M. Kurkuri and D. Losic, *Int. J. Pharm.*, 2013, **443**, 230–241.
73. T. Kumeria, M. Bariana, T. Altalhi, M. Kurkuri, C. T. Gibson, W. Yang and D. Losic, *J. Mater. Chem. B*, 2013, **1**, 6302–6311.
74. R. B. Vasani, D. Losic, A. Cavallaro and N. H. Voelcker, *J. Mater. Chem. B*, 2015, **3**, 4325–4329.
75. C. Lauritano, J. H. Andersen, E. Hansen, M. Albrigtsen, L. Escalera, F. Esposito, K. Helland, K. Ø. Hanssen, G. Romano and A. Ianora, *Front. Mar. Sci.*, 2016, **3**, 68.
76. A. Saxena, A. Dutta, N. Kapoor, A. Kumar and A. Tiwari, *J. Drug Delivery Sci. Technol.*, 2022, **68**, 102983.
77. J. Esfandyari, B. Shojaedin-Givi, H. Hashemzadeh, M. Mozafari-Nia, Z. Vaezi and H. Naderi-Manesh, *Photodiagn. Photodyn. Ther.*, 2020, **30**, 101753.
78. J. Delasoie, J. Rossier, L. Haeni, B. Rothen-Rutishauser and F. Zobi, *Dalton Trans.*, 2018, **47**, 17221–17232.
79. M. Terracciano, M.-A. Shahbazi, A. Correia, I. Rea, A. Lamberti, L. De Stefano and H. A. Santos, *Nanoscale*, 2015, **7**, 20063–20074.
80. N. Rabiee, M. Khatami, G. Jamalipour Soufi, Y. Fatahi, S. Iravani and R. S. Varma, *ACS Biomater. Sci. Eng.*, 2021, **7**, 3053–3068.
81. U. T. Uthappa, V. Brahmkhatri, G. Sriram, H.-Y. Jung, J. Yu, N. Kurkuri, T. M. Aminabhavi, T. Altalhi, G. M. Neelgund and M. D. Kurkuri, *J. Controlled Release*, 2018, **281**, 70–83.
82. M. Mishra, K. Devi, D. Pinheiro and A. Nizam, *Russ. J. Org. Chem.*, 2020, **56**, 1784–1789.
83. M. Rezaeian, H. Afjoul, A. Shamloo, A. Maleki and N. Afjoul, *Nanomedicine*, 2021, **16**, 1581–1593.
84. J. R. Kalluri, J. West, G. R. Akkaraju, L. T. Canham and J. L. Coffer, *ACS Omega*, 2019, **4**, 8359–8364.
85. J. R. Costa, T. Neto, S. S. Pedrosa, S. C. Sousa, J. Azevedo-Silva, D. Tavares-Valente, A. Mendes, M. E. Pintado, J. C. Fernandes and A. L. S. Oliveira, *Colloids Surf., B*, 2023, **226**, 113305.
86. H. Ramphul, F. Gimié, J. Andries, D. Jhurry and A. Bhaw-Luximon, *Int. J. Biol. Macromol.*, 2020, **157**, 296–310.
87. N. Tritean, B. Trică, A. R. Gabor, A. Cimpean, F. Oancea and D. Constantinescu-Aruxandei, *Proceedings*, 2020, **57**, 15.

88. S. Tamburaci and F. Tihminlioglu, *Mater. Sci. Eng., C*, 2018, **91**, 274–291.
89. A. C. Özarslan and S. Yücel, *Mater. Sci. Eng., C*, 2016, **68**, 350–357.
90. A. Alhadhrami, G. G. Mohamed, A. H. Sadek, S. H. Ismail, A. A. Ebnalwaled and A. S. A. Almalki, *Materials*, 2022, **15**, 8211.
91. M. Abbasi, R. Gholizadeh, S. R. Kasaee, A. Vaez, S. Chelliapan, F. Fadhil Al-Qaim, I. F. Deyab, M. Shafiee, Z. Zareshahrabadi and A. M. Amani, *Sci. Rep.*, 2023, **13**, 5987.
92. R. H. Babu, P. Yugandhar and N. Savithramma, *Bull. Mater. Sci.*, 2018, **41**, 1–8.
93. R. Periakaruppan, P. Ranjitha and J. Danaraj, *Appl. Biochem. Biotechnol.*, 2022, **194**, 5594–5605.
94. M. C. Rodrigues, W. R. Rolim, M. M. Viana, T. R. Souza, F. Gonçalves, C. J. Tanaka, B. Bueno-Silva and A. B. Seabra, *J. Dent.*, 2020, **96**, 103327.
95. M. Noushad, I. Ab Rahman, A. Husein and D. Mohamad, *Powder Technol.*, 2016, **299**, 19–25.
96. J. R. Kalluri, R. Gonzalez-Rodriguez, P. S. Hartman, A. Loni, L. T. Canham and J. L. Coffer, *PLoS One*, 2016, **11**, e0163270.
97. T. Todd, Z. Zhen, W. Tang, H. Chen, G. Wang, Y.-J. Chuang, K. Deaton, Z. Pan and J. Xie, *Nanoscale*, 2014, **6**, 2073–2076.
98. S. Mura, J. Nicolas and P. Couvreur, *Nat. Mater.*, 2013, **12**, 991–1003.
99. P. Araichimani, K. M. Prabu, G. S. Kumar, G. Karunakaran, N. Van Minh, S. Karthi, E. K. Girija and E. Kolesnikov, *Ceram. Int.*, 2020, **46**, 18366–18372.
100. D. Losic, Y. Yu, M. S. Aw, S. Simovic, B. Thierry and J. Addai-Mensah, *Chem. Commun.*, 2010, **46**, 6323–6325.
101. V. S. Javalkote, A. P. Pandey, P. R. Puranik and P. K. Deshmukh, *Mater. Sci. Eng., C*, 2015, **50**, 107–116.
102. H. Yu, R. Zheng, F. Lei, W. Wang, W. Guo, L. Zhang, Y. Liu, X. Chen and Y. Wang, *Am. J. Transl. Res.*, 2022, **14**, 1518.
103. S. Managò, N. Migliaccio, M. Terracciano, M. Napolitano, N. M. Martucci, L. De Stefano, I. Rendina, A. C. De Luca, A. Lamberti and I. Rea, *J. Biophotonics*, 2018, **11**, e201700207.
104. C. Feng, J. Li, G. S. Wu, Y. Z. Mu, M. Kong, C. Q. Jiang, X. J. Cheng, Y. Liu and X. G. Chen, *ACS Appl. Mater. Interfaces*, 2016, **8**, 34234–34243.
105. C. Yu-Wai-Man, A. D. Tagalakis, M. D. Manunta, S. L. Hart and P. T. Khaw, *Sci. Rep.*, 2016, **6**, 21881.
106. M. J. Gomes, P. J. Kennedy, S. Martins and B. Sarmento, *Nanomedicine*, 2017, **12**, 1385–1399.
107. Y. Zhou, G. Quan, Q. Wu, X. Zhang, B. Niu, B. Wu, Y. Huang, X. Pan and C. Wu, *Acta Pharm. Sin. B*, 2018, **8**, 165–177.
108. M.-H. Kim, H.-K. Na, Y.-K. Kim, S.-R. Ryoo, H. S. Cho, K. E. Lee, H. Jeon, R. Ryoo and D.-H. Min, *ACS Nano*, 2011, **5**, 3568–3576.
109. N. A. Keasberry, C. W. Yapp and A. Idris, *Biochemistry*, 2017, **82**, 655–662.
110. J. R. Kalluri, *Sustainable Routes to Porous Silicon Nanostructures as Drug and Gene Delivery Vehicles*, Texas Christian University, 2018.

111. S. Kwon, R. K. Singh, R. A. Perez, E. A. Abou Neel, H.-W. Kim and W. Chrzanowski, *J. Tissue Eng.*, 2013, **4**, 2041731413503357.
112. A. A. Alshatwi, J. Athinarayanan and V. S. Periasamy, *Mater. Sci. Eng., C*, 2015, **47**, 8–16.
113. A. Bertucci, K.-H. Kim, J. Kang, J. M. Zuidema, S. H. Lee, E. J. Kwon, D. Kim, S. B. Howell, F. Ricci and E. Ruoslahti, *ACS Appl. Mater. Interfaces*, 2019, **11**, 23926–23937.
114. I. Rea, N. M. Martucci, L. De Stefano, I. Ruggiero, M. Terracciano, P. Dardano, N. Migliaccio, P. Arcari, R. Taté and I. Rendina, *Biochim. Biophys. Acta, Gen. Subj.*, 2014, **1840**, 3393–3403.
115. V. De Riso, R. Raniello, F. Maumus, A. Rogato, C. Bowler and A. Falciatore, *Nucleic Acids Res.*, 2009, **37**, e96.
116. J. Lavaud, A. C. Materna, S. Sturm, S. Vugrinec and P. G. Kroth, *PLoS One*, 2012, 7, e36806.
117. B. Delalat, V. C. Sheppard, S. Rasi Ghaemi, S. Rao, C. A. Prestidge, G. McPhee, M.-L. Rogers, J. F. Donoghue, V. Pillay and T. G. Johns, *Nat. Commun.*, 2015, **6**, 8791.
118. P. A. Siver, *J. Phycol.*, 2005, **41**, 720.
119. T. Fuhrmann, S. Landwehr, M. El Rharbi-Kucki and M. Sumper, *Appl. Phys. B: Lasers Opt.*, 2004, **78**, 257–260.
120. S. Yamanaka, R. Yano, H. Usami, N. Hayashida, M. Ohguchi, H. Takeda and K. Yoshino, *J. Appl. Phys.*, 2008, **103**, 074701.
121. M. L. Presti, R. Ragni, D. Vona, G. Leone, S. Cicco and G. M. Farinola, *MRS Adv.*, 2018, **3**, 1509–1517.
122. S. Briceño, E. A. Chavez-Chico and G. González, *Mater. Sci. Eng., C*, 2021, **123**, 112018.
123. J. Noyes, M. Sumper and P. Vukusic, *J. Mater. Res.*, 2008, **23**, 3229–3235.
124. T. Furukawa, M. Watanabe and I. Shihira-Ishikawa, *Protoplasma*, 1998, **203**, 214–220.
125. G. Di Caprio, G. Coppola, L. D. Stefano, M. D. Stefano, A. Antonucci, R. Congestri and E. D. Tommasi, *J. Biophotonics*, 2014, 7, 341–350.
126. J. Pan, J. Cai, D. Zhang, Y. Jiang, Y. Wang, M. Chen and A. Li, Diatom based biosensor for high sensitive fluorescence detection based on a spin-on glass bonding technique, *SENSORS, IEEE*, 2012, DOI: 10.1109/ICSENS.2012.6411218.
127. A. Setaro, S. Lettieri, P. Maddalena and L. De Stefano, *Appl. Phys. Lett.*, 2007, **91**, 051921.
128. L. De Stefano, P. Maddalena, L. Moretti, I. Rea, I. Rendina, E. De Tommasi, V. Mocella and M. De Stefano, *Superlattices Microstruct.*, 2009, **46**, 84–89.
129. A. Bismuto, A. Setaro, P. Maddalena, L. De Stefano and M. De Stefano, *Sens. Actuators, B*, 2008, **130**, 396–399.
130. S. Lettieri, A. Setaro, L. De Stefano, M. De Stefano and P. Maddalena, *Adv. Funct. Mater.*, 2008, **18**, 1257–1264.
131. F. Ren, J. Campbell, G. L. Rorrer and A. X. Wang, *IEEE J. Sel. Top. Quantum Electron.*, 2014, **20**, 127–132.

132. F. Ren, J. Campbell, X. Wang, G. L. Rorrer and A. X. Wang, *Opt. Express*, 2013, **21**, 15308–15313.
133. X. Kong, K. Squire, E. Li, P. LeDuff, G. L. Rorrer, S. Tang, B. Chen, C. P. McKay, R. Navarro-Gonzalez and A. X. Wang, *IEEE Trans. NanoBiosci.*, 2016, **15**, 828–834.
134. K.-C. Lin, V. Kunduru, M. Bothara, K. Rege, S. Prasad and B. L. Ramakrishna, *Biosens. Bioelectron.*, 2010, **25**, 2336–2342.
135. S. Kabiri, M. D. Kurkuri, T. Kumeria and D. Losic, *RSC Adv.*, 2014, **4**, 15276–15280.
136. D. Losic, J. G. Mitchell, R. Lal and N. H. Voelcker, *Adv. Funct. Mater.*, 2007, **17**, 2439–2446.
137. A. Kamińska, M. Sprynskyy, K. Winkler and T. Szymborski, *Anal. Bioanal. Chem.*, 2017, **409**, 6337–6347.
138. S. Viji, M. Anbazhagi, N. Ponpandian, D. Mangalaraj, S. Jeyanthi, P. Santhanam, A. S. Devi and C. Viswanathan, *Appl. Biochem. Biotechnol.*, 2014, **174**, 1166–1173.
139. Y. Huang, P. Li, R. Zhao, L. Zhao, J. Liu, S. Peng, X. Fu, X. Wang, R. Luo and R. Wang, *Biomed. Pharmacother.*, 2022, **151**, 113053.
140. S. Li, Y. Zhang, X.-W. He, W.-Y. Li and Y.-K. Zhang, *Talanta*, 2020, **209**, 120552.
141. R. Sasirekha, T. S. Sheena, M. S. Deepika, P. Santhanam, H. E. Townley, K. Jeganathan, S. D. Kumar and K. Premkumar, *Mater. Sci. Eng., C*, 2019, **94**, 56–64.
142. F. F. Razura-Carmona, A. Perez-Larios, S. G. Sáyago-Ayerdi, M. Herrera-Martínez and J. A. Sánchez-Burgos, *Polysaccharides*, 2022, **3**, 411–425.
143. E. I. Mancera-Andrade, A. Parsaeimehr, F. Ruiz-Ruiz, G. L. Rorrer, J. González-Valdez, H. M. N. Iqbal and R. Parra-Saldivar, *Biocatal. Agric. Biotechnol.*, 2019, **19**, 101175.
144. A. Kabir, N. Nazeer, R. Bissessur and M. Ahmed, *Int. J. Pharm.*, 2020, **573**, 118887.
145. S. Managò, C. Tramontano, D. Delle Cave, G. Chianese, G. Zito, L. De Stefano, M. Terracciano, E. Lonardo, A. C. De Luca and I. Rea, *Small*, 2021, **17**, 2101711.
146. U. T. Uthappa, M. Kigga, G. Sriram, K. V. Ajeya, H.-Y. Jung, G. M. Neelgund and M. D. Kurkuri, *Microporous Mesoporous Mater.*, 2019, **288**, 109572.
147. J. Delasoie and F. Zobi, *Pharmaceutics*, 2019, **11**, 537.
148. B. D. Mattos, B. L. Tardy, M. Pezhman, T. Kämäräinen, M. Linder, W. H. Schreiner, W. L. E. Magalhães and O. J. Rojas, *Sci. Rep.*, 2018, **8**, 5555.
149. A. M. Carvalho, R. A. Cordeiro and H. Faneca, *Pharmaceutics*, 2020, **12**, 649.
150. D. Napierska, L. C. J. Thomassen, D. Lison, J. A. Martens and P. H. Hoet, *Part. Fibre Toxicol.*, 2010, **7**, 1–32.
151. S. Rangaraj and R. Venkatachalam, *Artif. Cells, Nanomed., Biotechnol.*, 2018, **46**, 1415–1424.
152. J. Athinarayanan, V. S. Periasamy, M. Alhazmi and A. A. Alshatwi, *J. Biomed. Mater. Res., Part B*, 2017, **105**, 340–349.

CHAPTER 9

Biogenic Silica as Filler Material for Elastomers

TRINATH BISWAL

Department of Chemistry, VSS University of Technology,
Burla-768018, India
Email: biswal.trinath@gmail.com

9.1 Introduction

Although biogenic silica can be extracted from plant resources, the marine environment is the major source of biogenic silica. Biogenic silica is also termed amorphous opaline silica or biogenic opal, which is the most widely available biogenic mineral on the globe. The microscopic particles present in silica are termed phytoliths and can be detected in grasses and other plant species. Silica is a metal oxide of amorphous structure produced by the complex inorganic polymerization method. Biogenic silica is normally a hydrated form of silica ($SiO_2 \cdot nH_2O$), which is crucial for many animals and plants.[1,2] Airborne biogenic amorphous silica (BAS) fibers derived from rice enter the human digestive system through inhalation at the time of incineration or combustion. The nonfibrous BAS can adsorb poisonous organic materials and can easily enter the lung, causing a potential health hazard. Bio-silica has many advantages over nanoporous silica for its use in nanoscience and nanotechnology. The BAS fibers obtained from food crops can be ingested by humans and cause an adverse health hazard. A neglected component of the coupled global continental biogeochemical cycles of silicon and carbon is called biogenic silica. Biogenic silica is a universal and important component of soils produced under a wide range of tropological conditions, whereas inorganic pedogenic opal is produced only under a particular physico-chemical condition of the soil. Based on geologic

Biogenic Silica: Fundamentals and Applications
Edited by Ashok Kumar Nadda and Vijai Kumar Gupta

Published by the Royal Society of Chemistry, www.rsc.org

implications, it was found that some specific plants possess the capability of accumulating some significant metals such as Fe, Si, Ca, Al, Cr, Hg, *etc.* The presence of biogenic silica in groundwater increases with time and depth in association with the rocks present under the Earth's surface. Under suitable conditions, a high percentage of organically derived silica in soluble form can be mixed in the groundwater or recycled. The use of appropriate inorganic filler materials of various kinds drastically increases the mechanical properties of elastomers. The efficiency of reinforcing the biogenic silica particles usually depends on the complex interaction between the macromolecules of the elastomer and the filler particles.[3,4] The shape and size of the particles of biogenic silica, their reactivity, specific surface area, structure, and uniformity of distribution in the matrix phase of the elastomer play a key role in the interaction between the reinforcing phase and matrix phase. Today, the use of technologies and materials associated with bio-sources is the cause of a substantial decrease in petroleum-derived substances and feedstock, which is now a vital trend in the rubber industry. The use of biobased silica as filler materials in elastomers enhances sustainable development with green chemistry approaches. Not only biogenic silica, but also the use of other different inorganic filler materials drastically enhances the mechanical properties and other kinds of essential properties of elastomers. The major objective of this study is to use biogenic silica as a filler or reinforcing material during the development of composites based on elastomers. Rice husk is a potential low-cost, renewable raw material source of carbon and bio-silica.[5,6] Biogenic silica (BSi) is a chemically defined proxy, that measures the concentration of amorphous Si within the sediments. It is observed to be a suitable proxy for diatom abundance, and some other siliceous microfossils such as phytoliths and sponges are highly important as filler material for the rubber industry. Hence, as a single proxy, BSi offers an index related to diatom richness and most of the systems of diatom productivity.[7,8] The techniques used to measure the content of BSi in sediments are of five kinds, such as:

(1) X-ray diffraction (XRD)
(2) Point count of diatoms
(3) Infrared analysis (IR analysis)
(4) A technique of normative calculation whereby BSi is measured by the difference from the silicates present in the minerals
(5) Techniques of wet-alkaline digestion

The methodology of wet-alkaline is frequently used due to its ease of applicability and reliability. Suitable wet chemical techniques were not found to estimate BSi until the early 1980s. Therefore, prevalent paleolimnological applications started in the 1980s, and the measurement of BSi in sediments is currently one of the usual estimations carried out in paleolimnological studies. Figure 9.1 reflects the uses of biogenic silica in different important fields.[9,10]

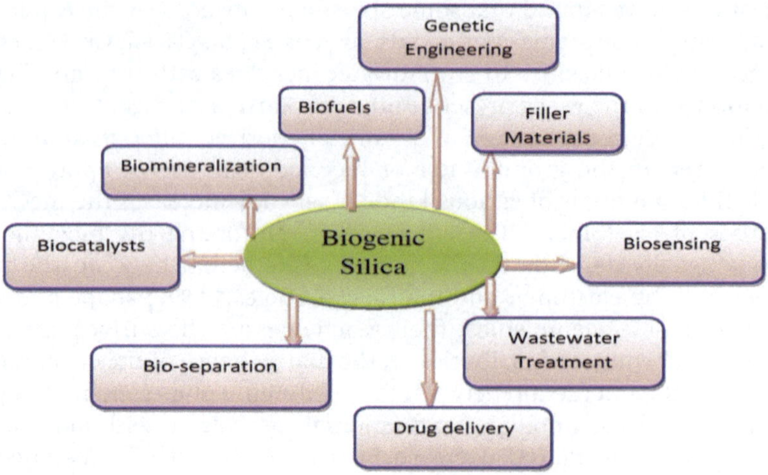

Figure 9.1 Use of biogenic silica in various important fields.

9.2 Sources of Biogenic Silica

Biogenic silica is a natural biomaterial and is found to be one of the most readily available biomaterials found in biogenic minerals. The various sources of biogenic silica are as follows:

(1) The marine sediments contain biogenic silica particles with an average of 14%, globally.
(2) Sugarcane biomass and rice husk are the two major sources of biogenic silica.
(3) Biogenic amorphous silica (BAS) is present in the soil normally in the range of 0–6% that protects the plants from drought.

Biogenic silica can be extracted from the plant biomass materials. The various steps of extraction of biogenic silica from biomass materials are shown in Figure 9.2.[11]

Rice husk ash and pineapple peel are important sources of biogenic silica. Various agricultural residues are a potential source for the extraction of biogenic silica. Various agro-wastes are potential sources of biogenic silica. The nanoparticles (NPs) of biogenic silica are highly significant in the application of some important fields such as textile, electrical, electronic, and elastomer or rubber. Biogenic silica was also extracted from wheat husk ash. Biogenic silica is also present in grasses and other plant materials.[12,13]

9.2.1 Biogenic Amorphous Silica Used for Elastomers

The insertion of different kinds of specific inorganic filler materials into elastomers or rubber potentially enhances the mechanical characteristics of

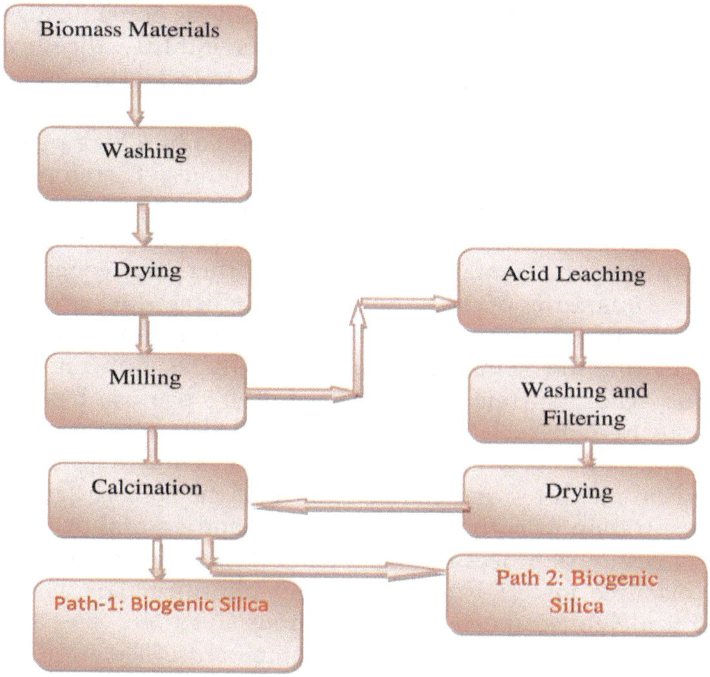

Figure 9.2 Extraction of biogenic silica from biomass agricultural residue products.

the elastomers. The efficiency of the reinforcement of biogenic silica as filler material in the matrix phase of elastomer normally depends upon two factors, such as:

(1) The complex interaction between the macromolecules of the elastomer and molecules of the filler materials (biogenic silica particles)
(2) The interaction between the particles of the filler materials themselves

In both of the cases, the shape, size, reactivity, structure, specific surface area, and distribution of the filler particles (uniform or non-uniform) play a significant role in the process of interaction. In the present decade, the application of materials from renewable bio-resources and their use in appropriate modern technologies has facilitated the green pathway of synthesis and a significant decrease in petroleum-derived substances and feedstock materials in the rubber or elastomer industries. The increase in effort for the development and use of biopolymeric materials and bio-fillers (as reinforcing materials) in the present decade improves the concept of green and sustainable development in the industrial sector. The use of bio-based materials as filler materials for the synthesis of numerous polymer composites has been reported by several researchers. A part of the green approach to synthesis is the use of huge quantities of agricultural waste biomass products, which is an innovative pathway for utilizing environmentally friendly substances in the elastomeric industry. Rice husk, sugar cane wastes, and wheat husk deserve special attention for production of biogenic silica particles.

These waste biomass products are characterized by a high percentage of ash, low nutritional values, and low calorific values. Therefore, these are unsuitable feeds for cattle and cause serious environmental issues. The open burning of rice and wheat waste biomass products is the cause of serious environmental contamination, and the flue gases generated from it contain huge quantities of crystalline forms of silica, which are highly toxic to human health. These wastes contain some basic components such as cellulose, lignin, and hemicellulose, along with a significant quantity of ash residue, whose sustainable solution for effective use has yet to be a challenge to human society. The quantity of SiO_2 present in rice husk normally ranges from 18–23% by mass, which mainly depends upon the agro-technical measures and characteristics of the soil. The presence of a high concentration of carbon (mass% of 38) and amorphous SiO_2 in the biomass of rice husk proves it to be a potential ecological, renewable, and low-cost resource material for carbon and bio-silica. The white ash obtained after incineration contains amorphous silica >96 wt%. The ratio of carbon to SiO_2 depends on the pyrolysis temperature. The heat-treated rice husk used as filler materials for elastomers such as white ash is a good reinforcing material for some important polymers such as polyethylene (PE), polystyrene (PS), and polypropylene (PP). Rice husk ash (RHA) is found to contain an appreciable content of bio-silica and is used as an excellent filler material for elastomers, particularly as an appropriate substitute for the synthetically developed commercially obtainable silica. RHA increases the grip on snow and ice in a better way as compared to standard silica and, therefore, can be suitably used in the formulations of shoe soles designed for extreme hiking footwear, winter tyre protectors, winter sports, ski boot soles, *etc.*[14–16]

9.2.2 Use of Biogenic Silica as Filler Material in Rubber Blend Composites

In industry, especially in tyre and tube industries, synthetic and natural rubbers are frequently used by modifying rubber on bleeding with silica or some other suitable reinforcing agent. The elastomeric substances associated with polyisoprene rubber possess quite improved crack resistance, weather resistance, high strength, and wet grip. An elastomeric material like chlorosulphonated polyethylene rubber shows improved properties of oxygen and ultraviolet stability. The rubber is blended to balance properties such as cost, workability, *etc.* Although synthetic rubber materials can be blended with each other to improve the properties of any one component of the blend, the insertion of inorganic filler material in the binary blended material can influence the separation of phases owing to the interaction of the individual components present in the blend with the surface of the solid phase. It was reported that inorganic filler materials like silica are compatible with the elastomer for blending and reinforcement. For the polymethylmethacrylate/polyvinyl acetate [(PMMA/PVAc)] blend, the biogenic silica (BSi) is found to be highly compatible, which is confirmed by identifying the rate of phase separation of the filler materials present in the blend. The addition of the filler surface compatibilizer to the matrix phase of the elastomer or its blend can change the shape and position of the phase separation curve. This is because the capitalizing effect of the inorganic filler material (biogenic silica) reduces the molecular mobility, which is the cause of the phase separation of the components. A new mechanism has been suggested where the filler materials are of high aspect ratio (more than 1 nm in dimension) and can produce *in situ* grafts *via* adsorbing huge quantities of polymers, that are highly suitable for retarding the compatibilization induced in immiscible blends and interfacial tension. The main aim of this study is associated with the impact of biogenic silica (BSi) as a filler material on the mechanical properties and deformation mechanisms of the composite of elastomer matrix.[17,18]

Biogenic silica (BSi) was incorporated at various ratios into some of the polymer blends of chlorosulphonated polyethylene rubber (CSM) and polyisoprene rubber (NR) cured by the usual sulphur system. It was observed that there is a drastic reduction in scorch time and cure time, which is due to the enhancement of the cure rate index (CRI) and the filler loading of about 30 phr in the various blend ratios. The hardness and tensile strength were 4–5 Sh-A higher for the compositions of various blends, whereas the swelling resistance in toluene is higher. The NR/CSM/BSi composites show improved thermal stability.[19]

9.2.3 Rice Husk as Filler Material for Rubber Blend

Rice is one of the major sources of food for most people in the world, and almost 90% of the global need for rice, which is mainly cultivated in Asian countries. According to the 2020 data of the Food and Agriculture Organization (United States), nearly 760 million tonnes of rice are produced globally. In rice, an average of about 20% is husk, which protects the grain. Therefore, the total husk production throughout the globe is approximately 152 million tonnes. The rice husk (RH) is a vital source of fuel to produce thermal and electrical energy. It is a renewable and reliable source for the extraction of biogenic silica. The RH used for extraction of bio-silica is an eco-friendly and significant source of material, that is widely found in paddy cultivation countries. In most of the situations it is considered that the waste solid materials are mostly insignificant, and their use is limited. The composition of the RH includes 80–90% silica, 1–2.5% alumina and the rest are oxides of Ca, Fe^{2+}, Na, Mg, and K. The higher percentage of biosilica in RH ash makes it a suitable material for concrete and cement production. The bio-silica can be extracted from the biomass ash (rice husk ash) using various methodologies, such as treatment with a base or acid on association with pyrolysis, sol–gel technique, biological treatment by using enzymes, and thermal treatment. Among these, the sol–gel treatment method is the most useful. The bio-silica obtained from rice husk ash (RHA) can be used as an effective filler material for the manufacturing of rubber composites.[20,21]

Ramasamy *et al.* used powdered forms of RHA and RH in the compound of polyisoprene (natural rubber) as filler material in the composite, but they have not reinforced it in the rubber material. The NR inserted with bio-silica, which is obtained from RHA, exhibits improved mechanical characteristics owing to its capability of reinforcing as carbon black. Later on, it was reported that the material NR filled with bio-silica showed better thermal stability and ozone and oil resistance properties.[22] But in the rubber industry, as per requirement, two or more different kinds of rubber are combined (rubber blend) to achieve improved properties, that can be successfully used in various important applications such as tyres, seals, bearings, and other important engineering product materials. Bio-silica has been successfully used as a filler material in blends of polar rubber like NBR and non-polar rubber like NR to make the material compatible and prevent phase separation owing to the difference in polarities of such rubbers. Hence, RHA can be potentially used as filler material in the NR/NBR blends to enhance the useful properties of the rubber blend to an appreciable extent. Since silica is an inorganic substance, it is not compatible with rubbers (organic compounds); hence, compatibilizers of a special kind have to be mixed to enhance the compatibility between silica filler materials and rubber.[23]

Example: Bis(triethoxysilylpropyl) tetrasulfide is a silane compound that can be used as a compatibilizer in the material of NR filled with silica. The compound developed exhibits improved dynamic mechanical properties and physico-mechanical properties.[24]

The use of silica as filler material in the NR/NBR blends is highly amazing because it offers much improved physico-mechanical properties in the product materials of rubber and overcomes the demerits of using carbon or carbon black as filler material, which is widely used as filler material in rubber-based compounds for commercialization. The use of bio-silica extracted from RHA in place of carbon powder is cost-effective, beneficial because of improved properties, and eco-friendly for the environment. Particularly, the percentage of equilibrium swelling of NR/NBR blends in engine oil, brake oil, toluene, and hydraulic oil is highly interesting, therefore can be used for applications of such kinds of solvents and oils for sealing, which promotes the chances of substituting carbon black as filler material in RHA. Although RH after burning is discarded on the paddy fields without any valuable use, it is a potential source of extraction of bio-silica, which can be effectively used as filler materials in rubber blends to improve their physical and mechanical properties.[25]

9.2.4 Bio-silica Extraction from Rice Husk Ash

At first, the RHA was collected from the factory or industry of clay bricks, perfectly washed with distilled water, and then stored at a constant temperature of 105 °C in an air-circulating oven until a constant weight was achieved. Now, after drying, the RHA is burned for about 6 hours at a temperature of 850 °C in the Hobersol Muffle Furnace. After that, it is kept overnight for cooling until room temperature (25 °C) is achieved. The sample of RHA was digested by boiling in a solution of 3.0 M NaOH for about three hours in a round bottom flask, maintaining the ratio of NaOH to RHA of 8 : 1, then cooling up to room temperature. Now the pH of the mixture must be decreased to pH = 2 by the addition of 2.5 M H_2SO_4 followed by mixing NH_3 to get the precipitate of silica at pH = 8.5 at normal room temperature. After that, the mixture of the solution was filtered and dried perfectly at 120 °C for about 12 hours in an air-circulating oven. Now the sample was refluxed by using 6.0 M HCl for approximately 4 hours and repeatedly washed with deionized water until the mixture of the solution was free from acidic components. Then the purified silica obtained was centrifuged at 8000 rpm for about 30 minutes and finally dried at 105 °C for about 2 hours.[26,27]

9.2.5 Short Fiber Biogenic Silica as Reinforcing or Filler Materials for Silicone Rubber

The siloxane polymeric materials, especially poly(dimethyl siloxane) (PDMS) of high molecular weight, show some valuable properties like excellent flexibility and high mobility of chains, which fit this material for application in the rubber industry. But such rubbers (PDMS-based rubbers) usually exhibit low mechanical properties such as tensile strength, tenacity, *etc.* The required mechanical properties of this elastomer can be achieved by incorporating suitable filler materials such as particulates or powdered or short-fiber forms of silica, specifically fumed silica, quartz, and precipitated silica. This kind of filler material is more abundantly used as a filler material for reinforcing the silicone rubber matrix. In addition to the short-fiber biogenic silica, other reinforcing materials such as fibers of sisal and cellulose are incorporated in the rubber matrix to increase or modify the mechanical properties while maintaining the cost of the composite material. Figure 9.3 represents the various potential renewable sources of biogenic silica.[28]

The mechanical performance of a polymer-based composite material depends on the kind of matrix phase, filler materials, its properties, and adhesive forces of attraction at the matrix–filler interface. The adhesive forces of attraction can be improved by surface treatment with an appropriate coupling agent, which can produce covalent linkages between the filler surface and the reactive functional groups present in the polymer matrix. The efficiency of the stress transfer at the interface of the matrix–filler mechanical properties of the synthesized polymer composite increases.

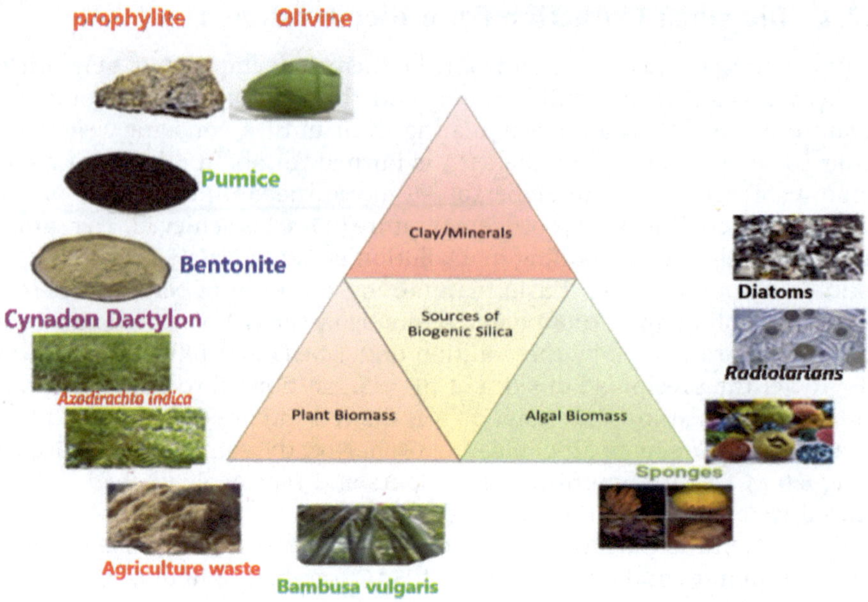

Figure 9.3 Various sources of biogenic silica.

Short fibers of biogenic silica are normally recovered in high quantities in the sediments present at lake beds, which contain clay materials, organic matter, and sand. This is an inorganic substance with the internal promoting structure of sponges (Demospongiae). Owing to its siliceous nature, the short fibers of biogenic silica can be used as a suitable, cost-effective reinforcing material in the case of elastomeric composite materials. The short fiber of biogenic silica is either a primary or secondary filler material for the elastomeric matrix of PDMS. The silica short fibers enhance the thermal stability of the matrix and prevent the swelling effect, which is caused by the interaction of the fibers with the rubber matrix. The short fiber of biogenic silica is highly effective as the primary filler material in the case of PDMS with respect to the tensile strength and Young's modulus, whereas as secondary filler materials, nanofiber (NF) and microfiber (MF) in association with Young's modulus and storage result in an improvement in the rubber stiffness. Hence, the short fibers of biogenic silica proposed the probability of using MF or NF as secondary or primary reinforcing filler materials in the case of PDMS, where the main aim for industrial application is to improve the thermal stability of the rubber matrix with the prevention of swelling behaviour in the solvent used, decreasing stiffness, and increasing the cost of the PDMS rubbers. The short fibers of modified biogenic silica or natural biogenic silica can be effectively used as secondary or primary filler materials in the case of PDMS matrix. The composite material obtained exhibits excellent mechanical, thermal, and anti-swelling properties.[29,30]

9.2.6 Composites of NR/CSM/Biogenic Silica Rubber Blend

Biogenic silica (BSi) was introduced in various proportions in some of the polymer blends associated with chlorosulphonated polyethylene rubber (CSM) and polyisoprene rubber (NR), then cured properly by the usual traditional system. It was found that the extraordinary reduction in the optimum scorch time, and cure time is observed in association with the increase in the rate index (CRI) with the filler loading up to 30 phr in various blend ratios. The hardness and tensile strength were 4–5 Sh-A higher for the various compositions of the blends, whereas the prevention of swelling in toluene comparatively increased. From the SEM images, it was found that the filler materials are filled at the interface between the polymers, indicating uniform distribution and compatibilization of the polymer blends. Particularly in the tyre industry, silica particles are used as highly suitable reinforcing agents in synthetic rubbers. The use of biogenic silica on polyisoprene rubber (NR) shows quite high strength, excellent resistance to cracks, good weather resistance, and wet grip. Chlorosulphonated polyethylene rubber (CSM) reinforced with biogenic silica shows remarkable oxygen and ultraviolet stability. Hence, along with gloves, this rubber blend can also be effectively used for coatings, linings, and adhesives. The silicon-based blended materials provide balanced useful properties at a low-cost, which is not achieved for single elastomers. Hence, synthetic rubbers are mixed with each other in the form of a binary mixture to enhance the inferior properties of the desired component. The silica minerals can be used as filler materials to function as a good compatibilizer for a binary mixture of polymethacrylates, polymethylmethacrylate/polyvinyl acetate, polymethacrylates, and polyacrylates. Blending with biogenic silica (BSi) drastically enhances the compatibility of the matrix and filler in between the polymers. The insertion of biogenic silica as a filler material in the immiscible blend of polymers can influence the phase separation owing to the cohesion of the separate components of the blended material with the solid surface.[31,32]

9.2.7 Biocomposite Based on Phyto-ash Rich BSi

The porous structured material obtained from plant biomass products is predicted to exhibit better performances in various fields, including sensors in different chemical reactions, cosmetics adsorption, pharmaceuticals, drug delivery, and catalysis. The porous substances of specific categories can be effectively used as additive materials in the case of polymer-based composites. The phyto-ashes obtained naturally from the biomass of plant species contain a high concentration of biogenic silica. The nanobiogenic silica is highly useful in medical science, especially because it is biocompatible with our body tissues and, in modified form, can be used for the treatment of cancer. The sol–gel technique is mostly used for the synthesis of silica nanoparticles. The biogenic silica is obtained from nature, but no biochemical reactions are carried out, which is necessary for silicon. A good example of an important source of silica is horsetail, which contains about 25 wt% of silica. Most of the silica of this kind is present in the secondary cell wall of horsetail, which is found to be a good vascular plant. In addition to silica, horsetail contains other substances such as lipids, proteins, cellulose, and some metal ions (K^+, Ca^{2+}, Na^+, and Al^{3+}). The properties of biogenic silica are highly dependent on the pre-treatment and processing of the crop biomass. Another class of crop that contains a high concentration of silica in its cell tissues is the Gramineae family. Wheat straw, rice straw, and rice husk are also important sources of biogenic silica, where biogenic silica can be extracted in a sustainable and economical way. Besides biogenic silica, the nanofibers of cellulose can also be extracted from wheat straw. The composition of natural rubber is 94% *cis*-polyisoprene, with the remaining 4% being other materials including proteins, carbohydrates, and lipids. The rubber core contains lipids and proteins on its surface. The lipids and some other chemicals present in the rubber are responsible for the overall performance, quality, and some essential properties of the vulcanized rubber. Although numerous kinds of filler materials and additives are added to the vulcanized natural rubber for improving the required properties, the incorporation of silica particles modifies the various properties such as tear strength, and rolling resistance, which are highly essential for the manufacturing of tyres. Besides silica, the second most effective filler material for rubber composites is carbon black. It is the cause of the improvement in strength, abrasion resistance, and modulus of the composite material. Natural rubber reinforced by biogenic silica is quite sustainable because both are found from natural renewable sources.[33,34]

9.2.8 Rubber Composite Reinforced by Nanostructured Biogenic Silica

Silica is a valuable material of commercial importance owing to its wide application in the development of various products. Synthetic amorphous silica (SAS) is normally SiO_2 that is deliberately produced without any change in its physico-chemical properties. The production of nanostructured SiO_2 on an industrial scale is a challenge because of the requirement for costly raw materials and increased energy consumption. But non-petroleum-based filler materials such as nanostructured biogenic silica are eco-friendly, cost-effective, efficient, and renewable, which makes them suitable filler materials today. The significant alternate precursors include synthetic chemicals, minerals, and biogenic resources. The use of biogenic silica in nanostructure as a reinforcing material for the synthesis of rubber composites remarkably improves some of the essential properties, such as mechanical, dynamic, and thermal stability. The efficacy of reinforcement of biogenic silica at the nanoscale is associated with various factors such as rubber–filler interactions, structural features of the filler material, surface area, and filler–filler interactions. Silica is an amazing material with great commercial importance owing to its wide application in the manufacturing of different product materials. Synthetic amorphous silica (SAS) is nothing but a simple form of SiO_2 that has been deliberately synthesized industrially and marketed for many years without appreciable change in its chemical and physical properties. The white dry powder form of SAS is used as a good filler or additive material for rubber and in a variety of industrial uses. Normally, three kinds of SAS are manufactured industrially, such as:

(1) Fumed SiO_2
(2) Gel form of SiO_2
(3) Precipitated SiO_2

Among these three, precipitated SiO_2 and amorphous SiO_2 have more commercial importance. The precipitated SiO_2 is a white powder in a finely divided state consisting of aggregates up to about 0.5 mm in radius and is little soluble in water. Some of the researchers throughout the world emphasized replacing costly synthetic silica with comparatively renewable, cost-effective biogenic silica as the resource material. Furthermore, many researchers studied extensively to develop suitable synthesis techniques, specifically the sol–gel method with optimization of synthesis parameters, to develop the product materials based on nanostructured SiO_2 having the physico-chemical properties suitable for reinforcement in rubber or elastomer matrix. Suitable filler materials are highly important in the rubber industry to enhance the ease of processing and service efficiency, and their addition leads to some basic changes in some valuable properties of rubber. Considering the kinetic theory of elasticity, rubber modulus increases with an increase in temperature (without the filler material), but the introduction of filler materials remarkably varies the temperature coefficient of modulus and can even change the sign of the coefficient, leading to a decrease in modulus with an increase in temperature. Filler materials are classified according to their chemical composition, compatibility with the rubber matrix, and influence on the properties of rubber. Again, the compounding of rubber is classified basically into three kinds based on their reinforcing effect, such as very active, inactive, and semi-active. The term active indicates the effect of filler material on the viscosity of the rubber matrix (degree of reinforcement), resulting in an improvement in mechanical properties. The major properties that measure the effect of reinforcing the filler materials include surface properties and the structure of the material. Biogenic silica is an active filler material that exhibits a huge relative surface area with adequate structure, offering strong chemical and physical forces of interaction between the rubber or elastomer matrix and the filler materials. But an extremely active filler surface is the cause of very high interparticle forces, which negatively impacts the processing properties and leads to the agglomeration of the particles of filler materials at the time of storage and mixing. There are a few properties that affect the property profiles of pure rubber matrix, filler materials, and processability. These properties include geometry, volume fraction, interaction between filler materials and elastomer or rubber, particle dimension, quality of dispersion, and the degree of orientation of the filler materials due to their anisotropic properties. All these factors influence each other and create synergy if an optimum balance is obtained. Besides silica and biogenic silica, the other most commonly used filler materials in the rubber and elastomer industries are basically particles of carbon materials such as graphene, carbon black, graphite, carbon fiber (CF), and carbon nanotubes (CNTs), particles of inorganic materials such as halloysite, nanoclays, alumina (Al_2O_3,), talc, polyhedral oligomeric silsesquioxane (POSS), titanium oxide (TiO_2), calcium carbonate ($CaCO_3$), zinc oxide (ZnO), and some bio-filler materials such as cellulose, coir, husk, and wood. Huge volumes and categories of rubber possessing excellent properties of elasticity and mechanical properties such as stiffness, abrasion resistance, adequate properties of anti-scratching, high strength, toughness, friction resistance behaviour, *etc.* are used as matrix phases, and biogenic silica, precipitated SiO_2, or carbon black are predominately used as filler materials. Nanostructured silica or biogenic silica is the predominant reinforcing material for elastomers or rubbers, but the filler materials of non-petroleum bases are also showing promising efficiency for decreasing the rolling resistance of rubber nanocomposites. But nanostructured silica, or biogenic silica, is a significant, cost-effective filler material for the elastomer and rubber industries. The significant reduction in rolling resistance of the rubber makes biogenic silica a sustainable and economical filler material for manufacturing green tyres. The new modified tyre treads are fuel-saving and specifically reinforced by biogenic silica, owing to the decrease in rolling resistance and improved wet grip compared to the carbon black filler materials. The composite materials based on rubber or elastomer reinforced with nanostructured biogenic silica exhibit excellent mechanical, thermal, and dynamic properties and are associated with various important factors such as the structure of the filler materials, interactions of rubber–SiO_2 filler, and impact of biogenic silica as a filler material for the synthesis of rubber nanocomposites.[35–38]

9.3 Extraction of Amorphous Biogenic SiO_2

Most of the global agricultural waste material byproducts contain SiO_2 or biogenic silica, which is particularly a third group of biomass products, *i.e.*, biowaste product materials, which truly generates another category of useful resource biomaterials. Today, extensive research work has been carried out to determine the methodology for getting a source of SiO_2, which is a completely safe, economical, and eco-friendly alternative to the commercial production of rubber-based composite materials. According to the reports of various researchers, the various residues of agricultural products such as rice husk, wheat husk, rice straw, cane husk, coffee waste biomass, *etc.* are the major precursors for the extraction of pure SiO_2 in both crystalline and

Table 9.1 Percentage of SiO_2 in some plant species (ref. 41 and 42).

Plant species	Part of the plant species	% of biogenic silica
Cane	Husk	8.0
Sorghum	Leaf sheath epidermis	88.75
Coffee	Husk	12.0
Rice	Straw	82.5
Rice	Husk	93.0
Bagasse	Biomass	73.0
Lantana	Leaf and stem	23.38
Bamboo	Nodes (inner portion)	57.4
Corn	Leaf sheath	64.32
Bread fruit tree	Stem	81.8

amorphous states. The most important biogenic sources of SiO_2 are rice husk, wheat husk, and ash. The rice husk is a comparatively hard substance containing lignin and SiO_2, which protect seeds from damage. During milling, one kilogram of milled white rice produces almost 0.28 kg of rice husk. Table 9.1 indicates the percentage of ash in some important plant species.[39,40]

9.4 Biogenic Silica Present in Terrestrial Ecosystems

Among the biological living beings of the land, only horsetails (Equisitaceae) are the conventional source of silicon. In the 19th century some plant physiologists identified many families of plant species present in terrestrial ecosystems containing silicon in appreciable amounts, like Ca, P, Mg, and S, but not an essential plant nutrient. Silicon is the second most abundant element present in the Earth's crust and larger sized plant species in terrestrial environments consume $Si(OH)_4$ from the soil *via* their root systems. It is then carried by the stream, and due to transpiration, it reaches and is deposited irreversibly in the form of amorphous, hydrated silica in various parts of the plants, such as roots, stems, fruits, leaves, and inflorescences, which is nothing but biogenic silica. Biogenic silica has the general chemical formula $[SiO_n{\cdot}OH(4-2n)]m$, where m is large and n is <2 indicating the range of composition in the formula with the end members $m\mathrm{Si(OH)_4}$ and $m\mathrm{SiO_2}$. The silica parts that discretely form in plant tissues are basically termed opal phytoliths. It was reported that the content of SiO_2 present varies from <0.5% to 19% dry weight, but a wide variation in silica is observed at different parts of the same plant species. According to the rule of thumb, the highest percentage of biogenic silica (BSi) is deposited at the organs or tissues, where the percentage of loss of water is maximum. The percentage of silica also exhibits remarkable phylogenetic variability. It was found that the concentration of BSi increases in the order of liverworts < horsetails < clubmosses < mosses < angiosperms < gymnosperms < ferns.

Among the angiosperms, normally the maximum percentage of BSi is reported in monocots, such as grasses, preferably *Phragmites,* bamboo, and rice. The other biomass products such as sedges, bananas, palms, and some dicots, such as many temperate-forest trees like beech, oak, sugar maple, elm, cannabis, sunflower, and stinging nettles also contains an appreciable percentage of biogenic silica. Less coniferous trees, such as gymnosperms and larch tree are also noticeable Si accumulators. The wood of many hardwood trees of the rainforest contains silica in appreciable concentration.

The category of higher plant species is specified into three categories, such as non-, high-, and intermediate accumulators of silicon.

(1) The high accumulator's kind basically contains concentrations of $SiO_2 > 1\%$ dry weight and consumes dissolved silica (DSi) faster as compared to water, this indicates an active-uptake mechanism. Recently, the Si-transporter gene was recognized in rice.
(2) The intermediate group of plant species contains almost 1% SiO_2, which basically absorbs DSi inactively with water.
(3) The non-accumulator plant species contain a negligible percentage of BSi and successfully exclude it.

For any kind of individual class, the quantity of BSi accumulated changes as a function of the kind of soil, pH, existence of ectomycorrhizal fungi on the roots of the plant species, concentration of DSi, rate of transpiration, plant maturation, and presence of plant nutrients (like P and N). The different benefits of biogenic silica have been successfully recognized in higher plant species. They can be classified as structural, protective, or physiological. Biogenic silica serves as an energetically economical stiffening material that stimulates the oxygen supply to roots, resistance to lodging, which is flattened by rain or wind, favourable exposure of leaves towards incident light, and upright stature. It improves yield and growth, particularly in the accumulator species, and retards vulnerability to pathogens like grazing insects, mammals, mildew, and molluscs. The consumption of biogenic silica enhances the nutrient (N and P) supply and enriches the heavy metal toxicity through Mn, Al, Zn, and Fe. In forests in acidic environments, including conifer forests, swamp forests, and tropical rainforests, substantial quantities of Al can co-precipitate with BSi. Silicification facilitates protecting against temperature extremes, thereby decreasing transpiration and reducing the influence of salinity stress and drought. The higher plant species enhance the weathering of minerals based on silicate by enhancing the presence of organic material and moisture status in soil, which is done by incorporation of H^+ ions (organic acids, CO_2, and detritus) in the proximity of the mineral surfaces with exclusion of potent chelating agents like malic acid, citric acid, phytosiderophores, and rhizosphere.[43–45]

9.5 Biogenic Silica Present in Aquatic Ecosystems

The organisms present in freshwater vigorously exhibit cycle leakage of DSi from terrestrial ecosystems. Although limited floating or submerged vascular plants have been found to contain remarkable quantities of BSi, the number is gradually increasing because of the existence of plants like *Trapa natans, Ruppia maritima* that colonize salt lakes. But Si is a vital nutrient for a small number of highly important aquatic silica bioengineers such as diatoms, flagellates (chrysophytes) and sponges. Chrysophytes and diatoms can use inorganic C (dissolved state) for photosynthesis. Diatoms are normally assumed to be the most significant Si sink in rivers and lakes because of their proliferation across a huge variety of habitats. They form DSi into nanospheres along with proteins, which are commonly termed silaffins. These spheres basically serve as the building blocks for the creation of frustule. A membrane of organic materials prevents the frustule from dissolving, and the gels of thick polysaccharides related to some specific species add to this coating. The membrane created supports the adsorption of dissolution-retarding cations such as Fe^{2+} and Al^{3+} and offers a good physical barrier. It decays very quickly after death, which is partially owing to the activity of the bacteria that accelerates the process of dissolution *via* a bacterial loop within the Si cycle. Siliceous sponges are found to be distributed globally, both in marine environments and freshwater ecosystems. They embellish deep lakes with hydrothermal Si, including Baikal, Taupo, and Tanganyika, however, it is most widely available in shallower waterbodies, which contribute 40% of the BSi in the surface sediments of the lakes in Florida. Biosilicification through sponges is also like diatoms. They discharge spicules that are needle-like, which offer structural support and serve against predation. The spicules encircle an axial organic filament, which contains protein silicate that stimulates the condensation process of amorphous silica. Some researchers claimed that the role of sponges present in the biogeochemical Si cycle worldwide has been undervalued, particularly as sponge spicules are huge and have a less specific surface area as compared to diatom frustules. The high concentration of Al is also the cause of remarkable resistance towards dissolution. The chrysophyte microalgae are effective competitors for nutrients, but they have more resistance in distribution as compared to sponges or diatoms. BSi is usually more widely distributed in the Antarctic lakes, the cold oligotrophic waters of the Arctic, and some specific tropical swamps. Like diatoms, chrysophytes are also comparatively rare in eutrophic and alkaline waters, but there are a few exceptions. These exist in numerous life forms and are particularly widespread in the plankton of lakes of low trophic status that uses their flagella to break into the pelagic zone and compete for nutrients. The siliceous components of these aquatic species are cysts and scales, which are usually formed in a similar manner to those of diatoms, and sponges. The lake ecosystem is normally efficient at silica trapping.

Example: In Lake Malawi, phytoplankton is dominated by diatoms at the time of the dry season, at the time when upwelling transports nutrients towards the surface from anoxic hypolimnion.

It was reported that deep lake sediments in unglaciated areas shows permanent sinks in the biogeochemical Si cycle. This also depends on plate-tectonic timescales despite leaks in different quantities of DSi.

Example: 35% of the inputs pass *via* Lake Superior, whereas 20% *via* Lake Michigan. Both lakes possess higher outflows as compared to Lake Malawi. Shallower lakes are comparatively less likely to release and release greedy silica if they desiccate. Under the dry and stormier climatic conditions of the Last Glacial Maximum (LGM), it was found that the aeolian flux of BSi was highly strengthened. In all these examples, it was recommended that lakes store appreciable quantities of BSi on timescales ranging from 1–10 years in hydrological open systems to about 106 years for deep closed basins. The fluvial transport of DSi from tropical rivers to the oceanic ecosystem is comparatively higher. The siliceous organisms remove huge quantities of DSi, although recycling of Si is comparatively quicker and BSi storage in floodplains and river channels is less permanent relative to lakes.

Example: In the flooded forest areas, sponges live on trees along the Rio Negro, which is a tributary of extremely limited transport in the Amazon region that changes huge quantities of DSi into spicules that accumulate in alluvial deposits. High diatom growth on the submerged wood is the cause of gradual silicification *via* silica gel, creating extra Si storage. Globally, riverine ecosystems convert 16% of the gross fluvial DSi load to BSi (suspended), leading to the declination of the introduction of DSi to the oceans. Human activity, particularly eutrophication, manipulation of macrophyte communities, and changes in the river channels, also influence the transportation of silica particles. The manmade water reservoirs are a good example of human activity enhancing fluvial filtering of Si on the continents. The availability of DSi and decline in river flow rate impact negatively most of the lake ecosystems and many coastal water bodies like the Baltic and Black Seas by facilitating the flagellates or noxious blooms of N-fixing cyanobacteria.[46–49]

9.6 Conclusions

Since the late 19th century, geochemists have identified that the silicon and carbon cycles of the globe are coupled with each other based on geological timescales through the chemical weathering of minerals based on silicate. Biologists recognized that many freshwater organisms and accumulator plant species consume huge amounts of Si from lakes, rocks, rivers, and soils. All molecules of silica in the dissolved state spontaneously drain into the water body and then flow through the ecosystem, although only a microbial community survives in the weathering peels of boulders. The concentration of biogenic silica is comparatively high in rice husk, wheat husk, and wheat straw and is therefore used as a potential filler material for rubber composites, particularly as a replacement for synthetic silica. The condition of extraction of biogenic silica is highly significant to get the filler material for elastomers. It was advised not to exceed the temperature of 800 °C at the time of extraction during incineration. Especially the vulcanizates containing biogenic silica, if used as filler material, show enhanced modulus at 300% of elongation, less residual elongation, and a greater mechanical loss angle tangent at 0 °C and less at 60 °C than those of commercially available vulcanizate filler materials for the rubber industry. It was ultimately concluded that the composite of elastomer containing biogenic silica, rice husk, wheat husk, wheat straw, and phyto-ashes containing a higher percentage of biogenic silica derived from field horsetail and wheat straw used as filler materials shows outstanding grip in tyres on ice, and the composite material can be effectively used for the synthesis of footwear, winter shoe soles, ski boot soles, *etc.*

References

1. N. Dishovsky, P. Malinova and I. Uzunov, Biogenic amorphous silica as filler for elastomers, *J. Renewable Mater.*, 2018, **6**(4), 402–414.
2. N. Llopis Monferrer, T. Biard, M. M. Sandin, F. Lombard, M. Picheral, A. Elineau, L. Guidi, A. Leynaert, P. J. Treguer and F. Not, Siliceous Rhizaria abundances and diversity in the Mediterranean Sea assessed by combined imaging and metabarcoding approaches, *Front. Mar. Sci.*, 2022, **9**, 1–4.
3. N. Llopis Monferrer, D. Boltovskoy, P. Tréguer, M. M. Sandin, F. Not and A. Leynaert, Estimating biogenic silica production of Rhizaria in the global ocean, *Global Biogeochem. Cycles*, 2020, **34**(3), 1–13.
4. M. Masłowski, J. Miedzianowska, M. Delekta, A. Czylkowska and K. Strzelec, Natural rubber biocomposites filled with phyto-ashes rich in biogenic silica obtained from wheat straw and field horsetail, *Polymers*, 2021, **13**(7), 1177–1200.
5. M. L. Kaneko and I. V. Yoshida, Effect of natural and organically modified montmorillonite clays on the properties of polydimethylsiloxane rubber, *J. Appl. Polym. Sci.*, 2008, **108**(4), 2587–2596.
6. E. Phumnok, P. Khongprom and S. Ratanawilai, Preparation of natural rubber composites with high silica contents using a wet mixing process, *ACS Omega*, 2022, 7(10), 8364–8376.
7. H. G. Wijesinghe, T. N. Etampawala, D. G. Edirisinghe, G. R. Gamlath, R. R. Wadugodapitiya and T. A. Bandara, Properties of Rice Husk Ash Silica Filled Natural Rubber and Acrylonitrilebutadiene Rubber Blends, *J. Agric. Sci.*, 2022, **17**(1), 1–18.
8. P. R. Leavitt and D. Hodgson, Tracking environmental change using lake sediments, *Sedimentary pigments*, Springer Netherlands, Dordrecht, 2002, pp. 295–325.
9. H. O. Lijun, L. I. Min, X. U. Shiyuan, Y. A. Huimin, O. U. Dongni, S. Cheng and L. I. Xiao, Distribution, and accumulation of biogenic

silica in the intertidal sediments of the Yangtze Estuary, *J. Environ. Sci.*, 2008, **20**(5), 543–550.

10. L. Xuegang, S. Jinming, D. Jicui, Y. Huamao, L. Ning, L. Fengye and S. Song, Biogenic silicate accumulation in sediments, Jiazhou Bay, *Chin. J. Oceanol. Limnol.*, 2006, **24**(3), 270–277.
11. O. Ragueneau, N. Savoye, Y. Del Amo, J. Cotten, B. Tardiveau and A. Leynaert, A new method for the measurement of biogenic silica in suspended matter of coastal waters: using Si:Al ratios to correct for the mineral interference, *Cont. Shelf Res.*, 2005, **25**(5–6), 697–710.
12. M. Yadav, V. Dwibedi, S. Sharma and N. George, Biogenic silica nanoparticles from agro-waste: Properties, mechanism of extraction and applications in environmental sustainability, *J. Environ. Chem. Eng.*, 2022, **10**(6), 108550.
13. H. Ahmad-Alyosef, H. Uhlig, T. Münster, G. Kloess, W. D. Einicke, R. Gläser and D. Enke, Biogenic silica from rice husk ash–Sustainable sources for the synthesis of value-added silica, *Chem. Eng.*, 2014, **37**, 667–672.
14. Y. R. Corrales-Ureña, C. Villalobos-Bermúdez, R. Pereira, M. Camacho, E. Estrada, O. Argüello-Miranda and J. R. Vega-Baudrit, Biogenic silica-based microparticles obtained as a sub-product of the nanocellulose extraction process from pineapple peels, *Sci. Rep.*, 2018, **8**, 10417–10425.
15. N. Dishovski, F. Mitkova, M. Kandeva, Y. Angelov, I. Uzunov, M. Ivanov and D. Klissurski, Elastomer Composites with Enhanced Ice Grip Based on Renewable Resources, *Mater. Sci. Appl.*, 2018, **9**(4), 412–429.
16. A. M. Muhammud and N. K. Gupta, Nanostructured SiO_2 material: synthesis advances and applications in rubber reinforcement, *RSC Adv.*, 2022, **12**(29), 18524–18546.
17. V. P. Silva, M. C. Gonçalves and I. V. Yoshida, Biogenic silica short fibers as alternative reinforcing fillers of silicone rubbers, *J. Appl. Polym. Sci.*, 2006, **101**(1), 290–299.
18. Y. Zhou, M. Fan, L. Chen and J. Zhuang, Lignocellulosic fiber mediated rubber composites: An overview, *Composites, Part B*, 2015, **76**, 180–191.
19. M. Mohan Prasad, S. M. Sutharsan, K. Ganesan, N. R. Babu and T. Maridurai, Role of sugarcane bagasse biogenic silica on cellulosic Opuntia dillenii fibre-reinforced epoxy resin biocomposite: mechanical, thermal, and laminar shear strength properties, *Biomass Convers. Biorefin.*, 2023, **13**, 17027–17037.
20. M. G. Segatelli, I. V. Yoshida and M. do Carmo Gonçalves, Natural silica fiber as reinforcing filler of nylon 6, *Composites, Part B*, 2010, **41**(1), 98–105.
21. D. Kavitha, S. C. Murugavel and S. Thenmozhi, Flame retarding cardanol based novolac-epoxy/rice husk composites, *Mater. Chem. Phys.*, 2021, **263**, 124225.
22. S. Poomathi and S. S. Roji, Experimental investigations on Palmyra sprout fiber and biosilica-toughened epoxy bio composite, *Biomass Convers. Biorefin.*, 2022, 1–9.
23. S. Ramasamy, A. Bouaissi, M. Saidani and K. Mbarki, The Physico-Mechanical Properties of Natural Rubber Latex Foam when Loaded by

Different Sizes of Rice Husk Powder, *Knowledge-Based Engineering and Sciences*, 2022, **3**(3), 83–92.
24. M. C. Li, Y. Zhang and U. R. Cho, Mechanical, thermal and friction properties of rice bran carbon/nitrile rubber composites: Influence of particle size and loading, *Mater. Des.*, 2014, **63**, 565–574.
25. P. Saramolee, K. Sahakaro, N. Lopattananon, W. K. Dierkes and J. W. Noordermeer, Compatibilization of silica-filled natural rubber compounds by combined effects of functionalized low molecular weight rubber and silane, *J. Elastomers Plast.*, 2016, **48**(2), 145–163.
26. G. Liang and L. Zhang, An improved process for the preparation of high-quality bio-silica microparticles from rice husk ash, *Waste Biomass Valorization*, 2020, **11**(5), 2227–2233.
27. S. Steven, E. Restiawaty and Y. Bindar, Routes for energy and bio-silica production from rice husk: A comprehensive review and emerging prospect, *Renewable Sustainable Energy Rev.*, 2021, **149**, 1–18.
28. S. Steven, E. Restiawaty and Y. Bindar, Operating Variables on Production of High Purity Biosilica from Rice Hull Ash by Extraction Process, *J. Eng. Technol. Sci.*, 2022, **54**, 220304.
29. Q. Xu, M. Pang, L. Zhu, Y. Zhang and S. Feng, Mechanical properties of silicone rubber composed of diverse vinyl content silicone gums blending, *Mater. Des.*, 2010, **31**(9), 4083–4087.
30. H. Li, B. Cheng, W. Gao, C. Feng, C. Huang, Y. Liu, P. Lu and H. Zhao, Recent research progress and advanced applications of silica/polymer nanocomposites, *Nanotechnol. Rev.*, 2022, **11**(1), 2928–2964.
31. G. Marković, M. Marinović-Cincović, V. Jovanović, S. Samaržija-Jovanović and J. Budinski-Simendić, NR/CSM/biogenic silica rubber blend composites, *Composites, Part B*, 2013, **55**, 368–373.
32. S. Bhattacharyya, C. Sinturel, O. Bahloul, M. L. Saboungi, S. Thomas and J. P. Salvetat, Improving reinforcement of natural rubber by networking of activated carbon nanotubes, *Carbon*, 2008, **46**(7), 1037–1045.
33. J. Miedzianowska, M. Masłowski, P. Rybiński and K. Strzelec, Modified nanoclays/straw fillers as functional additives of natural rubber biocomposites, *Polymers*, 2021, **13**(5), 799–828.
34. G. Raju, M. Khalid, M. M. Shaban and B. Azahari, Preparation, and characterization of eco-friendly spent coffee/ENR50 biocomposite in comparison to carbon black, *Polymers*, 2021, **13**(16), 2796–2807.
35. R. Mohammadinejad, S. Karimi, S. Iravani and R. S. Varma, Plant-derived nanostructures: types and applications, *Green Chem.*, 2016, **18**(1), 20–52.
36. K. Adach, D. Kroisova and M. Fijalkowski, Biogenic silicon dioxide nanoparticles processed from natural sources, *Part. Sci. Technol.*, 2021, **39**(4), 481–489.
37. M. Yadav, V. Dwibedi, S. Sharma and N. George, Biogenic silica nanoparticles from agro-waste: Properties, mechanism of extraction and applications in environmental sustainability, *J. Environ. Chem. Eng.*, 2022, **10**(6), 108550.

38. S. Samarzija-Jovanovic, V. Jovanovic, G. Markovic, I. Zekovic and M. Marinovic-Cincovic, Properties of vulcanized polyisoprene rubber composites filled with opalized white tuff and precipitated silica, *Sci. World J.*, 2014, **2014**, 1–9.
39. A. Pandey, M. Nguyen-Vu and P. Schwab, Novel extraction protocol for evaluating abundances and structural features of amorphous SiO_2, *Icarus*, 2022, **384**, 115096.
40. N. Meftah, A. Hani and A. Merdas, Extraction and Physicochemical Characterization of Highly-pure Amorphous Silica Nanoparticles from Locally Available Dunes Sand, *Chem. Afr.*, 2023, **6**(6), 3039–3048.
41. J. T. Cornelis, B. Delvaux, R. B. Georg, Y. Lucas, J. Ranger and S. Opfergelt, Tracing the origin of dissolved silicon transferred from various soil-plant systems towards rivers: a review, *Biogeosciences*, 2011, **8**(1), 89–112.
42. O. Katz, D. Puppe, D. Kaczorek, N. B. Prakash and J. Schaller, Silicon in the soil–plant continuum: Intricate feedback mechanisms within ecosystems, *Plants*, 2021, **10**(4), 652–687.
43. D. Puppe, Review on protozoic silica and its role in silicon cycling, *Geoderma*, 2020, **365**, 114224.
44. R. J. Haynes, The nature of biogenic Si and its potential role in Si supply in agricultural soils, *Agric., Ecosyst. Environ.*, 2017, **245**, 100–111.
45. A. Tischer, E. Struyf, K. Potthast, M. Bremer, P. Meire and W. Clymans, Fire enhances solubility of biogenic silica, *Sci. Total Environ.*, 2016, **572**, 1289–1296.
46. X. Ran, J. Liu, J. Zang, B. Xu, S. Zhao, W. Wu, H. Wang and S. Liu, Export and dissolution of biogenic silica in the Yellow River (Huanghe) and implications for the estuarine ecosystem, *Mar. Chem.*, 2018, **200**, 14–21.
47. Y. Churakova, A. Aguilera, E. Charalampous, D. J. Conley, D. Lundin, J. Pinhassi and H. Farnelid, Biogenic silica accumulation in picoeukaryotes: Novel players in the marine silica cycle, *Environ. Microbiol. Rep.*, 2023, **15**(4), 282–290.
48. Y. Churakova, A. Aguilera, E. Charalampous, D. J. Conley, D. Lundin, J. Pinhassi and H. Farnelid, Biogenic silica accumulation in picoeukaryotes: Novel players in the marine silica cycle, *Environ. Microbiol. Rep.*, 2023, **15**(4), 282–290.
49. N. V. Lobus and M. S. Kulikovskiy, The co-evolutional aspects of the biogeochemical role of phytoplankton in aquatic ecosystems: A review, *Biology*, 2023, **12**(1), 92–112.

Subject Index